History of Analytic Philosophy
Series Editor: **Michael Beaney, University of York, UK**

Titles include:

Graham Stevens
THE THEORY OF DESCRIPTIONS

Mark Textor (editor)
JUDGEMENT AND TRUTH IN EARLY ANALYTIC PHILOSOPHY AND
PHENOMENOLOGY

Maria van der Schaar
G.F. STOUT AND THE PSYCHOLOGICAL ORIGINS OF ANALYTIC PHILOSOPHY

Nuno Venturinha (editor)
WITTGENSTEIN AFTER HIS NACHLASS

Pierre Wagner (editor)
CARNAP'S LOGICAL SYNTAX OF LANGUAGE

Pierre Wagner (editor)
CARNAP'S IDEAL OF EXPLICATION AND NATURALISM

Forthcoming:

Andrew Arana and Carlos Alvarez (editors)
ANALYTIC PHILOSOPHY AND THE FOUNDATIONS OF MATHEMATICS

Rosalind Carey
RUSSELL ON MEANING
The Emergence of Scientific Philosophy from the 1920s to the 1940s

Sandra Lapointe (translator)
Franz Prihonsky
THE NEW ANTI-KANT

Consuelo Preti
THE METAPHYSICAL BASIS OF ETHICS
The Early Philosophical Development of G.E.Moore

History of Analytic Philosophy
Series Standing Order ISBN 978–0–230–55409–2 (hardcover)
Series Standing Order ISBN 978–0–230–55410–8 (paperback)
(outside North America only)

You can receive future titles in this series as they are published by placing a standing order. Please contact your bookseller or, in case of difficulty, write to us at the address below with your name and address, the title of the series and one of the ISBNs quoted above.

Customer Services Department, Macmillan Distribution Ltd, Houndmills, Basingstoke, Hampshire RG21 6XS, England

The History and Philosophy of Polish Logic

Essays in Honour of Jan Woleński

Edited by

Kevin Mulligan
University of Geneva, Switzerland

Katarzyna Kijania-Placek
Jagiellonian University, Poland

and

Tomasz Placek
Jagiellonian University, Poland

First published 2014 by
PALGRAVE MACMILLAN

Palgrave Macmillan in the UK is an imprint of Macmillan Publishers Limited, registered in England, company number 785998, of Houndmills, Basingstoke, Hampshire RG21 6XS.

Palgrave Macmillan in the US is a division of St Martin's Press LLC, 175 Fifth Avenue, New York, NY 10010.

Palgrave Macmillan is the global academic imprint of the above companies and has companies and representatives throughout the world.

Palgrave® and Macmillan® are registered trademarks in the United States, the United Kingdom, Europe and other countries

ISBN 978-1-349-44063-4   ISBN 978-1-137-03089-4 (eBook)

DOI 10.1057/9781137030894

This book is printed on paper suitable for recycling and made from fully managed and sustained forest sources. Logging, pulping and manufacturing processes are expected to conform to the environmental regulations of the country of origin.

A catalogue record for this book is available from the British Library.

A catalog record for this book is available from the Library of Congress.

Contents

Part III Ontology, Mereology and the Philosophy of Mathematics

Series Editor's Foreword

During the first half of the twentieth century analytic philosophy gradually established itself as the dominant tradition in the English-speaking world, and over the last few decades it has taken firm root in many other parts of the world. There has been increasing debate over just what 'analytic philosophy' means, as the movement has ramified into the complex tradition that we know today, but the influence of the concerns, ideas and methods of early analytic philosophy on contemporary thought is indisputable. All this has led to greater self-consciousness among analytic philosophers about the nature and origins of their tradition, and scholarly interest in its historical development and philosophical foundations has blossomed in recent years, with the result that history of analytic philosophy is now recognized as a major field of philosophy in its own right.

The main aim of the series in which the present book appears, the first series of its kind, is to create a venue for work on the history of analytic philosophy, consolidating the area as a major field of philosophy and promoting further research and debate. The 'history of analytic philosophy' is understood broadly, as covering the period from the last three decades of the nineteenth century to the start of the twenty-first century, beginning with the work of Frege, Russell, Moore and Wittgenstein, who are generally regarded as its main founders, and the influences upon them, and going right up to the most recent developments. In allowing the 'history' to extend to the present, the aim is to encourage engagement with contemporary debates in philosophy, for example, in showing how the concerns of early analytic philosophy relate to current concerns. In focusing on analytic philosophy, the aim is not to exclude comparisons with other – earlier or contemporary – traditions, or consideration of figures or themes that some might regard as marginal to the analytic tradition but which also throw light on analytic philosophy. Indeed, a further aim of the series is to deepen our understanding of the broader context in which analytic philosophy developed, by looking, for example, at the roots of analytic philosophy in neo-Kantianism or British idealism, or the connections between analytic philosophy and phenomenology, or discussing the work of philosophers who were important in the development of analytic philosophy but who are now often forgotten.

The present volume, edited by Kevin Mulligan, Katarzyna Kijania-Placek and Tomasz Placek, is concerned with the history and philosophy of Polish logic, and in particular, the Lvov–Warsaw School, founded in 1895 by Kazimierz Twardowski (1866–1938). Twardowski had studied with Franz Brentano (1838–1917) in Vienna, and his *Habilitationsschrift, On the Content and Object of Presentations* (1894), had reworked Bolzano's distinction between content and object in a Brentanian framework. The work had been reviewed by G. F. Stout (1860–1944) in *Mind* in 1894, and it was through Stout that Brentano's and Twardowski's ideas influenced Moore and Russell. (On Stout's influence, see the book by Maria van der Schaar in the present series.) More significantly, Twardowski's work influenced a whole generation of Polish logicians and philosophers. Twardowski became Professor in Lvov in 1895, and remained there until he retired in 1930. Among his students were Jan Łukasiewicz (1878–1956), Stanisław Leśniewski (1886–1939), Tadeusz Kotarbiński (1886–1981) and Kazimierz Ajdukiewicz (1890–1963). Łukasiewicz became Professor in Warsaw in 1915, marking the point at which the Lvov School became the Lvov–Warsaw School, and he was joined there in 1919 by Leśniewski and Kotarbiński. All three taught Alfred Tarski (1901–83), who was to become the most famous Polish logician and analytic philosopher.

As amply demonstrated by the papers in the present volume, written by analytic philosophers and historians of analytic philosophy who have themselves made substantial contributions to both analytic philosophy and history of analytic philosophy, the Lvov–Warsaw School constitutes an important tradition in twentieth-century philosophy, and indeed, an important subtradition of analytic philosophy. It built on the pioneering logical work of Frege and Russell, while drawing inspiration, too, from Bolzano and Brentano. Like early British analytic philosophy and logical empiricism, it stressed the virtues of clarity and rigorous argumentation. It is worth noting that when Ernest Nagel visited Europe in 1934–5 to report on the new philosophical scene for the *Journal of Philosophy*, he singled out for attention 'the philosophy professed at Cambridge, Vienna, Prague, Warsaw and Lwów'. His report was published in 1936 as 'Impressions and Appraisals of Analytic Philosophy in Europe' (the first article, in fact, to contain 'analytic philosophy' in its title; the quotation is from p. 6). Despite this mention of Polish philosophy, though, Nagel focused his discussion on the work of Moore, Wittgenstein and Carnap; and until recently, the Lvov–Warsaw School has not been accorded the recognition it deserves in the history of analytic philosophy. That it does at all is due to the advocacy of one man, in particular, Jan Woleński, whose pioneering book, *Logic and Philosophy in the Lvov-Warsaw School* (1988), has done much to raise the profile of Polish logic and philosophy. This volume is dedicated to Jan Woleński, and I am delighted that it is appearing in this series. It will not only help secure the position of the Lvov–Warsaw School in the history of analytic philosophy but also deepen our understanding of its central ideas and influence.

Michael Beaney

Notes on Contributors

Joseph Agassi is an emeritus professor at Tel-Aviv University and York University, Toronto. His prime interests lie in science, metaphysics and politics. He has made contributions to the history of sciences and scientific method.

Arianna Betti is Professor of Philosophy of Language at the University of Amsterdam (UvA). Holder of two ERC grants, she leads a group investigating the development of ideas in logic as a methodology of the sciences from Kant to Tarski with the aid of computational techniques. She has worked on historical and systematic aspects of philosophical concepts such as the concepts of axiom, explanation, fact and truth.

Jaakko Hintikka is a professor of at the Boston University. He has made pioneering contributions to game-theoretical semantics, the interrogative approach to inquiry, and independence friendly logic. He has heavily influenced the development of possible-worlds semantics, tree methods, infinitely deep logics and the theory of inductive generalization.

Alexander Karpenko is a professor at the Moscow State University, a member of the Russian Academy of Natural Sciences as well as chair of the Department of Logic in the Institute for Philosophy of the Russian Academy of Sciences. His logic expertise lies in particular in the functional properties of finite-valued logics. His interests are in the foundations of logic. In philosophy he has published on logical fatalism and Lovejoy's plenitude principle.

Wolfgang Künne is an emeritus professor at the University of Hamburg and member of the Göttingen Academy of Sciences. He has worked on the ontology of abstract entities and on theories of truth and of meaning. He also made contributions to the history of analytic philosophy and phenomenology (Bolzano, Frege and Husserl in particular) as well as to the philosophy of mind.

Iris Loeb is an assistant professor in Mathematics at VU Univerisity Amsterdam. As a postdoctoral fellow in Philosophy she has worked in the ERC-funded project *Tarski's Revolution: A New History*.

Kevin Mulligan is Ordinary Professor of Analytic Philosophy at the University of Geneva. His publications are in the areas of analytic metaphysics, the philosophy of mind and the history of Austro-German philosophy from Bolzano to Musil and Wittgenstein.

Roman Murawski is a professor in the Faculty of Mathematics and Computer Science of Adam Mickiewicz University in Poznań. His areas of expertise include models of first and higher order arithmetic, nonstandard satisfaction classes, and computation theory. He has worked as well on the history and philosophy of logic and mathematics.

Ilkka Niiniluoto is a professor of theoretical philosophy at the University of Helsinki and the Chancellor of this university. He has made contributions to the philosophy of science, particularly to the topic of verisimilitude or truth approximation. He works as well on probability, induction, theory change, scientific progress and philosophy of law.

David Pearce is a professor at the Universidad Politécnica de Madrid. His philosophical interests focus on applications of logic to problems of knowledge representation and reasoning. He has contributed to semantics of logic programs.

Gabriel Sandu is a professor of philosophy at the University of Helsinki. He has been interested in games in logic, in particular in the relations between games in economic sciences and semantic notions such as truth or logical consequence. His works contributed to the foundations of mathematics, to theories of truth and the theory of quantification. His research concerns also theories of reference as well as the logic of modalities.

Maria van der Schaar is an assistant professor of at Leiden University. Her research concerns the theory of judgement and its history. Her project contributes to the philosophical foundations of mathematical constructivism through work on assertion, judgement, meaning and knowledge.

Peter Simons is a professor of at the Trinity College Dublin. He has made contributions to mereology. He has worked on various topics in metaphysics, philosophy of language and logic, and philosophy of mathematics. He has strong interests in history of philosophy and logic in Central Europe (Bolzano, Brentano, Frege, Meinong, Husserl, Twardowski, Łukasiewicz, Leśniewski and Tarski).

Introduction: The History and Philosophy of Polish Logic: Some Basic Thoughts

Kevin Mulligan, Katarzyna Kijania-Placek, and Tomasz Placek

Modern Polish logic is the fruit of the Lvov-Warsaw School of philosophy, which existed between 1895 and World War II. From today's perspective, the emergence of the Lvov-Warsaw School appears to be a result of the intentional and pre-meditated efforts of one philosopher in particular, namely Kazimierz Twardowski. In 1895 Twardowski, then 29 years old but already an accomplished philosopher who was based in Vienna and a former student of Brentano and Zimmerman, was appointed to the post of extraordinary professor at the University of Lvov, one of the two universities in which Polish was the language of instruction. He saw his coming to Lvov as part of a mission to export Brentano's philosophy and conception of philosophy to Poland and, more generally, to reform Polish philosophy by making it rigorous and oriented towards detailed analyses of philosophical problems. Clarity of thought was by no means a priority of the dominant forces in Polish philosophy at the time. For almost 150 years the Polish state had ceased to exist, its former lands being divided between Russia, Prussia and the Austro-Hungarian Empire. After a few unsuccessful uprisings aimed at resurrecting the Polish state, the trend of Polish Messianism, a peculiar mixture of Hegelian themes, Catholicism and national myths, had gained popularity, as it attempted to explain, or at least provide, an eschatological sanction for the country's national suffering. Kantianism was another source of inspiration: philosophy professors like Struve or Mahrburg disputed Kant's *Critiques* and aimed to restore metaphysics. At some institutions, in particular those located in the territory under Austrian administration, Catholic philosophy was taught as well. There was also a positivist movement in Warsaw but its orientation was largely political and social and it had no clear stance on epistemological issues, in contrast to its French or American namesakes. (For more, see Simons, 2009.)

In order to fully comprehend the intellectual forces behind the Lvov-Warsaw School, it is important to reflect on the philosophical interests of its founding father, Twardowski. Twardowski's main work before coming to Lvov, his *Habilitationsschrift, Zur Lehre vom Inhalt und Gegenstand der Vorstellungen,* clarified a subtle question in Brentanian philosophy. An important thesis of Brentano's distinguishes between mental and non-mental phenomena. A distinctive feature of the former is intentionality: a mental phenomenon always has an object towards which it is oriented. This immediately raises the question: What are the objects of mental acts, in particular, are they in our heads? The question is made even more pressing by Brentano's account of judgements as expressible by existential claims such as '*A* exists/does not exist', where *A* is the so-called immanent object of judgement, what is presented-by-a-presentation. This immediately leads to a conundrum concerning our positive and negative judgements. Clearly, the aim and meaning of the assertion that black matter exists/does not exist is not to claim the existence/non-existence of some mental object. This kind of problem led to criticism of Brentano's theory of judgements, and of defences of the theory by Brentano and his followers. It is to this context that Twardowski's *Zur Lehre vom Inhalt und Gegenstand* ... belongs. The work's central claim is that every presentation has both a content and an object. The distinction is similar to that drawn by Frege in the philosophy of language between an expression's sense and its reference. Although it might have been 'in the air' in the German-speaking philosophical world of the 1890s, no Brentanian had accepted such a distinction before Twardowski. Furthermore, it is not perhaps the distinction alone, but Twardowski's lucid argumentation, his persuasive everyday examples, and a reliance on linguistic analysis that make it an extraordinary piece of philosophical prose. The work influenced Husserl's and Meinong's theories of intentionality and, thanks to a review in *Mind* (1894) by G.F. Stout it was to stir up some thoughts in anglophone analytic philosophy as well. We will not, however, present or discuss it here, as this job has already been done perfectly in a number of accounts. We merely seek to bring it to the reader's attention here, as it casts some unexpected light on the background of the logically and scientifically minded Lvov-Warsaw School. That background was late nineteenth-century theories of intentionality, and, connected with this, the dispute in the foundation of psychology between Brentanian descriptive psychology and Wundt's experimental psychology. The main auxiliary discipline for philosophy according to Twardowski was psychology, descriptive and genetic or causal. (He lectured extensively on psychology in his Lvov years and set up the first psychological laboratory in Poland.) Twardowski's infatuation with all matters psychological is worth bearing in mind, in particular if one recalls Tarski's or Łukasiewicz's mathematical inspirations, or the role physics played in the thought of Wiener Kreis or the Berliner Kreis, the major contemporaries of Twardowski's school.

The philosophical interests of the members of the Lvov-Warsaw School were multifarious; two achievements, Tarski's definition of truth and Łukasiewicz's multivalued logics became known world-wide, a fact which was probably due in part to the fact that these two scholars ended up in Western academic institutions as a result of World War II (Tarski at UC Berkeley, Łukasiewicz at University College Dublin). Since most other members of the school either did not survive the war, or stayed in communist Poland after it had become a satellite of the Soviet Union, and thus cut off intellectually from the West, their work was condemned to relative obscurity. The task of reclaiming their work is a part of the large project which Jan Woleński initiated with his *Logic and Philosophy in the Lvov-Warsaw School*, and to which (we hope) the present volume contributes.

Uncontroversially, the School's best-known contributions to the world of learning lie in the field of logic. Was there some driving force which motivated the research conducted in the School, no matter whether it was research in logic, the philosophy of science, ethics, descriptive psychology (the philosophy of mind), the philosophy of language or the history of philosophy? The factor was, we believe, an Austro-Polish obsession with truth, to use the apt expression of Woleński and Simons (1989). The topic of truth passed to Twardowski from his teacher, F. Brentano,[1] and dominated Twardowski's efforts to bring greater rigour to Polish philosophy as well. In 1900 he published 'On the so-called relative truths', in which he first surveyed the putative examples of relative truths, and then argued that the relativity of truth is illusory: it disappears if one pays close attention to the quantifiers or if one removes indexical expressions from (what he calls) incomplete utterances. He thus argues that in every case one can de-indexicalize indexical utterances without compromising their content, a claim later commonly accepted in the Lvov-Warsaw School (see Betti, 2006).

The topic of truth receives a more logic-oriented treatment in Łukasiewicz's (1910) 'On the principle of contradiction in Aristotle' where he draws a distinction between three senses of the principle, a psychological, a logical, and a metaphysical sense, critically examines Aristotle's arguments in favour of the principle, in all three senses, and envisions a discourse in which the principle fails.[2] The book is a prelude to Łukasiewicz's construction of multi-valued logics: there is evidence that he toyed with a three-valued calculus already at the time of writing the book (see the quote in Woleński, 1988, p. 119).

A related and fascinating debate about the eternity of truth broke out with Kotarbiński's (1913) 'The problem of the existence of the future'. As for sentences about the past, he adopts a line similar to his teacher, Twardowski, by arguing that temporal indexicals should be eliminated. Thus, he translates 'A meteor seen yesterday in Spain was red', asserted on 2 October 1912, into 'A meteor seen on 1 October 1912 in Spain is red', and argues that if a sentence (with temporal indexicals removed) is once true at one time, it is always true. This is the doctrine of sempiternal truth. He then contrasts sentences about the past with sentences about the future. Concerning the latter, he argues that even with temporal indexicals removed, it cannot be that one of the two, a sentence and its negation, is true. He argues that if one of two such sentences is true, then it is true at any later time, leaving no room for freedom of action. He suggests that some sentences about the future are neither true nor false, yet, they acquire truth values over the course of time. Truth is thus sempiternal, but not eternal.

Leśniewski (1913) objected to this position and, in the crucial part of this paper, he undertakes an analysis of what 'free' creativity means, a vital concept in Kotarbiński's and Łukasiewicz's arguments and discussions. If I do something because I have to do so, because I cannot do otherwise, he writes, that action cannot be considered the product of free creativity. Following this intuition, he puts forward the following explication of free action: 'if at a time t an object p "must" possess a property c and if it "cannot fail" to possess it, – then the possession by object p of property c at time t cannot be a product of free creativity'.[3] From this explication, he then reconstructs Kotarbiński's argument, the conclusion of which is:

> If it is already now true that the object p will possess property c at time t, then the object p cannot not possess property c at time t.

Accordingly, if it is now already true that the object p will possess property c at time t, an agent's action before time t to bring about property c in object p at time t is superfluous; this property is, so to speak, guaranteed to materialize at t.

Leśniewski finds the reasoning here faulty, however. Interestingly, he accepts the following claim: if it is now true that it will be the case that *q*, then it must already be false that it will not be the case that *q*. (This claim is as sound as the principle of contradiction, he writes.) He identifies a faulty move in the implication of the form:

If (a) I may not write this section, then (b) it may be true that I shall not write this section.

If parsed in a more up-to-date manner, the spurious implication is from (a') 'It is possible that it will not be the case that ψ' to (b') 'It is possibly true that it will not be the case that ψ', where ψ abbreviates 'I write this section'. If we eliminate the modifier 'true' in 'possibly true' in (b') above, however, the two sentences (a') and (b') appear to be synonymous, and hence must form a correct implication. But for Leśniewski this modifier is significant; moreover, his theory of assertion makes him oppose the parsing of (a) and of (b) as (a') and (b'), respectively. The two sentences, (a) and (b) ascribe properties to different objects, the former to an agent (the referent of the pronoun 'I' above), the latter to the assertion (the property of being true). Accordingly (a) and non-(b) are not contradictory, as they do not ascribe opposing properties to one and the same object. And clearly, in order for the implication from (a) to (b) to be correct, (a) and non-(b) would have to be contradictory. Since there is no proof that they are contradictory, the implication is not established. Thus, he concludes, a crucial step is wanting in the alleged derivation of the superfluity of free creativity from the eternity of truth.

As we mentioned, a conflict (illusory or not) between free creativity and eternal bivalent truth also lies behind Łukasiewicz's project. In his essay 'On Determinism' (1961), which is based on his inaugural lecture for the academic year 1922/3, when he was Rector of the University of Warsaw, he follows Kotarbiński in arguing that eternality of truth entails determinism, that it makes the future as much determined as the past. In a beautiful passage, he compares a deterministic scenario to 'a film drama produced in some cinematographic studio'. The 'ending is there, it exists from the beginning of the performance, for the whole picture is completed from the eternity. In it all our parts, all our adventures, both good and bad, are fixed in advance'.

How does he explicate determinism, however? Oddly enough, he formulates determinism as follows:

(DT) If *A* is *b* at instant *t*, it is true at any instant earlier than *t* that *A* is *b* at instant *t*.

The oddity becomes apparent when we observe that (DT) is likely to be innocent of all deterministic connotations. If read in natural tensed language, it is rendered as:

> If A is b at instant t, then it was true at any instant earlier than t that A would be b at instant t.

If we further replace 'it was true at any instant earlier than' by the past tense sentential operator 'it has always been the case that', abbreviated by *Has*, and replace the reference to instants by tenses, we get the formula $\varphi \rightarrow Has : Will : \varphi$, which is typically assumed to be correct in tense logics. Tense logicians argue that the formula is correct and does not have deterministic underpinnings. Such underpinnings are present in the sentence:

> 'It was settled as true that it would be the case that φ'

but this sentence does not follow from:

> 'It was true that it would be the case that φ'.

If we test this against natural language, we see that although it was true yesterday that it would rain today, it was not inevitable (settled as true) yesterday that it would rain today. The same intuition appears to be present in Polish as well.[4]

Thus, although the rejection of Łukasiewicz's (DT) would lead to a much stronger position than indeterminism, the criticism of it motivated Łukasiewicz to construct a three-valued logic, with the basic idea that future contingents have an intermediate (third) value assigned to them.[5]

Lack of space prohibits us from tracing any further the debate about truth in Polish philosophy. It seems fair to say, however, that in the 1920s and 1930s the debate moved from general philosophy to logic. Starting from 1920 Łukasiewicz produced a number of systems of many-valued logics. Beginning with his (1930) paper, Tarski started publishing on truth in formalized languages, and although his results are negative with respect to defining truth in natural languages, the work blazed a trail for later work on truth in ordinary languages. The obsession with truth, which was brought by Twardowski from Vienna to Lvov, and which in the early years of the twentieth century was transmitted to most of the prominent scholars in the Lvov-Warsaw School, continued until much later, up to the 1970s, and in many different fields, such as aesthetics and general epistemology.

The present volume aims to present the state of the art of research into the history and philosophy of Polish logic. It contains 13 essays written by philosophers which exemplify different approaches to the history of philosophy. One approach focuses on some little-known aspect of Polish philosophy (for example, Leśniewski's arithmetic, Tarski's geometry, philosophy of mathematics in interwar Cracow), analysing it in great detail, sometimes by using current formal techniques. Another group of papers looks at the inspiration which Poles obtained from the founding fathers of analytic philosophy (Frege, Husserl and Wittgenstein), and locates Polish philosophy in the larger landscape of European analytic philosophy. Finally, some contributors pick a topic from the Polish school (sometimes only mentioned, but not developed by Poles), and construct an alternative account which is then compared with the earlier account. Most of the papers were presented at a symposium celebrating the seventieth birthday of Jan Woleński, whose book *Logic and Philosophy of the Lvov-Warsaw School* has played a substantial role in sparking contemporary interest in Polish analytic philosophy.

The topic of Part 1 of the book is logic, proofs, and models. It begins with A. Karpenko's account of the development of many-valued logics in the Lvov-Warsaw School. Next, G. Sandu's paper sketches the basis of the independence-friendly logic that allows for interdependencies between quantifiers. Although this new system has greater expressive power than first order logic, it introduces indeterminacy. The author then discusses how to overcome indeterminacy by means of a probabilistic account of the quantifiers. A concept of a model that is an alternative to the one developed in the Tarskian tradition, is investigated in D. Pearce's study. M. van der Schaar develops an analysis of adjectives, the starting point of which is K. Twardowski's paper 'On the logic of adjectives'. The (classical) concept of rationality (which Polish analytic philosophers accepted) is critically discussed in J. Agassi's paper. He claims that analytic philosophy traditionally advocated rational belief where rationality is understood in terms of proof or proof-surrogates, and argues that this classical theory of rationality must give way to a view of rationality as open-mindedness, or as openness to criticism.

Part II deals with truth and concepts; it opens with J. Hintikka's essay which points to a circularity in generalizations of Tarski's T-schema to a truth definition, and shows how this flaw can be corrected by using Independence-Friendly logic. In the spirit of critical realism, I. Niiniluoto defends the objective and absolute nature of the concept of material truth. He examines and rejects attempts to define truth in epistemic or doxastic terms. W. Künne takes up a question first raised by Prior: Should we understand truth as a truth connective ('It is true that ...') or as a truth predicate ('...is true')? He argues for the latter against recent attempts to revive Prior's view by Mulligan. He then considers a similar choice which arises in the analysis of propositional attitude reports and rejects Prior's view, the gist of which is that 'NN φs that p' is to be parsed, not as 'NN/φs / that p', but as 'NN/φs that/p'. K. Mulligan examines attempts to distinguish between formal and non-formal concepts by Husserl and Wittgenstein, in particular their accounts of operations and form. These attempts and accounts, he suggests, are the immediate predecessors of Tarski's attempt to distinguish between logical and non-logical terms and the Jaśkowski-Gentzen theory of inferences from suppositions.

Part III is devoted to ontology, mereology and the philosophy of mathematics. P. Simons's essay investigates and reconstructs a system of arithmetic in Leśniewski's ontology. The Leśniewskian theme is continued by A. Betti, who probes Tarski's (1929) 'mereological' foundations of the geometry of solids. Although Tarski's approach looks Leśniewskian, its logical basis is Russellian (simple) type theory. Betti argues that Tarski's choice of background logic is important since it naturally reinforces the distinction between the domain of discourse, the range of quantifiers and the individuals of the underlying type theory. She indicates the relevance of this finding to the current debate about the correct interpretation of Tarski's notion of a model. Another Tarskian theme is developed in I. Loeb's reflections on Tarski's manuscript on atomless Boolean algebras, a manuscript which Tarski never published. A different historical subject is tackled in R. Murawski's paper which discusses the relations between logic and mathematics in interwar Cracow. This period saw the intensive development of the so-called Polish Mathematical School alongside the Lvov-Warsaw School of logic. The author investigates how these parallel developments were reflected in the philosophical and methodological views of some of the luminaries of the two schools.

The book is dedicated to Jan Woleński on the occasion of his seventieth birthday. Jan Woleński has done his utmost to revive the memory of the Lvov-Warsaw School in Poland and to bring the School's significance to the attention of analytic philosophers all over the world. In many cases, the research reported in this book has been inspired by his work on the history and philosophy of Polish logic.

Notes

1. For Brentano's theory of truth and its influence on Twardowski see (Woleński & Simons, 1989 or Albertazzi, 1993).
2. He points to the Athanasian Creed and says that outside a theological context it appears to be contradictory.
3. The single quotation marks are Leśniewski's.
4. Poles say: 'Wypadnie reszka, ale nie jest przesądzone, że wypadnie reszka', and in the past tense: 'Było prawdą, że wypadnie reszka, ale nie było przesądzone, że wypadnie reszka'.
5. Note that in Łukasiewicz's three-valued logic, DT with true antecedent and intermediate consequent is intermediate rather than true.

References

Albertazzi, L. (1993) 'Brentano, Twardowski, and the Polish Scientific Philosophy', in F. Coniglione, R. Poli, and J. Woleński (eds) *Polish Scientific Philosophy: The Lvov-Warsaw School* (Poznań: Studies in Philosophy of the Sciences), 11–39.

Betti, A. (2006) 'Sempiternal Truth: The Bolzano-Twardowski-Leśniewski Axis', in J. J. Jadacki, and J. Paśniczek (eds) *The Lvov-Warsaw School – The New Generation* (Poznań: Studies in the Philosophy of the Sciences), 371–99.

Kotarbiński, T. (1913) 'Zagadnienie istnienia przyszłości', *Przegląd filozoficzny*, 16(1), 74–92; translated as: T. Kotarbiński (1968) 'The Problem of the Existence of the Future', *The Polish Review*, 13(3), 7–22.

Leśniewski, S. (1913) 'Czy prawda jest tylko wieczna czy też i wieczna i odwieczna?', *Nowe Tory*, 8, 493–528; translated as: S. Leśniewski (1991) 'Is All Truth Only True Eternally or Is It Also True without a Beginning', in S. J. Surma *et al.* (eds) *Stanisław Leśniewski. Collected Works* (Dordrecht: Kluwer), 86–114.

Łukasiewicz, J. (1910) *O zasadzie sprzeczności u Arystotelesa* (Cracow: Polska Akademia Umiejętności).

—— (1920) 'O logice trójwartościowej', *Ruch filozoficzny*, 5, 170–1. English translation: 'On Three-Valued Logic', in L. Borkowski (ed.) *Selected Works by Jan Łukasiewicz* (Amsterdam: North-Holland), 1970, 87–8.

—— (1961) 'O determinizmie', in T. Borkowski (ed.) *Z zagadnień logiki i filozofii. Pisma wybrane* (Warsaw: PWN), 114–26; translated as: J. Łukasiewicz (1967) 'On Determinism', in S. McCall (ed.) *Polish Logic 1920–1939* (Oxford: Oxford University Press).

Simons, P. (2009) 'Twardowski on Truth', *The Baltic International Yearbook of Cognition, Logic, and Communication*, 4, 1–14.

Tarski, A. (1929) 'Les fondements de la géometrie des corps (résumé)', *Annales de la Société Polonaise de Mathématiques*, Lwów 7–10.IX.1927 (Cracow), 29–33.

—— (1930) 'O pojęciu prawdy w odniesieniu do sformalizowanych nauk dedukcyjnych', *Ruch Filozoficzny*, 12, 210–11.

Twardowski, K. (1900) 'O tak zwanych prawdach względnych', *Księga Pamiątkowa Uniwersytetu Lwowskiego ku uczczeniu 500 rocznicy fundacji Jagiellońskiej Uniwersytetu Krakowskiego*, 25–37; translated as: K. Twardowski (1999) 'On So-Called

Relative Truths', in J. Brandl, and J. Woleński (eds) *On Actions, Products and Other Topics in Philosophy* (Amsterdam: Rodopi), 147–70.

Woleński, J. (1988) *Logic and Philosophy in the Lvov-Warsaw School* (Dordrecht: Kluwer).

Woleński, J., and Simons, P. (1989) 'De Veritate: Austro-Polish Contributions to the Theory of Truth from Brentano to Tarski', in K. Szaniawski (ed.) *The Vienna Circle and the Lvov-Warsaw School* (Dordrecht: Kluwer), 391–443.

Part I

Logic, Proof and Models

1
Many-valued Logic in Poland: The Golden Age

Alexander S. Karpenko

1.1 Introduction

The origin of many-valued logic was closely connected with the Lvov-Warsaw School (LWS) (see Woleński, 1985). Here is an amazing case of the emergence of one of the most influential scientific schools in the world, which is famous for its research in philosophy and especially in logic.

1895 is considered to be the year of foundation of the LWS, when K. Twardowski (1866–1938) got a chair at Lvov University. Poland as an independent state did not exist at that point in history, for after three partitions it was completely torn into three parts: an Austrian one, a Prussian one and a Russian one (1772–95). The state was restored in 1919 by the Versailles agreement. The second period in the development of the school began just after that restoration. Quite paradoxically, despite the *non existence* of a Polish state, the conditions for the emergence of the above-mentioned scientific school were in place and thus the emergence of the unique Warsaw school of logic headed by J. Łukasiewicz (1878–1956) and S. Leśniewski (1886–1939) was possible. J. Woleński, the main specialist on the LWS, writes: 'The Second World War and the events that followed, first of all the presence of Poland in the "socialist camp" where *Marxism* was the official state philosophy, had led to the disappearance of Lvov-Warsaw school as an organized philosophical movement.' (Woleński, 2004a, p. 456) Nevertheless it should be noted that the level of development of logic in Poland remains very high. In another work Woleński (2010) writes:

> The logical achievements of the LWS became the most famous. Doubtless, the Warsaw school of logic contributed very much to the development of logic in the 20th century.[1]

13

Z. Jordan described the phenomenon of many-valued logic in the following manner (see the chapter V 'The Discovery of Many-Valued systems of Logic' in Jordan, 1945/1967, p. 389):

> Whatever value may be attached to the above-mentioned results, the discovery by Łukasiewicz of many-valued systems of logic stands out against all of them. Without any doubt it is a discovery of the first order, eclipsing everything done in the field of logical research in Poland.

And in Woleński (1985, p. 119), see the whole chapter 6.2 'Many-Valued Logics', we read: 'The construction of many-valued logical systems is commonly believed to have been one of the major achievements of the Warsaw School, and specifically of Łukasiewicz.'

The early development of many-valued logic in Poland is considered in detail in Woleński (2001) especially from 1910 to 1920. Woleński (2001, p. 196) writes that 'Kotarbiński[2] introduced into Poland the idea that there are sentences which are indefinite, that is, neither true nor false', (see also Woleński, 1990). It is interesting that Kotarbiński didn't mention it in chapter XXI, 'Many-Valued Propositional Calculus' of his book (1957) but criticized Łukasiewicz's three-valued logics for conceptual reasons.

The first mention of three-valued logic can be found in Łukasiewicz's lecture delivered in 1918. There, Łukasiewicz says: 'that new logic [...] destroys the former concept of science'; moreover, Łukasiewicz makes a connection between the 'new logic' and the 'struggle for the liberation of the human spirit' (see Łukasiewicz, 1918/1970, p. 86). One can say that passion for freedom led Łukasiewicz to the discovery of three-valued logic.

Philosophical ideas underlying the third truth-value are discussed in Łukasiewicz's seminal paper 'On determinism' (see Łukasiewicz, 1922).[3] There, Łukasiewicz states that Aristotle's solution of the problem of future contingency (the problem of logical fatalism) destroys one of the main principles of our logic, namely that *every proposition is either true or false*. Łukasiewicz calls this principle *the principle of bivalence*. According to him, it is an underlying principle of logic that cannot be proved – one can only believe in it. Łukasiewicz claims that the principle of bivalence does not seem self-evident to him. Therefore, he claims to have the right not to accept it and to stipulate that, along with truth and falsity, there should be at least one more truth-value, which Łukasiewicz considers to be intermediate between the other two. Łukasiewicz concludes that

If this third value is introduced into logic we change its very foundations. A trivalent system of logic [...] differs from ordinary bivalent logic, the only one known so far, as much as non-Euclidean systems of geometry differ from Euclidean geometry. (Łukasiewicz, 1922/1970, p. 126)

Similar passages occur in other papers of Łukasiewicz (up to 1951). Those claims foreshadowed a radical revision of the classical logic. Łukasiewicz's main philosophical conclusion is that determinism could be avoided by rejecting the principle of bivalence.[4]

1.2 Łukasiewicz's three-valued logic Ł₃

The first system of three-valued logic appeared in Łukasiewicz (1920). It means that for the first time three-valued logical connectives were defined and combined to form a logical system.

Adhering to the classical way of defining implication $p \to q$ and negation $\sim p$ wherever their arguments are the classical truth-values 0 and 1, Łukasiewicz defines the meaning of those connectives for the cases featuring his new truth-value in the following way:

$$(1 \to {}^1/_2) = ({}^1/_2 \to 0) = {}^1/_2,$$

$$(0 \to {}^1/_2) = ({}^1/_2 \to {}^1/_2) = ({}^1/_2 \to 1) = 1,$$

$$\sim {}^1/_2 = {}^1/_2.$$

The other propositional connectives are defined by means of the primary connectives:

$p \lor q =: (p \to q) \to q$ (disjunction).

$p \land q =: \sim (\sim p \lor \sim q)$ (conjunction).

$p \leftrightarrow q =: (p \to q) \land (q \to p)$ (equivalence).

Thus, the truth-tables for the logical connectives look as in Table 1:
A *valuation* is a function v from the set of formulae S to the set $\{0, {}^1/_2, 1\}$ of truth-values, 'compatible' with the above truth-tables. A formula α is a *tautology* if and only if $v(a) = 1$ for every valuation v, where 1 is the *designated value*. The set of tautologies thus defined is *Łukasiewicz's three-valued logic* Ł₃.

Wajsberg (1931) showed that Łukasiewicz's three-valued logic can be axiomatized in the following way:

Table 1

p	$\sim p$
1	0
½	½
0	1

$\to$	1	½	0
1	1	½	0
½	1	1	½
0	1	1	1

$\vee$	1	½	0
1	1	1	1
½	1	½	½
0	1	½	0

$\wedge$	1	½	0
1	1	½	0
½	½	½	0
0	0	0	0

$\leftrightarrow$	1	½	0
1	1	½	0
½	½	1	½
0	0	½	1

1. $(p \to q) \to ((q \to r) \to (p \to r))$.
2. $p \to (q \to p)$.
3. $(\sim p \to \sim q) \to (q \to p)$.
4. $((p \to \sim p) \to p) \to p$.

The rules of inference are as in the classical propositional logic C_2.

R1. *Modus ponens*: If α and $\alpha \to \beta$, then β.
R2. *Substitution*: If $\vdash \alpha(p)$, then $\vdash \alpha(\beta)$, where β, is a well-formed formula that is substituted uniformly for p.

Wajsberg's axiomatization means that for $Ł_3$ a formula α is a tautology if and only if α is a theorem. Thus, $Ł_3$ is, like C_2, deductively complete and consistent.

1.3   Differences between $Ł_3$ and C_2

It's worth pointing out once again that the behaviour of the connectives of $Ł_3$ over the set $\{1,0\}$ coincides with that of the connectives of C_2. Thus, logic $Ł_2$ (the two-valued version of $Ł_3$) is nothing else but the classical propositional logic C_2. It is evident that any tautology of $Ł_3$ is a tautology of C_2, but not *vice versa*.

In fact, $Ł_3$ is radically different from C_2 – some important laws of the classical logic, such as

$p \vee \sim p$ (the law of the excluded middle)

$\sim (p \wedge \sim p)$ (the law of non-contradiction),

Table 2

p	Tp
1	$1/2$
$1/2$	$1/2$
0	$1/2$

fail in Ł$_3$ (as well as in any Ł$_n$): these formulae get the value $1/2$ when p is $1/2$. Also the *contraction law*

$$(p \to (p \to q)) \to (p \to q)$$

is not a tautology of Ł$_3$ (assign $1/2$ to p and 0 to q, to get a counter-example).[5]

The most important difference between Ł$_3$ and C$_2$ is the following. Słupecki (1936) showed that Ł$_3$, in contrast with C$_2$, is not truth-functionally complete; that is, not every three-valued truth-function can be defined in Ł$_3$. To see this, take Słupecki's function Tp as in Table 2.

It is obvious that Tp cannot be defined in Ł$_3$; furthermore, Słupecki showed that by adding Tp to Ł$_3$, we get a *truth-functionally complete*, that is, *full* system and Post-complete system (when adding axioms $Tp \to T \sim p$ and $T \sim p \to Tp$ to Wajsberg's axiom system for Ł$_3$).[6] Słupecki also proved that this axiomatization is independent.[7]

We see that, on the one hand, Ł$_3$ is weaker than C$_2$ – it has not got some classical laws and it is not a functionally complete system. But on the other hand, unlike classical logic, Ł$_3$ is rich enough to provide the means for defining two *non-trivial* truth-functional modal operators M (possibility) and L (necessity). The following definition of *possibility* was suggested by A. Tarski (1921) (see Łukasiewicz, 1930/1970, p. 167):

$$Mp =: \sim p \to p.$$

Necessity is then defined as usual:

$$Lp =: \sim M \sim p.$$

The truth-tables for these operators look then as in Table 3.

It is worth noting that we cannot define Łukasiewicz's implication $\to$ with $\sim$, $\vee$, and $\wedge$ only,[8] as we can in classical logic, but if we add Tarski's modal operators to those connectives, then the following definition due to Słupecki (1964) does the trick:

$$p \to q =: (\sim p \vee q) \vee M(\sim p \wedge q).$$

Table 3

p	Mp	Lp
1	1	1
$1/2$	1	0
0	0	0

As remarked by Słupecki, the above interpretation of the implication complies with our intuition.

But we can define another implication in a classical way:

$$p \supset q =: \sim p \vee q \quad \text{or} \quad p \rightarrow_1 q =: \sim (Lp) \vee q.$$

Using the last definition J. Słupecki, J. Bryll and T. Prucnal (1967)[9] gave an axiomatization of Ł$_3$ with primitive connectives $\vee$, $\sim$, and L.

As a result, in three-valued logic we have many implications with completely different properties. For example, the last implication possesses all the good properties for proving the deduction theorem for Ł$_3$.[10]

1.4   Logical matrices and the consequence operation

The generalization of truth-tables led to the notion of a *logical matrix*. This is the central concept required for the construction of many-valued logics. The technical results concerning logical matrices appeared in the famous compendium (Łukasiewicz & Tarski, 1930).[11] Here for the first time the authors summed up the development of many-valued logic in the LWS. Besides J. Łukasiewicz & A. Tarski, A. Lindenbaum, B. Sobociński, and M. Wajsberg are also among the authors.

Let S be the set of all formulas formed in the usual manner by means of propositional variables $\{p, q, r, \ldots, p_1, q_1, r_1, \ldots\}$ and connectives $F_1, \ldots, F_m$. An algebra of formulas $\mathcal{L} = \langle S, F_1, \ldots, F_m \rangle$ is called a (denumerable) *sentential language*. The idea of treating the set of formulas as an abstract algebra is due to Lindenbaum. By a (finitary) inference rule over $\mathcal{L}$ we mean any pair $\langle X, \alpha \rangle$, where $X \subseteq S$ is a finite set of formulas and α is a single formula. Let R be a certain set of rules over $\mathcal{L}$ and let $X \subseteq S$ be a set closed under each rule $r \in R$. Every such pair $L = (X, R)$ is called a 'Hilbert-style' *propositional logic* (see above Wajsberg's axiomatization of Ł$_3$).

In current terminology a logical matrix $\mathfrak{M}$ for L is a pair $\mathfrak{M} = \langle \mathcal{A}, D \rangle$, where $\mathcal{A}$ is an algebra similar to $\mathcal{L}$, with universe A, and $D \subseteq A$. The set D is called the set of *designated values* of $\mathfrak{M}$. Since algebras $\mathcal{L}$ and $\mathcal{A}$ are the same type this enables us to define valuations h of $\mathcal{L}$ into $\mathfrak{M}$ as homomorphisms from $\mathcal{L}$ into $\mathfrak{M}$ (strictly speaking into $\mathcal{A}$). Let $\alpha \in S$. When $h(\alpha) \in D$ for each valuation h of $\mathcal{L}$ into $\mathcal{A}$, we say that α is tautology over $\mathfrak{M}$. By $E(\mathfrak{M})$ we denote the content of a matrix $\mathfrak{M}$, that is, the set of all tautologies over $\mathfrak{M}$.

Matrix $\mathfrak{M}$ is said to be (*weakly*) *adequate* for a system L if $E(\mathfrak{M}) = L$.[12] One of the most important and general results is the Lindenbaum theorem (see Łukasiewicz & Tarski, 1930/1970, p. 135):

For every system L there exists a denumerable matrix $\mathfrak{M}$ adequate for it.

Indeed, the matrix $\langle \mathcal{L}, X \rangle$, where $X \subseteq S$ is of the required kind. It is called a *Lindenbaum matrix*.

The problem is to find the *minimal* adequate matrix of (X, R). For example, a two-valued one is the minimal adequate matrix for $\mathbf{C_2}$. The following surprising result is due to Jaśkowski (1936). He constructed an infinite sequence of finite matrices *adequate for the intuitionistic propositional calculus* **Int**.[13] This sequence was obtained by Jaśkowski as a result of alternate application of $\tilde{A}$-operation (addition of the new greatest element) and the operation of direct product of logical matrices.[14] The sequence of matrices: $\{J_n | n = 0, 1, 2, \ldots\}$ is given by conditions: J_0 is a matrix of classical propositional logic $\mathbf{C_2}$, $J_{n+1} = \Gamma((J_n)^n)$. Then J_1 is *the first Jaśkowski matrix* which exactly coincides with three-valued Heyting-Gödel's logic $\mathbf{G_3}$.[15]

It was Tarski who first considered logical matrices as a general method of constructing logical systems (in addition to Łukasiewicz and Tarski, 1930, p. 133, see also Tarski, 1938).

The general theory of logical matrices is the contribution of Polish logicians mainly, starting with Łoś (1949).[16] In this book the proof of the above-mentioned Lindenbaum's theorem is provided. This book can serve as the foundation for a number of later developments in the study of finite models of sentential calculi (see Ulrich, 1986). It is an interesting coincidence that almost at that very time J. Kalicki published six papers on logical matrices (Kalicki, 1950a; 1950b; 1950c; 1952; 1954a; 1954b). Zygmunt (1981) writes that, from the technical point of view, Łoś and Kalicki investigated the same or similar questions. But the method used by Łoś, called by him 'the Lindenbaum method' is entirely algebraic in nature. In his turn, Kalicki preferred 'model-theoretic' techniques.

Kalicki defines the direct product (a generalization of the concept of a product used earlier by Wajsberg & Jaśkowski), the direct sum and a partial ordering of logical matrices. We distinguish only one result which is important now for studying the interrelation and classification of propositional calculi (Kalicki, 1952): 'There is an effective method to decide whether the set of tautologies of two given finite matrices are equal or not'. Note that Łoś gave criteria for equalities of two finite matrices, too.

Development of a general theory of logical matrices was continued in the fundamental paper by Łoś and Suszko (1958), where the general notion of a logic defined by a class of matrices is first given. This work marked a completely new stage in the development of Polish logic. Here we must remember Tarski's theory of the *consequence operation*, (Tarski, 1930b) and some notions of Tarski's methodology of deductive systems (Czelakowski & Malinowski, 1985). This apparatus will be useful for us in Section 10.

Nowadays an operation C defined on the power set $\mathcal{P}(S)$ of all subsets of $\mathcal{P}(S) \to \mathcal{P}(S)$ is said to be a *consequence operation* if and only if it satisfies the following conditions for every $X, Y \subseteq S$:

(T1) *Reflexivity*: $X \subseteq C(X)$,
(T2) *Idempotence*: $C(C(X)) = C(X)$,
(T3) *Monotonicity*: if $X \subseteq Y$, then $C(X) \subseteq C(Y)$.

If X is closed under a consequence operation C, that is, if $X = C(X)$, then X is called a *theory* of C. $C(X)$ is the least theory of C containing X taken as premises or hypothesis and $C(\varnothing)$ is the system of all provable or logically valid sentences of C. $\alpha \in C(X)$ reads: 'α *is a consequence of* X' or else 'α *is deducible from* X'.[17]

The following new notions were introduced in Łoś and Suszko (1958). C is *a structural consequence* if for every substitution ε of S and for every $X \subseteq S$,

$$\varepsilon C(X) \subseteq C(\varepsilon X).$$

Then $L = (\mathcal{L}, C)$ is said to be *a structural (Tarskian) logic*. It is important that in the propositional case the class of substitutions coincides with the class of all endomorphisms of S. The case of predicate logic is treated in Pogorzelski & Prucnal (1974). Łoś and Suszko also introduced the notion 'strong adequacy':

A matrix $\mathfrak{M}$ is strongly adequate for a propositional logic L iff for every $\alpha \in S$, every $X \subseteq S, \alpha \in C(X)$ iff for each homomorphism h, h$\alpha \in D$, whenever $hX \subseteq D$.

If $\mathfrak{M}$ is strongly adequate for $(\mathcal{L}, C)$, C is said to be a *matrix consequence* defined by $\mathfrak{M}$ and denoted as $Cn_{\mathfrak{M}}$. One may verify that $Cn_{\mathfrak{M}}(\varnothing) = E(\mathfrak{M})$. If instead of a single matrix $\mathfrak{M}$ we take any class K of matrices, all similar to $\mathcal{L}$, and every $\mathfrak{M} \in K$ is strongly adequate for L, then it is also said that K is a *matrix semantics* for L.

The main problem is to find a matrix representation of structural consequence in the form of some strong adequacy theorems. A general criterion on the existence of a strongly adequate matrix for a given logic L was provided in Łoś and Suszko (1958) and corrected and improved by Wójcicki (1969). In this work very important generalizations of the idea of a logical matrix were suggested. A generalized matrix (later replaced by a Lindenbaum bundle; see below) has many designated sets and its consequence operation is defined as the intersection of all matrix consequences with a single designated sets from these appearing in the signature. This matrix is a pair $\mathfrak{M} = \langle \mathbf{A}, D \rangle$, where $\mathbf{A}$ is an algebra similar to $\mathcal{L}$ and $D \subseteq \mathcal{P}(\mathbf{A})$. Each generalized matrix $\mathfrak{M} = \langle \mathbf{A}, \{D_i : i \in I\} \rangle$ is treated as semantically equivalent to the set $K = \{\langle \mathbf{A}, D_i \rangle : i \in 1\}$ of logical matrices in the usual sense. As noted in Font (2003, p. 62): 'Most model-theoretic or universal-algebraic ideas and constructions on matrices can be reproduced or adjusted for generalized matrices'.[18]

The great value of this generalization is to expand the notion of 'strong adequacy'. Wójcicki (1970) proved the adequacy (completeness) theorem for structural logics (generalization of the above-mentioned Lindenbaum theorem):

For every structural Tarskian logic $\langle S, C \rangle$ there is a generalized matrix adequate for it.

Note that $\mathbb{L}_C = \{\langle \mathcal{L}, C(X) \rangle : X \subseteq S\}$ is called a *Lindenbaum bundle* (of matrices). Then every logic is complete relative to the class of all its Lindenbaum matrices. In this sense every logic has a matrix semantics. So what this approach says about the benefits of matrix semantics is universal. Of course it uses many of the techniques and results offered by universal algebra and model theory for the semantic study of propositional logics. Font, Jansana and Pigozzi (2003, p. 18), in a beautiful survey, stress that the formal concept of logic (propositional language as algebra of formulas together with structural consequence operation) 'has been extensively studied since Tarski's work mainly by logicians in Poland [...][19] and has become one of the standard frameworks of contemporary AAL', where the abbreviation 'AAL' stands for 'Abstract Algebraic Logic', which is defined as 'the general theory of the algebraization of logical systems'.

Specific versions of the adequacy problem were worked on by many authors. Almost all of them came out of R. Wójcicki's group (J. Czelakowski, W. Dziobiak, G. Malinowski, J. Malinowski, M. Tokarz and J. Zygmunt),[20] their work resulted in numerous papers with technically difficult theorems. In paper (1973a) Wójcicki develops the matrix approach in the methodology of propositional logic, where the main problem is that of matrix adequacy when $\mathfrak{M}$ fulfils some special conditions. The following books are very important: Czelakowski (1980; 1992; 2001), Dziobiak (1980; 1981), Wójcicki (1984; 1988) and Zygmunt (1984). And, of course, see the books Pogorzelski and Wojtylak (1982; 2008).

Returning to the question of the equality of logical matrices the following result of Zygmunt (1983) should be noted: there exists an effective procedure for comparing $Cn_{\mathfrak{M}}$ and $Cn_{\mathfrak{N}}$ for any finite generalized matrices $\mathfrak{M}$ and $\mathfrak{N}$.

1.5   Łukasiewicz n-valued matrix logic

The generalization of his three-valued logic led Łukasiewicz, in 1922, to consider n-valued logic (see Łukasiewicz, 1923; Łukasiewicz & Tarski, 1930).

A logical matrix of the form

$$\mathfrak{M}_n = \langle V_n, \sim, \rightarrow, \{1\} \rangle$$

is called a *Łukasiewicz n-valued matrix* ($n \in N, n \geq 2$) provided that

$$V_n = \{0, {}^1\!/_{n-1}, \ldots, {}^{n-2}\!/_{n-1}, 1\};$$

$\sim$ (negation) is a unary operation, and $\rightarrow$ (implication) is a binary operation; these are defined on V_n as follows:

$$\sim x = 1 - x,$$

$$x \rightarrow y = min(1, 1 - x + y);$$

$\{1\}$ is the set of designated elements of $\mathfrak{M}_n$.

Operations $\vee$ (disjunction), $\wedge$ (conjunction), and $\equiv$ (equivalence) are defined through the above-mentioned functions as follows:

$$x \vee y = (x \rightarrow y) \rightarrow y = max(x, y)$$

$$x \wedge y = \sim (\sim x \vee \sim y) = min(x, y)$$

$$x \leftrightarrow y = (x \rightarrow y) \wedge (y \rightarrow x).$$

An algebra of formulas of the propositional language of Łukasiewicz's many-valued logics is an algebra freely generated on a denumerable set of propositional variables by connectives $\sim$ (negation) and $\rightarrow$ (implication). As usual we denote logical connectives and matrix operation by the same symbols.

Łukasiewicz n-valued matrix logic $Ł_n$ is the set of all tautologies of the matrix $\mathfrak{M}_n$, that is, the set of all such formulas α that $h(\alpha) = 1$ for each valuation h of $\mathcal{L}$ into $\mathfrak{M}_n$.

The problem of interrelations among the finite-valued systems $Ł_n$ was settled by Lindenbaum (see Łukasiewicz & Tarski, 1930/1970, p. 142) in the following way (relative to the classes of tautologies):

$Ł_m \subseteq Ł_n$ *if and only if* $n-1$ *is a divisor of* $m-1$.

1.5.1   Ordinal and Cardinal degrees of completeness of $Ł_n$

In 1930 in order to obtain a classification and characterization of incomplete sets of sentences, Tarski (1930a) introduced the concept of degree of completeness of a set of sentences. Subsequently he distinguished between ordinal and cardinal degrees of completeness (see Tarski 1930b).

We say that a set of axioms is absolutely consistent if there is some formula which is not a consequence of the axioms. Then *the ordinal degree of completeness* of a set L of axioms, in symbols $\gamma(L)$, is the smallest ordinal $\pi \neq 0$ such that there is no increasing sequence of type π of absolutely consistent nonequivalent sets of axioms which begins with L. According to Łukasiewicz and Tarski (1930), Lindenbaum proved that the ordinal degree of completeness of $Ł_3$ is 3, and Tarski showed that this takes place for every $n-1$ which is a prime number in $Ł_n$. Then in May 1930 'the problem of the degree of completeness was solved for systems $Ł_n$ with an arbitrary natural n; this was the joint result of members of a proseminar conducted by Łukasiewicz and Tarski in the University of Warsaw' (see the footnote on p. 142 in (Łukasiewicz & Tarski, 1930/1970)). Rose (1952) provides a proof of this result, namely that for any $n > 2$ the ordinal degree of completeness of $Ł_n$ is $d(n-1)+1$ where $d(n)$ is the number of distinct divisors of n including 1 and n. A different proof (and much shorter as well) may be found in Tokarz (1974a).

The cardinal degree of completeness of L, in symbols $g(L)$, is the number of logics containing the theorems of L.

Tarski proved that

$$\gamma(Ł_n) = g(Ł_n) = 3 \text{ for } n-1 \text{ being a prime number.}$$

Only in Tokarz (1974a) was a solution to the problem of the cardinal degree of completeness of $Ł_n$ for all finite n published. A shorter proof may be found in Tokarz (1977).

Let $C = \langle a_1, \ldots, a_n \rangle$ be an arbitrary sequence of natural numbers. We denote by $N_C(a_i)$ $(1 \leq i \leq n)$ the number of subsequences D of C which satisfy the following condition:

$a_i \in D$ and for every $b \in D$, $a_i \geq b$,
if $j \neq k$ and $a_j, a_k \in D$, then $a_j - 1$ is not a divisor of $a_k - 1$.

Let $c(n) = \langle a_1, \ldots, a_k \rangle$ be the sequence that has the following properties:

(i) $a_1 = n$,
(ii) $a_1 > \cdots > a_k > 1$,
(iii) for every i, $1 \leq i \leq k$, $a_i - 1$ is a divisor of $n - 1$.

Then for finite n,

$$g(Ł_n) = \left(\sum_{a_i \in c(n)} N_{c(n)}(a_i) \right) + 1.$$

The interesting development of these topics (including infinite-valued Łukasiewicz logic) is in Beavers (1993).

In 2000, M. N. Rybakov (Tver State University, Russia) wrote a computer program for calculating $\gamma(Ł_n)$. Table 1, which appears in Karpenko (2006) and is of great interest, contains the values of $\gamma(Ł_n)$ for $n \leq 1000$. Apparently, some natural numbers are not values of $\gamma(Ł_n)$ for any n. So, for the first ten thousand n, the values of $\gamma(Ł_n)$ contain the following natural numbers from the first one hundred:

2, 3, 4, 5, 6, 7, 8, 9, 10, 11, 12, 14, 15, 20, 21
28, 35, 36, 45, 50, 55, 56, 66, 70, 78, 84, 91.

1.5.2   Axiomatization of $Ł_n$

It is not straightforward to find a finite axiomatization of the set of tautologies of $Ł_n$ for arbitrary n. Although the axiom system given by Wajsberg for $Ł_3$ was simple, there was no hint of how to extend this method to the other systems. It is claimed that Wajsberg found an axiomatization of $Ł_n$ for all n such that $n - 1$ is a prime number and that Lindenbaum generalized it onto all natural numbers (see Łukasiewicz & Tarski, 1930/1970, p. 142), however this result has never been published.

Later, a general theorem on finite axiomatizability of some class of finite-valued logics, including all Ł$_n$, was given by Wajsberg (1935). But this method has no practical value. Another approach was offered for a wide class of finite-valued logics by Rosser and Turquette (1952). Here for the first time, it was pointed out that J_i-operators (see below Section 8) must be definable from the initial connectives. This is the case for Ł$_n$, and also Ł$_n$ fulfils other conditions. So axiomatization of Ł$_n$ can be obtained, but the axiomatization becomes very complicated.

The problem of axiomatization directly for Ł$_n$ was solved by M. Tokarz (1974b) but the resulting formulas in his axiomatization are very long and the completeness for finite-valued logics Ł$_n$ is derived from the completeness of the infinite-valued Łukasiewicz logic Ł$_\infty$ (see below). Surma (1974a; 1974b) presented a simple Gentzen-style characterization of every finite logic. As an illustration he gave a rather detailed Gentzen-style characterization of Ł$_n$. A very simple way to use axiomatization of Ł$_n$ was suggested by Tuziak (1988).[21]

The comparison between operations $C(X)$ and $Cn_{\mathfrak{M}}$ determined by Łukasiewicz calculi was made in Wójcicki (1973b); (see also Malinowski and Zygmunt (1978)). The paper (Wójcicki, 1977) gathers the results concerning strongly finite propositional calculi (including Ł$_n$). A logic $(\mathcal{L}, C)$ is strongly finite if and only if there exists a finite set of finite matrices strongly adequate for $(\mathcal{L}, C)$. Here we find a criterion of strong finiteness.[22] Also Wójcicki (1973a) proved that the degree of completeness of such logics (the cardinality of all logical, that is, closed under substitutions, theories of C) is always finite. About axiomatization of strongly finite logics see (Wojtylak, 1979).

In connection with the axiomatization of finite-valued logic the following question arises: is there a nonfinitely axiomatizable finite matrix? For the history of the subject in detail see (Pałasińska, 1994). Here we note only that Wroński (1979) considered an algebraic three-valued matrix (that is, a matrix with one designated element) and showed that the consequence operation of this matrix is not finitely based. Wojtylak (1984) constructed a 5-element matrix with two designated elements that is not finitely axiomatizable. Dziobiak (1991) presented a 4-element nonfinitely axiomatizable algebraic matrix. Finally, Pałasińska (1994) found two three-valued algebraic matrices with the same property. Since every 2-element matrix can be finitely axiomatized (see Rautenberg, 1981), the matrices presented here in some sense are the best possible. One of them is as follows:

$$\mathfrak{M} = \langle \{0, 1, 2\}, \otimes, \{2\} \rangle,$$

where $x \otimes y$ takes 2 except in the case when $2 \otimes 0 = 1$.

Thus, we have an essential difference between two-valued logic and three-valued logic. But the real difference between two-valuedness and three-valuedness lies in their functional properties (see Section 10).

1.6 Unusual functional properties of $Ł_n$

Neither the axiomatic nor the algebraic (nor, for that matter, any other semantic) approach can bring out the uniqueness and peculiarity of Łukasiewicz finite-valued logics $Ł_n$. Only the consideration of $Ł_n$ as a functional system can help us to decipher the essence of $Ł_n$-s. It was exactly this approach that allowed the discovery that the functional properties of $Ł_n$ are extremely unusual.

A function $f(x_1, \ldots, x_s)$ with a finite number of arguments is called an n-valued function, or a function of n-valued logic, if f is a map from the power-set V_n^s into V_n, where $V_n = \{0, 1, 2, \ldots, n-1\}$. Let P_n be the set of all n-valued functions defined on the set V_n.[23] Then, a pair (P_n, C), where C is the operation of *superposition* of functions, is a functional system. Roughly speaking, the result of superposition of functions $f_1, \ldots, f_k$ is the function obtained from $f_1, \ldots, f_k$ either (1) by substituting some of these functions for arguments of $f_1, \ldots, f_k$ or (2) by renaming arguments of $f_1, \ldots, f_k$ or by both (1) and (2).

We will employ the following terminology, introduced by A. V. Kuznetsov (see Janowskaja, 1959). Let $F \subseteq P_n$. The operation of superposition leads to a closure operator [ ] on the power-set of P_n; intuitively, $[F]$ is the set of all superpositions of functions from F. It is worth noting that the properties of a closure operator [ ] are the same as Tarski's consequence operation Cn (see above the three conditions for Cn in Section 4). A set F of functions is said to be closed if $F = [F]$. A set F of functions is functionally complete in P_n, if $[F] = P_n$. A set F is called *precomplete* in P_n, if $[F] \neq P_n$, and $[F \cup \{f\}] = P_n$, where $f \in P_n$, and $f \notin F$ (in other terminology, a precomplete class of functions is called *maximal clone*).

The first investigations of the functional properties of finite-valued logics were due to Słupecki (1939). He obtained the following fundamental result. A function $f \in P_n$ is called a *Słupecki function*, if f depends on at least two variables essentially and takes n different values. Then *F is functionally complete if and only F contains all unary functions and a Słupecki function.*[24]

Returning to the three-valued Słupecki's T-function[25] (see Section 2 above) let's note that in order to make $Ł_3$ functionally complete it is sufficient to add any function from P_3 that is not definable in $Ł_3$. This means that the set of functions of $Ł_3$ is functionally *precomplete* in P_3 (see Finn, 1969). Therefore, there raises the question of the criterion of functional precompleteness for the set of functions $Ł_n$ for arbitrary n. The solution was given by Finn (1970) (see in detail Bočvar and Finn, 1972): 'The set of functions of the logic $Ł_n$ is functionally precomplete in P_n if and only if $n-1$ is a prime number'.[26]

So, we have a direct connection between Tarski's theorem on cardinal degrees of completeness of $Ł_n$ (see above) and Finn's theorem on functional precompleteness of $Ł_n$:

If $Ł_n$ is functionally precomplete in P_n, where $n \geq 3$, then $g(Ł_n) = 3$.

Finn's theorem led to the discovery of some interesting properties of prime numbers. There is an algorithm which for every prime number n constructs a rooted tree where nodes are natural numbers and n is a root. Thus, each prime number is given a structure, which proves to be an algebraic structure of p-abelian groups. Finite-valued logics K_n are specified that have tautologies if and only if $n-1$ is a prime number. It is discovered that K_n have the same functional properties as $Ł_n$ whenever $n-1$ is a prime number. Thus, K_n is the 'logic' of prime numbers. Amazingly, the combination of different logics of prime numbers led to uncovering *the law of generation of classes of prime numbers*. All prime numbers can be generated in such a way. Along with the characterization of prime numbers there is characterization, in terms of Łukasiewicz logical matrices, of the powers of primes, odd numbers, and even numbers. All these results, as well as other related ones, can be found in Karpenko (2006).

1.7   Infinite-valued logic $Ł_\infty$

The first infinite-valued logic was introduced by Łukasiewicz (1929, §6), but the main results were published in Łukasiewicz & Tarski (1930). Infinite-valued matrix M_∞ is defined – just like the finite ones – (since the symbol n does not enter into the definition of matrix operations $\sim$ and $\rightarrow$) on the rational numbers in the range 0 to 1, or on the real interval $[0, 1]$. Matrix logic $Ł_\infty$ is the set of all tautologies of the matrix $\mathfrak{M}_\infty = ([0, 1], \sim, \rightarrow, \{1\})$. Lindenbaum showed that the sets of tautologies of the countable-valued logic and the continual-valued logic are the same. The following fact was established by Łukasiewicz:

$$Ł_\infty = \bigcap_{n=2}^{\infty} Ł_n.$$

It was Łukasiewicz's hypothesis that $Ł_\infty$ is axiomatizable with the rules of substitution and modus ponens and the following axioms (Łukasiewicz & Tarski, 1930/1970, p. 143):

Ł1 $(p \to q) \to ((q \to r) \to (p \to r))$,

Ł2 $p \to (q \to p)$,

Ł3 $((p \to q) \to q) \to ((q \to p) \to p)$,

Ł4 $(\sim p \to \sim q) \to (q \to p)$,

Ł5 $((p \to q) \to (q \to p)) \to (q \to p)$.[27]

A proof of this conjecture was announced by Wajsberg (1935/1977, p. 105), but never published. The proof of the completeness result was only published in Rose and Rosser (1958) and in Chang (1959).

The infinite-valued Łukasiewicz logic $Ł_\infty$ is among the most important and widely studied of all non-classical logics. According to the theorem of R. McNaughton (1951), it is the logic characterized by 'piecewise continuous' linear functions on [0,1]. This property is very important for proving many fundamental metatheorems (for example, in Rose & Rosser, 1958) related to $Ł_\infty$ (and of course to $Ł_n$)[28]. Also, along with Gödel logic and Product logic, it is viewed as one of the fundamental 't-norm based' fuzzy logics, where t-norm in $Ł_\infty$ is the strong conjunction $x \otimes y = max(0, x + y - 1)$, and Łukasiewicz's implication $\to$ is the residual implication (see Hájek, 1988 for details).

But the best thing is that $Ł_\infty$ is the logic characterized by the class of MV-algebras. In order to give a purely algebraic proof of completeness of $Ł_\infty$ with respect to semantics of the interval [0,1], Chang introduced MV-algebras, which play a huge role in modern algebra-logical research and which may be defined in the following way:

An algebraic structure $\mathcal{A} = (A, \otimes, \neg, 0)$ with a binary operation $\oplus$, a unary operation $\neg$ and constant 0 is MV-algebra iff $(A, \otimes, 0)$ is abelian monoid, and if furthermore for all $x, y \in A$ the following is true:

(i) $\neg\neg x = x$,
(ii) $x \oplus \neg 0 = \neg 0$,
(iii) $\neg(\neg x \oplus y) \oplus y = \neg(\neg y \oplus x) \oplus x$.

The simplest example of an MV-algebra is an arbitrary (finite or not) Łukasiewicz matrix, where $x \oplus y = \sim x \to y = min(1, x + 1)$ and $\neg x = 1 - x$. It was proven by Chang that Boolean algebras coincide with MV-algebras satisfying the additional equation $x \oplus x = x$ (idempotency). Note that

MV-algebras characterize $Ł_\infty$ in a manner analogous to the way that Boolean algebra characterizes C_2.

It soon became clear that this class of structures is very important in mathematics. Since 1986 D. Mundici, in a series of works, has discovered a relation between MV-algebras and other mathematical structures, for example to abelian *l*-groups with strong unit, to approximately finite-dimensional (AF) C^*-algebras (AF C^*-algebras are the standard tool to give a mathematical description of spin systems in quantum mechanics), to De Finetti's subjective probability theory. He also discovered that deduction in $Ł_\infty$ is related to desingularizations of toric varieties. These examples could easily be multiplied.

All basic results concerning in $Ł_\infty$ and MV-algebra are given in Cignoli, D'Ottaviano & Mundici (2000), but especially see the book Mundici (2011). In the 'Introduction' to this book we read:

> The book embodies the viewpoint that modern Łukasiewicz logic and *MV*-algebras provide a benchmark for the study of several deep mathematical problems, such as Rényi conditionals of continuously valued events, the many-valued generalization of Carathéodory algebraic probability theory, morphisms and invariant measures of rational polyhedra, bases and Schauder bases as jointly refinable partitions of unity, and first-order logic with [0,1]-valued identity on Hilbert space.

1.8 Łukasiewicz's rejection of his many-valued logic

Łukasiewicz was not to know of the amazing developments of the logics created by him. It is worth mentioning that the problem of the intuitive interpretation of $Ł_3$ led to a strong objection to this very logic, and $Ł_3$ has been much criticized for failing to observe the law of non-contradiction. In this regard $Ł_3$ was seriously criticized at the conference in Zurich in 1938 by F. Gonseth in discussing Łukasiewicz's report. Gonseth's objection[29] is considered to be an irrefutable argument against any informal interpretation of $Ł_3$. He observed that the proposition 'In a year I shall be in Warsaw and I shall not be in Warsaw' has the truth-value $1/2$ according to Łukasiewicz since the proposition "In a year I shall be in Warsaw" has the truth-value $1/2$ in Łukasiewicz's interpretation and the conjunction and negation operations do not change this value. However, as Gonseth notes, it is quite reasonable that such a conjunctive proposition must be false now. Among many other works we shall single out only the review of Rasiowa's paper (Rasiowa, 1950) on Łukasiewicz many-valued logic by Mostowski (1950).[30] He notes that the author ignores Gonseth's quite important remark though she gives some examples corresponding to the interpretation of '$1/2$' as 'possibility'. Mostowski states that 'this remark destroys any hope that it will ever be possible to find a reasonable interpretation of the three-valued logic of Łukasiewicz in terms of everyday language'. Moreover, in the early 1950s, Łukasiewicz discovered a nasty thing, the fact that the formula

$$p \to (p \to Lp)$$

is a tautology in his modal three-valued logic (see Section 2 above).

Without replying to any of his critics, Łukasiewicz rejected $Ł_3$ and consequently all its generalizations:

> When I had discovered in 1920 a three-valued system of logic, I called the third value, which I denoted by $1/2$, 'possibility'. Later on, after having found my n-valued modal systems, I thought that only two of them may be of philosophical importance, viz., the 3-valued and the $\aleph_0$-valued system.[31] [...] This opinion, as I see it today, was wrong.
>
> (Łukasiewicz, 1953/1970, pp. 370–371)

But the main thing is the following:

> I am of the opinion that in any modal logic the classical calculus of propositions should be preserved. This calculus has hitherto manifested solidity and usefulness, and should not be set aside without weighty reasons.
>
> (see Łukasiewicz, 1957, p. 167)

In 1953 Łukasiewicz constructed four-valued Ł-modal logic (it is denoted by Ł) in the base of which lay the direct product of the two-element matrix for C_2 (classical propositional calculus) with itself.[32] From this fact it follows that modality-free formulas are tautologies of Ł if and only if they are tautologies of C_2. So Ł is an expansion of C_2. The irony is that $Ł_3$ and all Łukasiewicz finite-valued logics $Ł_n$ are expansions of C_2.

In a series of works, beginning from the 1980s, Anshakov and Rychkov (1985) suggested a general, effective method for constructing Hilbert-type first-order finite-valued calculi extending the C_2, and for the limited version to propositional calculi only see Anshakov and Rychkov (1995). It is applicable to logics with arbitrary sets of designated values. But matrix logic should have the following properties:

(I) We say that a logic L_n is *truth-complete* iff all J_i-operators ($i \in V_n$), are functionally definable in the signature σ (σ consists of operations on V_n), where

$$J_i(x) = \begin{cases} 1, & \text{if } x = i \\ 0, & \text{if } x \neq i. \end{cases}$$

Note that not every finite-valued logic enjoys this property. However, J_i-operators are definable through $\sim$ and $\rightarrow$ in Łukasiewicz logic $Ł_n$. For the proof, see Rosser and Turquette (1952, pp. 18–22).[33]

(II) A logic L_n is said to be *C-extending* iff in L_n one can functionally express the binary operations $\supset, \vee, \wedge$ and the unary operation $\neg$, whose restrictions to the subset $\{0, 1\}$ of V_n coincide with the classical logical operations of implication, disjunction, conjunction, and negation. Note that L_n coincides with classical logic C_2 over the set $\{0, 1\}$. Thus, propositional Łukasiewicz logic $Ł_n$ is truth-complete C-extending logic. Note that in D'Ottaviano and Epstein (1988) an axiomatization of $Ł_3$ (with two designated truth values) is given as an expansion of C_2.

It should be said that only a man of genius like Łukasiewicz could reject his main scientific achievement.

1.9 Other resources

Of course it is impossible to review completely the contribution of the Polish logicians to the development of many-valued logic. But it is necessary to point out the following works.

First of all, there are A. Mostowski's investigations in the field of many-valued predicate logics, see Hájek (2006). Mostowski's paper (1948) is one of the pioneering papers on many-valued predicate logic. Here he proves some formulas to be unprovable in the intuitionistic logic using a many-valued predicate calculus. In another paper Mostowski (1961a) studies generalized quantifiers in many-valued logic, that is, $\forall$ is an infinitary generalization of the $\wedge$ conjunction, and $\exists$ is an infinitary generalization of the disjunction $\vee$. There is a section on finite-valued logics and a section on the continuous set of truth-values (for example, $[0,1]$). Mostowski obtained a very general result to the effect that the set of tautologies in the continuous matrix with the designated set D which is an open interval in $[0,1]$ (including also $[0, a]$ and $(a, 1)$) is axiomatizable (recursively enumerable). This general approach does not provide a corresponding logical calculus. But Mostowski (1961b) has the priority in giving a non-axiomatizable many-valued predicate logic. Note that in 1962 B. Scarpellini proved that tautologies of $\forall Ł_\infty$ are not axiomatizable. Also we must mention the paper Mostowski (1963) about generalizations to the many-valued case of two fundamental theorems concerning the classical predicate calculus, and Mostowski (1969), where, in Lecture V, he considers Boolean-valued models for the set theory as many-valued logics.

In connection with the last remark let us note the words of H. Rasiowa: 'We regard elements of any algebra associated with [a formalized language] L as truth values, in particular the element 1 as the value truth'. (1974a, p. 168) In this, the best-known of her books, Rasiowa develops in detail a general algebraization theory for propositional logics including the connection between Post algebras of order n and n-valued propositional logic P_n in detail.[34] One of the main results there is the completeness theorem for P_n w.r.t. the class of all Post algebras of order n. Of particular note is the nice work Rasiowa (1974b), where Post's algebras were taken as a semantic foundation of n-valued logic. Orlowska (1985) (a pupil of Rasiowa) develops a resolution style proof system for P_n.

An interesting problem which proved to be important was to find possible infinitely many-valued generalizations of these logics, or of these Post algebras. Gottwald (2001, p. 323) emphasizes that the most influential paper in this field was the paper of Rasiowa (1973) in which she introduced Post algebras of order[35] ω^+ and the corresponding systems of infinitely many-valued (first order) logic.

Another infinitely many-valued generalization of Post logics P_n is considered in Epstein and Rasiowa (1990; 1991) – Post algebras of order[36] $\omega + \omega^+$. An axiomatization of such logics requires infinitely many rules and one of them is (unfortunately) an infinitary one which needs infinitely many premises.[37]

Rasiowa relates algebraic methods of non-classical logics to applications in the foundations of computer science, in particular she investigates applications of various generalizations of Post algebras to the logics of programs and approximation logics (more than 30 papers and two lecture notes).[38] We refer only to Rasiowa (1977).

Now we should mention a few books. First, the very interesting volume devoted to Łukasiewicz propositional calculi which includes papers by G. Malinowski, M. Spasowski, W. Suchoń, S. Surma, M. Tokarz, R. Wójcicki (see Wójcicki & Malinowski, 1977).[39]

The book of Malinowski (1993) is an excellent introduction to many-valued logics, see also Malinowski (2001; 2006; 2007). There are other books by Polish authors on many-valued logics, (see Bolc & Borowik, 1992; 2000). A survey of works and of his own results in the field of nonsense logic may be found in the book (Piróg-Rzepecka, 1977).

Finally, I want to stress that the development of many-valued logic in the world is impossible to understand without comprehension of the phenomenon of the emergence of many-valued logic in Poland (see Rescher, 1969/1993; Gottwald, 2001; Karpenko, 2010).

But there is one question that remains unsolved and it is the central problem in modern many-valued logic.

1.10 Suszko's thesis and its functional refutation

To everyone's surprise, during a talk delivered at the 22nd Conference on the History of Logic held in 1976 (Cracow) one of the most prominent Polish logicians, Roman Suszko made the shocking claim that many-valued logics did not really exist and they could be reduced to logics with only two logical values. In his own words: 'After 50 years we still face an illogical paradise of many truths and falsehoods.' (1977, p. 379)

The philosophical standpoint according to which "every logic is (logically) two-valued" (Suszko, 1977, p. 378) is nowadays given the label of *Suszko's thesis* (cf. Malinowski, 1993, ch. 10.1; Da Costa *et al.*, 1996; Tsuji, 1998; Wansing & Shramko, 2008; Woleński, 2009).

Suszko's thesis states that every structural Tarskian logic (see Section 4 above) is two-valued.[40] To illustrate this Suszko (1975) shows how three-valued Łukasiewicz logic $Ł_3$ could be given a two-valued (non-truth functional) semantics.[41] Suszko does not consider the elements of the Łukasiewicz matrix, $0, 1/2, 1$, as *logical* values, but as *algebraic* values. For him, an algebraic valuation h is a homomorphism from an absolutely free algebra of formulas into an algebra of similar type, and a logical valuation is an arbitrary function from a set of formulas into a two-element set $\{0,1\}$. The very simple idea is to associate with the set of designated values D its characteristic function $\mu_D : A \rightarrow \{0,1\}$ defined by: $\mu_D(x) = 1$ if $x \in D$, and $\mu_D(x) = 0$ otherwise. So Suszko's thesis shifts logical values over the set of algebraic (matrix) values and it refers to the division of the matrix universe into two subsets of elements: designated and others using their characteristic functions as logical valuations V. For every propositional logic $L = \langle S, C \rangle$ one can select a set V of function mapping sentences of S into the set $\{0,1\}$of two truth-values with the following adequacy property:

$\alpha \in C(X)$ *iff for every* $h \in V$, *if* $h(X) \subseteq \{1\}$ *then* $h(\alpha) = 1$.

Thus every propositional logic can be regarded as two-valued (see Stachniak, 1988, p. 171). Malinowski (1993, pp. 72–73) also showed that

Suszko's reduction of many-valued logic to a two-valued one is general in nature and supported Wójcicki's adequacy theorem (see Section 4 above).

Deep discussion of Suszko's thesis and the construction of counter-examples to it are to be found in the paper by Malinowski (1990). He generalizes Tarski's concept of consequence operation (rejecting T1 and weakening T2), and introduces a *quasi*-consequence (*q*-consequence) relation. The fruitful idea of Malinowski is the *bi*-partition of the algebraic values into a set D of *designated values* and a set D^* of *antidesignated values*.[42] Then D and D^* are disjoint subsets of A such that $D \cap D^* = \varnothing$. This distinction leaves room for values that are *neither* designated *nor* antidesignated. To obtain a kind of consequence relation that does not admit Suszko's reduction method, Malinowski defines the *q*-consequence relation as depending on both sets D and D^*. The *q*-consequence coincides with standard concept of a matrix consequence only if $D \cup D^* = A$, that is, when D and D^* are complementary. Malinowski constructs a three-valued logic (with the help of the *q*-matrix) for which Suszko semantics could not be employed. So his counter-example is based upon the trivalent division of *q*-matrices and says that any structural *q*-consequence may have a bivalent or a three-valued semantics.

Moreover, Malinowski made uneasy steps towards logical many-valuedness for the case $n > 3$. He writes:

> It seems that the situation is somewhat like with geometrical intuition: man can easily accept up to three dimensions only. Even the fourth dimension is painful.
>
> (Malinowski, 2009, p. 206, fn. 1)

The first step in that direction is easy and it consists of a division of the matrix universe into more than three subsets. Malinowski introduces the notion of *s-dimensional q-matrix* and points out their kinship with 'generalized matrices' (see Section 4 above). However, the next step, that is, construction of an appropriate matrix $q(s)$-consequence, 'seems extremely difficult'.

Discussion of Suszko's method (semantics) has spread around the world and attracts more and more new authors (see especially Caleiro *et al.*, 2005; Font, 2009). In Caleiro *et al.* (2005) the authors obtain an effective method for constructing the two-valued semantics of any logic that has a truth-functional finite-valued semantics and a sufficiently expressive language. In the important paper of Font Suszko semantics is criticized very strongly: 'it is very difficult to accept Suszko semantics as a real semantics'[43] and it is suggested that attention should be paid to the notion of 'degrees of truth'. (Font, 2009, p. 387)

We must note that Suszko semantics belongs to a broad class of *bivalent semantics* which began to spread in the early 1970s (see some surveys in connection with Suszko's thesis in Da Costa *et al.*, 1996). But the main problem of many-valued logic pointed out in this paper is the following (Da Costa *et al.*, 1996, p. 281): "Undoubtedly, a fundamental problem concerning many-valuedness is to know what it really is." Jordan (1945, pp. 393–394) warned about this problem. Similarly, Dana Scott writes:

> Before you accept many-valued logic as a long-lost brother, try to think what these fractional truth-values could possibly mean. And do they have any use? What is the conceptual justification of intermediate values? (Scott, 1976, p. 66)

Note that before this Scott also suggested one of the variants of bivalent semantics (see Scott, 1974).

The great sensation provoked by Suszko's thesis and his discussion of many-valued logic (maybe the sensation was deliberate; it was the last work of Suszko) turned attention to the main question: *what are truth-values?* This question (problem) is crucial in the foundation of many-valued logics. Woleński (2009, p. 107) writes: 'almost every monograph or survey dealing with many-valued logic points out at this problem'. Recently this subject, due to Suszko, has been discussed very intensively. The international Polish journal *Studia Logica* (one of the most famous logical journals in the world) devoted two special issues to this theme, that is, 'Truth Values': Part I (2009, Vol. 91, N 3) and Part II (2009, Vol. 92, N 2). Also under the same title see the very important work (Shramko & Wansing, 2010).[44] The main value of this discussion is that the participants were dealing with the foundations of many-valued logic critically. Here it is worth recalling Łukasiewicz's definition of logic (Łukasiewicz, 1921/1970, p. 90): 'Logic is the science of objects of a specific kind, namely a science of *logical values*.'

At last we come to the fundamental difference between many-valued logic and two-valued logic and the explanation of the fact that Suszko's reduction (and others too) do not make sense.[45]

1. *Suszko's reduction*: any finite-valued logic is reduced to two-valued logic.
2. *Post's theorem*: the set of closed classes of P_2 (or all possible clones on $\{0,1\}$; see Section 6 above) is *countable*, where P_2 is the set of all Boolean functions defined on $\{0,1\}$ (see (Post, 1920; 1941)).

3. *Mučnik's theorem*: for every $n \geq 3$, P_n has a *continuum* of closed classes (see Janov & Mučnik, 1959).[46]

4. From (2) and (3) follows the *contradiction* with (1), that is, *even P_3 cannot be reduced to P_2*.

Note that Mučnik's theorem holds only for a functionally complete finite-valued logics starting with Post's three-valued logic P_3. So, it is interesting to ask what is the cardinality of the set of closed classes of other three-valued logics. First and foremost this question refers to $Ł_3$, having in view Suszko's bivalent semantics for it (see Suszko, 1975).

Let G_3 be the set of functions of Heyting' three-valued logic G_3 (the first matrix of Jaśkowski (1936), see Section 4 above). It was proven in Ratsa (1982)[47] that G_3 has a continuum of closed classes of functions. It is easy to check that G_3 is functionally enclosed in $Ł_3$. Then $Ł_3$ also has a continuum of closed classes of functions. The difficult problem is to find a criterion for three-valued logics which have a continuum of closed classes. For example, this problem is open for Bočvar's three-valued nonsense logic B_3 (see Karpenko, 2010a).

For over a century the development of many-valued logic has been strongly influenced by the ideas of the Polish school of logic. There are, without any doubt, great insights and discussions to come.

Notes

1. About the development of logic in Poland see also Woleński (1995; 2003a; 2004b) and Wójcicki and Zygmunt (2003).
2. See Kotarbiński (1913).
3. This paper is a revised version of the address that Łukasiewicz delivered as a Rector at the start of the academic year 1922–3 at Warsaw University. Later on Łukasiewicz revised the address, giving it the form of a paper without changing the main claims and arguments. It was published for the first time in Polish in 1961. The first version in English was published in McCall (1967, pp. 19–39).
4. There is considerable discussion around this bewildering topic and about the attempts at intuitive interpretations of Jan Łukasiewicz's three-valued logic. Here we refer only to the Polish authors: see Kłósak (1948), Jordan (1963), Słupecki (1964), Borkowski (1977), Trzęsicki (1993), Lechniak (2002), Woleński (2003b) and Łukasiewicz (2011). On the other hand, Zawirski (1934a; 1934b), starting from the idea of indeterminism in quantum physics came to see interrelations between many-valued logics and probability theory.
5. So $Ł_3$ is a '*resource conscious*' logic; it is historically the first *logic without contraction*. See Ono and Komori (1985) in which Łukasiewicz's many-valued logics are discussed from this point of view.

6. Ten years later Słupecki (1946) investigated another three-valued system with the same properties (full and Post-completeness).

7. A set of sentences is called *independent* if it is not equivalent to any of its proper subsets (Tarski, 1930b). Independence is proven by constructing a suitable model (in which each axiom could not be derived from the others). Apparently P. Bernays (1918) in his dissertation, partly published in 1926, first used many-valued truth-tables for independence proofs. But Łukasiewicz (1941/1970, p. 284) writes: 'The same method was also known to me even before it was published by Professor Bernays.' (See also Wajsberg, 1936)

8. The logic that has exactly these connectives as its initial connectives is *Kleene's three-valued logic* (see Kleene, 1952, §64).

9. This paper contains an English translation of Słupecki (1964).

10. About the problem of deduction theorem for Łukasiewicz many-valued logics see (Pogorzelski, 1964; Wójcicki, 1973b).

11. The first publication of this paper in English is in Tarski (1956, pp. 39–59).

12. Usually such a matrix is called *characteristic* for L. Wajsberg axiomatization of Ł$_3$ is of this type.

13. A detailed proof following closely Jaśkowski's strategy is presented in Surma (1971). See also Surma, Wroński and Zachorowski (1975). The following interesting result is due to Wroński (1974): it is proved that no denumerable matrix is strongly adequate for the intuitionistic propositional logic **Int**.

14. A theorem saying that the content of the product of matrices is equal to the intersection of the contents of these matrices was given there. The proof of this theorem can be found in Kalicki (1950a).

15. The first axiomatization of **G$_3$** is due to Łukasiewicz (1941/1970, p. 286): the axioms of intuitionistic logic **Int** plus the axiom $(\neg p \Rightarrow q) \Rightarrow (((q \Rightarrow p) \Rightarrow q) \Rightarrow q)$. In Bolc and Borowik (1992, p. 84) this axiomatization was mistakenly attributed to K. Gödel.

16. See the extensive review of Kalicki (1951).

17. A consequence operation C on a set S can be transformed into a usual *relation* $\vdash_C \subseteq \mathcal{P}(S) \times S$ between subsets of S and elements of S by postulating for every $X \subseteq S$ and every $\alpha \in S$ that *if and only if*

    $X \vdash_S \alpha$ *if and only if* $\alpha \in C(X)$.

18. Generalized matrices have a well-known dual presentation as pairs $\langle A, C \rangle$, where A is an abstract algebra and C is an abstract consequence operation without structurality (*a close operator*). These structures are called "abstract logics" and were developed by Suszko (see Suszko & Brown, 1973).

19. Here the authors refer to Czelakowski (2001) and Wójcicki (1988).

20. See the paper Malinowski and Woleński (2003) dedicated to R. Wójcicki on his eightieth birthday.

21. It is worth also mentioning the paper by Sobociński (1936) for axiomatization of certain many-valued systems, where every value except 'false' is a designated value, and Słupecki (1939) for axiomatization of full many-valued logics with an arbitrary set of designated values.

22. Stachniak (1988) extends Wójcicki's criterion of strong finiteness to all propositional logics.

23. The set of functions P_n corresponds to Post n-valued logic **P$_n$** (see Post, 1921). The standard definition of Post's n-valued matrix logics is as follows. A matrix

of the form $\mathfrak{M}_n = \langle V_n, \neg, V, \{n-1\}\rangle$ is called a *Post n-valued matrix* ($n \in N, n \geq 2$) provided that $V_n = \{0, 1, 2, \ldots, n-1\}$; $\neg x = x + 1 (mod\, n), x \vee y = max(x, y)$, and $\{n-1\}$ is the set of the designated elements of $\mathfrak{M}_n$.

24. Let's note another criteria for functional completeness based on this result, see Rosenberg (1980). About the results of Słupecki in logic see also (Woleński and Zygmunt, 1989).

25. Generalization of this function for $\mathbf{Ł_n}$ see (Evans & Schwartz, 1958).

26. It follows that $\mathbf{Ł_n}$ has *a pure number-theoretical nature* and all attempts to interpret it in any intuitive way are doomed to failure. However, $\mathbf{Ł_n}$ can be interpreted by Kripke-style semantics (see Urquhart, 1973) or by the two-valued Boolean algebras (see Karpenko, 1983). Note, that these are examples of bivalent semantics (see Section 10).

27. The deduction of axiom Ł5 from the rest of the axioms was first published by Łukasiewicz's pupil C. Meredith (1958).

28. Specifically, for the finite case the criterion of definability of functions in Łukasiewicz matrices is given in Prucnal (1969).

29. See p. 105 in Łukasiewicz (1941). There is an English translation in Łukasiewicz (1970) but without the text of the discussion.

30. However Mostowski (1957) appreciates the contribution of Łukasiewicz to mathematical logic.

31. See Łukasiewicz (1930/1970, p. 173).

32. After years of oblivion interesting consideration of this system is given in Font and Hájek (2002).

33. W. Suchoń (1974) showed that Boolean-valued endomorphisms

$$\delta_i(^j/_{n-1}) = \begin{cases} 1, & if\, i+j \geq n \\ 0, & if\, i+j < n \end{cases} \quad (1 \leq i \leq n-1)$$

are also definable through $\sim$ and $\rightarrow$ in Łukasiewicz matrix $\mathfrak{M}_n$. Incidentally, the δ_i are definable through J_i. Note that the endomorphisms δ_i play a special role in the different algebraic investigations of finite-valued logics.

34. Previously Rasiowa (1969; 1970; 1972) studied predicate calculi constructed on the basis of $\mathbf{P_n}$.

35. By ω^+, or also by $\omega + 1$, one denotes the order type of the set of natural numbers completed with an additional greatest element ∞, that is, of the set $\{0, 1, 2, \ldots, \infty\}$.

36. By $\omega + \omega^+$ one denotes the order type of the set $\{0, 1, 2, \ldots, -2, -1, 1\} (= \Sigma)$.

37. In connection with this we note that in 1985 $\omega + \omega^+$-generalization of Łukasiewicz logics $\mathbf{Ł_n}$ is considered (see Karpenko, 1988). Let's consider a matrix

$$\mathfrak{M}_\Sigma = \langle \Sigma, \sim^\Sigma, \rightarrow^\Sigma, \{1\}\rangle.$$

Logical operations are defined in the following way:

$$\sim^\Sigma x = -x, x \rightarrow^\Sigma y = \begin{cases} 1 & if\, x \leq y \\ y - x & if\, x > y. \end{cases}$$

One may verify that $\mathfrak{M}_\Sigma$ is a model for infinite-valued Łukasiewicz logics $\mathbf{Ł_\infty}$. Note that $\mathfrak{M}_\Sigma$ is a linearly-ordered MV-algebra which Chang denotes by C and gives it as an example of *not* representable MV-algebra (see Chang, 1958, p. 486). Now we can say that matrix $\mathfrak{M}_\Sigma$ is a characteristic matrix for the following propositional calculus:

$$Ł_\infty \text{ plus } [(p \to (p \to \sim p))] \to [(\sim p \to p) \to (\sim p \lor p)].$$

Let's denote this logic by $Ł_\Sigma$. In Karpenko (1988) there is a proof that $Ł_\Sigma$ has an adequate interpretation by the two-valued Boolean algebras.

38. See chapter 3, 'Mathematical Foundations of Computer Science' in Bartol, Orłowska and Skowron (1997).
39. Part 4 of this book prepared by G. Malinowski consists of two papers: 'Historical Note' and 'Bibliography of Łukasiewicz's logics'.
40. Note that Da Costa *et al.* (1996, p. 287) emphasize that the assumption of structurality is not necessary for Suszko's reduction to two-valuedness. See also Caleiro *et al.* (2005, p. 172).
41. Also Malinowski (1977) demonstrates this for $Ł_n$.
42. For the first time this concept was introduced by Rescher (1969, p. 67). However, a similar concept called 'super-designated truth-values' appeared in Kalicki (1952, p. 161).
43. Earlier Font (2009, p. 384) explains: 'Since arbitrary functions from a set into a two-element set can be identified with arbitrary subsets of the given set, it turns out that specifying a Suszko semantics amounts to specifying a family of subsets of the set of all formulas.' It is interesting to note the following remark of Malinowski (1993, p. 73): 'even for simple relations of inference the conditions defining valuations are illegible'.
44. It is worth mentioning a paper (Karpenko, 1989), entirely devoted to this topic. Also for the first time the question of structuralization (generalization) of truth-values is discussed here.
45. Reduction, but not the *interpretation*.
46. Independently and at the same time this remarkable result was proven by the Polish logicians Hulanicki and Świerczkowski (1960). See also the fundamental book (Lau, 2006, Section 8.1).
47. Since 1965 Ratsa devoted to the first matrix of Jaśkowski a series of papers.

References

Anshakov, O., and Rychkov, S. (1985) 'On the Axiomatization of Finite-Valued Logical Calculi', *Math. USSR Sbornik*, 51, 473–91.
—— 'On Finite-Valued Propositional Logical Calculi', *Notre Dame Journal of Formal Logic*, 36(4), 606–29.
Bartol, W., Orłowska, E., and Skowron, A. (1997) 'Helena Rasiowa, 1917–1994', *Bull. European Association for Theoretical Computer Science*, 62, 353–66.
Beavers, G. (1993) 'Extensions of the $\aleph_0$-Valued Lukasiewicz Propositional Logic', *Notre Dame Journal of Formal Logic*, 34(2), 251–62.
Bočvar, D. A., and Finn, V. K. (1972) 'On Many-Valued Logics that Permit the Formalization of Analysis of Antinomies. I', in D. A. Bochvar (ed.) *Researches on Mathematical Linguistics, Mathematical Logic and Information Languages* (Moscow: Nauka Publishers), 238–95 [in Russian].

Bolc, L., and Borowik, P. (1992) *Many-Valued Logics. Vol. 1: Theoretical Foundations* (Berlin: Springer).

—— (2000) *Many-Valued Logics. Vol. 2: Automated Reasoning and Practical Applications* (Berlin: Springer).

Borkowski, L. (1977) 'W sprawie intuicyjnej interpretacji logiki trójwartościowej Łukasiewicza', *Roczniki Filozoficzne*, 25, 61–8. Engl. trans: 'On the Intuitive Interpretation of Łukasiewicz's Three-Valued Logic', in L. Borkowski, A. B. Stępień (eds) *Studies in Logic and Theory of Knowledge 2* (Lublin: TNKUL), 1991, 25–30.

Brown, D. J., and Suszko, R. (1973) 'Abstract Logics', *Dissertationes Mathematicae*, 102, 7–41.

Caleiro, C. *et al.* (2005) 'Two's Company: "The Humburg of many-logical values", in J. Y. Béziau (ed.) *Logica Universalis* (Basel: Birkhäuser Verlag), 169–89.

Chang, C. C. (1958) 'Algebraic Analysis of Many-Valued Logics', *Transactions of the American Mathematical Society*, 88, 467–90.

—— (1959) 'A New Proof of the Completeness of the Łukasiewicz Axioms', *Transactions of the American Mathematical Society*, 93, 74–80.

Czelakowski, J. (1980) *Model-Theoretic Methods in Methodology of Propositional Calculi* (Warsaw: IFIS PAN).

—— (1992) *Consequence operations: Foundational Studies* (Warsaw: IFIS PAN).

—— (2001) *Protoalgebraic Logics* (Dordrecht: Kluwer).

Czelakowski, J., and Malinowski, G. (1985) 'Key Notions of Tarski's Methodology of Deductive Systems', *Studia Logica*, XLIV(4), 321–51.

Cignoli, R., D'Ottaviano, I. M. L., and Mundici, D. (2000) *Algebraic Foundations of Many-Valued Reasoning* (Dordrecht: Kluwer).

Da Costa, N. C. A., Béziau, J. Y., and Bueno, O. A. S. (1996) 'Malinowski and Suszko on Many-Valued Logics: On the Reduction of Many-Valuedness', *Modern Logic*, 6(3), 272–99.

D'Ottaviano, I. M. L., and Epstein, R. L. (1988) 'A Paraconsistent Many-Valued Propositional Logic: J_3', *Reports on Mathematical Logic*, 22, 89–103.

Dziobiak, W. (1980) 'Silnie finitystyczne operacje konsekwencji. Teorie równościowe skończonych algebr – pewne ich kratowe badania' [Strongly Finite Consequence Operations. Equational Theories of Finite Algebras – Their Lattice-Theoretic Investigation]. Doctoral dissertation.

—— (1981) 'The Lattice of Strengthenings of a Strongly Finite Consequences Operation', *Studia Logica*, 40(2), 177–94.

—— (1991) 'A Finite Matrix whose set of Tautologies is not Finitely Axiomatizable', *Reports on Mathematical Logic*, 25, 83–90.

Epstein, G., and Rasiowa, H. (1990) 'Theory and Uses of Post Algebras of Order $\omega + \omega^+$. I' in *20th International Symposium on Multiple-Valued Logic* (New York: IEEE Computer Society Press), 42–7.

—— (1991) 'Theory and Uses of Post Algebras of Order $\omega + \omega^+$. II' in *21st International Symposium on Multiple-Valued Logic* (New York: IEEE Computer Society Press), 248–54.

Evans, T., and Schwartz, P. B. (1958) 'On Słupecki *T*-functions', *The Journal of Symbolic Logic*, 23, 267–70.

Finn, V. K. (1969) 'The Precompleteness of a Class of Functions that Correspond to the Three-Valued Logic of J. Łukasiewicz', *Naucno-Tehnicheskaja Informacia* (*VINITI*), Ser. 2(10), 35–8 [in Russian].

Font, J. M. (2003) 'Generalized Matrices in Abstract Algebraic Logic', in V. F. Hendricks, and J. Malinowski (eds) *Trends in Logic: 50 Years of Studia Logica* (Dordrecht: Kluwer), 57–86.

—— (2009) 'Taking Degrees of Truth Seriously', *Studia Logica*, 91, 383–406.

Font, J. M., and Hájek, P. (2002) 'On Łukasiewicz's Four-Valued Modal Logic', *Studia Logica*, 70(2), 157–82.

Font, J. M., Jansana, R., and Pigozzi, D. (2003) 'A Survey of Abstract Algebraic Logic', *Studia Logica*, 74(1–2), 13–97.

Gottwald, S. (2001) *A Treatise on Many-Valued Logics* (Baldock: Research Studies Press).

Hájek, P. (1998) *Metamathematics of Fuzzy Logic* (Dordrecht: Kluwer).

—— (2006) 'Mathematical Fuzzy Logic – What it Can Learn from Mostowski and Rasiowa', *Studia Logica*, 84, 51–62.

Hulanicki, A., and Świerczkowski, S. (1960) 'Number of Algebras with a Given Set of Elements', *Bull. Acad. Polon., Sci. Ser. Sci. Math. Astronom. Phys.*, 8, 283–4.

Janov, Y. I., and Mučnik, A. A. (1959) 'On the Existence of k-Valued Closed Classes without a Finite Basis', *Dokl. Akad. Nauk SSSR*, 127(1), 44–6 [in Russian].

Janowskaja, S. A. (1959) 'Mathematical Logic, and Foundations of Mathematics', in *Mathematics in the USSR during 40 Years*, ch. 13 (Moscow: Nauka) [in Russian].

Jaśkowski, S. (1936) 'Recherches sur le système de la logique intuitioniste', *Actes du Congrès International de Philosophie Scientifique*, 6, 58–61. Engl. trans: 'Investigations into the System of Intuitionist Logic', *Studia Logica*, 34 (1975), 117–20.

Jordan, Z. (1945) *The Development of Mathematical Logic and of Logical Positivism in Poland between the Two Wars* (London: Oxford University Press). Reprinted shortened version under the title 'The Development of Mathematical Logic in Poland between the Two Wars', in (McCall, 1967), 346–97.

—— (1963) 'O determinizmie logicznym', *Studia Logica*, XIV. Engl. trans: 'Logical Determinism', *Notre Dame Journal of Formal Logic*, 4 (1963), 1–38.

Kalicki, J. (1950a) 'On Tarski's Matrix Method', *Comptes Rendus des Séances de la Société des Sciences et des Lettres de Varsovie III*, 41, 130–42.

—— (1950b) 'Note on Truth-Table', *The Journal of Symbolic Logic*, 15(3), 174–81.

—— (1950c) 'A Test for the Existence of Tautologies According to Many-Valued Truth-Tables', *The Journal of Symbolic Logic*, 15(3), 182–4.

—— (1951) 'Review of Łoś (1949)', *The Journal of Symbolic Logic*, 16, 59–61.

—— (1952) 'A Test for the Equality of Truth-Table', *The Journal of Symbolic Logic*, 17(3), 161–3.

—— (1954a) 'On Equivalent Truth-Tables of Many-Valued Logics', *Proceedings of the Edinburgh Mathematical Society*, 10, 56–61.

—— (1954b) 'An Undecidable Problem in the Algebra of Truth-Tables', *The Journal of Symbolic Logic*, 19, 172–6.

Karpenko, A. S. (1983) 'Factor-Semantics for n-Valued Logics', *Studia Logica*, XLII(2/3), 179–85.

—— (1988) 'T-F-sequence and their Sets as Truth-Values', in I. M. Bodnár, A. Máté, and L. Pólos (eds) *Intensional Logic, History of Philosophy and Methodology: To Imre Ruzsa on the occasion of his 65th Birthday*, 2 (Budapest: Eötvös University), 109–19.

—— (1989) 'Truth Values: What Does it Mean?', in V. A. Smirnov (ed.) *Investigations of Non-Classical Logics* (Moscow: Nauka), 38–53 [in Russian].

—— (2006) *Łukasiewicz Logics and Prime Numbers* (Beckington: Luniver Press).

—— (2010) *The Development of Many-Valued Logic* (Moscow: LKI Publishers) [in Russian].

—— (2010a) 'Continuity of Three-Valued Logics: Problems and Hypotheses', *Logicheskie Issledovanija*, 16, 127–33 [in Russian].

Kleene, S. C. (1952) *Introduction to Metamathematics* (New York: D. Van Nostrand Company).

Kłósak, K. (1948) 'Teoria indeterminizmu ontologicznego a trójwartościowa logika zdań prof. Jana Łukasiewicza', *Ateneum Kapłańskie*, 49, 209–230.

Kotarbiński, T. (1913) 'Zagadnienie istnienia przyszłości', *Przegląd Filozoficzny*, 16, 74–92. Engl. trans: 'The Problem of the Existence of the Future', *Polish Review*, 13 (1968), 7–22.

—— (1957) *Wykłady z dziejów logiki*, (Łódź: Ossolineum).

Lau, D. (2006) *Function Algebras on Finite Sets: A Basic Course on Many-Valued Logic and Clone Theory* (Berlin: Springer).

Lechniak, M. (2002) 'Some Remarks about Attempts of Intuitive Interpretation of Jan Łukasiewicz's Three-Valued Logic' in S. Kiczuk *et al.* (eds) *Studies in Logic and the Theory of Knowledge*, 5 (Lublin: TNKUL), 49–56.

Łoś, J. (1949) 'O matrycach logicznych', in *Prace Wrocławskiego Towarzystwa naukowego, Seria B*, vol. 19 (Wrocław: University of Wrocław), 1–141. Engl. trans. in: D. E. Ulrich *Matrices for Sentential Calculi*. Ph.D. (Detroit, MI: Wayne State University), 1967.

Łoś, J., and Suszko, R. (1958) 'Remarks on Sentential Logics', *Indagationes Mathematicae*, 20, 177–83.

Łukasiewicz, D. (2011) 'On Jan Łukasiewicz's Many-Valued Logic and his Criticism of Determinism', *Philosophia Scientiae*, 15(2), 7–20.

Łukasiewicz, J. (1918) 'Wykład pożegnalny, 7 III 1918', *Pro Arte et Studio*, 11, 3–4. Engl. trans: 'Farewell Lecture by Professor Jan Łukasiewicz, Delivered in the Warsaw University Lecture Hall on 7 March 1918' in Łukasiewicz (1970), 84–6.

—— (1920) 'O logice trójwartościowej', *Ruch Filozoficzny*, 5, 170–1. Engl. trans: 'On Three-Valued Logic', in Łukasiewicz (1970), 87–88.

—— (1921) 'Logika dwuwartościowa', *Przegląd Filozoficzny*, 13, 189–205. Engl. trans: 'Two-Valued Logic', in Łukasiewicz (1970), 89–109.

—— (1922) 'O determinizmie', *Z zagadnień logiki i filozofii. Pisma wybrane*, J. Słupecki (ed.) (Warsaw: PWN), 1961, 114–26. Engl. trans: 'On Determinism', in Łukasiewicz (1970), 110–28.

—— (1923) 'Interpretacja liczbowa teorii zdań', *Ruch Filozoficzny*, 7, 92–3. Engl. trans: 'A numerical interpretation of theory of propositions', in Łukasiewicz (1970), 129–30.

—— (1929) *Elementy logiki matematycznej*, (Warsaw: Koło Matematyczno-Fizyczne Słuchaczów Uniwersytetu Warszawskiego). Engl. trans: *Elements of Mathematical Logic* (Oxford: Pergamon Press), 1963.

—— (1930) 'Philosophische Bemerkungen zu mehrwertigen Systemen des Aussagenkalküls', in *Comptes Rendus des Séances de la Société des Sciences et des Lettres de Varsovie*, III(23), 51–77. Engl. trans: 'Philosophical Remarks on Many-Valued Systems of Propositional Logic', in Łukasiewicz (1970), 153–78.

—— (1941) 'Die Logik und das Grundlagenproblem', in *Les Entretiens de Zürich sur les fondements et la méthode des sciences mathématiques*, 6–9 decembre 1938

(Zürich: Leemann frères), 82–100. Engl. trans: 'Logic and the Problem of the Foundations of Mathematics', in Łukasiewicz (1970), 278–94.

—— (1953) 'A System of Modal Logic', *The Journal of Computing Systems*, 1, 111–149. Reprinted in Łukasiewicz (1970), 352–90.

—— (1957) *Aristotle's Syllogistic from the Standpoint of Modern Formal Logic* (Oxford: Oxford University Press), 2nd edn enlarged.

—— (1970) *Selected Works*, L. Borkowski (ed.) (Amsterdam and Warsaw: North-Holland and PWN).

Łukasiewicz, J., Tarski, A. (1930) 'Untersuchungen über den Aussagenkalkül', *Comptes Rendus des Séances de la Société des Sciences et des Lettres de Varsovie*, Cl. III, 23, 1–21. Engl. trans: 'Investigations into the sentential calculus', in Łukasiewicz (1970), 131–52.

Malinowski, G. (1977) 'Classical Characterization of *n*-valued Łukasiewicz Calculi', *Reports on Mathematical Logic*, 9, 41–5.

—— (1990) 'Q-consequents Operation', *Reports on Mathematical Logic*, 24, 49–59.

—— (1993) *Many-Valued Logics* (Oxford: The Clarendon Press).

—— (2001) 'Many-Valued logics', in L. Goble (ed.) *The Blackwell Guide to Philosophical Logic* (Oxford: Blackwell Publishers), 309–35.

—— (2006) *Logiki Wielowartościowe* (Warsaw: PWN).

—— (2007) 'Many-Valued Logic and its Philosophy', in D. M. Gabbay, and J. Woods (eds) *Handbook of the History of Logic, vol. 8: The Many Valued and Nonmonotonic Turn in Logic* (Amsterdam: North-Holland), 13–94.

—— (2009) 'Beyond Three Inferential Values', *Studia Logica*, 92(2), 203–13.

Malinowski, G., and Zygmunt, J. (1978) 'Results in General Theory of Matrices for Sentential Calculi with Applications to the Łukasiewicz Logics', *Zeszyty Naukowe Uniwersytetu Wrocławskiego, Prace Filozoficzne*, XX, 43–57.

Malinowski, G., and Woleński, J. (2003) 'Logic, Formal Methodology and Semantics in Works of Ryszard Wójcicki', *Studia Logica*, 99, 7–30.

McCall, S. (ed.) (1967) *Polish Logic 1920–1939* (New York: Oxford University Press).

McNaughton, R. (1951) 'A Theorem about Infinite-Valued Sentential Logic', *The Journal of Symbolic Logic*, 16, 1–13.

Meredith, C. A. (1958) 'The Dependence of an Axiom of Łukasiewicz', *Transactions of American Mathematical Society*, 87, 54.

Mostowski, A. (1948) 'Proofs of Non-Deducibility in Intuitionistic Functional Calculus', *The Journal of Symbolic Logic*, 13, 204–7.

—— (1950) 'Review of (Rasiowa 1950)', *The Journal of Symbolic Logic*, 15, 213.

—— (1957) 'L'oeuvre scientifique de Jan Łukasiewicz dans le domaine de la logique mathématique', *Fundamenta Mathematicae*, 44, 1–11.

—— (1961a) 'Axiomatizability of Some Many-Valued Predicate Calculi', *Fundamenta Mathematicae*, 50(2), 165–90.

—— (1961b) 'An Example of a Non-Axiomatizable Many-Valued Logic', *Zeitschrift für mathematische Logic und Grundlagen der Mathematik*, 7, 72–6.

—— (1963) 'The Hilbert Epsilon Function in Many-Valued Logic', *Acta Philosophica Fennica*, XVL, 169–88.

—— (1969) 'Models of Set Theory', in E. Casari (ed.) *Aspects of Mathematical Logic* (Rome: Edizione Cremonese), 65–179; repr. in 2010.

Mundici, D. (2011) *Advanced Łukasiewicz Calculus and MV-Algebras* (Dordrecht: Kluwer).

Ono, H., and Komori, Y. (1985) 'Logics without the Contraction Rule', *The Journal of Symbolic Logic*, 50, 169–201.

Orłowska, E. (1985) 'Mechanical Proof Methods for Post Logics', *Logique et Analyse*, 110–11, 173–192.

Pałasińska, K. (1994) 'Three-element Nonfinitely Axiomatizable Matrices', *Studia Logica*, 53, 361–72.

Piróg-Rzepecka, K. (1977) *Systemy Nonsense-Logics* (Warsaw-Wrocław: OTPN-PWN).

Pogorzelski, W. A. (1964) 'The Deduction Theorem for Łukasiewicz's Many-Valued Propositional Calculi', *Studia Logica*, XV, 7–19.

Pogorzelski, W. A., and Prucnal, T. (1974) 'Structural Completeness of the First-Order Predicate Calculus', *Bulletin de l'Académie Polonaise des Sciences, Série des Sciences, Mathematiques, Astronomiques et Physiques*, 22(3), 349–51.

Pogorzelski, W. A., and Wojtylak, P. (1982) *Elements of the Theory of Completeness in Propositional Logic* (Katowice: Silesian University).

—— (2008) *Completeness Theory for Propositional Logics* (Basel: Birkhäuser Verlag).

Post, E. L. (1920) 'Determination of all Closed Systems of Truth Table', *Bull. Amer. Math. Soc.*, 26, 427.

—— (1921) 'Introduction to a General Theory of Elementary Propositions', *American Journal of Mathematics*, 43(3), 163–85. Reprinted in J. van Heijenoort (ed.) *From Frege to Gödel: A Source Book in Mathematical Logic, 1879–1931* (Cambridge: Harvard University Press, 1967), 264–83.

—— (1941) 'Two-Valued Iterative Systems', *Annals of Mathematical Studies*, 5 (Princeton, NJ: Princeton University Press).

Prucnal, T. (1969) 'Kryterium definiowalnosci funkcji w matrycach Łukasiewicza', *Studia Logica*, 23, 71–7.

Rasiowa, H. (1950) 'Z dziedziny logiki matematycznej II. Logiki wielowartościowe Łukasiewicza', *Matematyka*, 3.

—— (1969) 'A Theorem on the Existence of Prime Filters in Post Algebras and the Completeness Theorem for some Many-Valued Predicate Calculi', *Bull. Acad. Polon. Sci., Serie Sci. Math. Astr. Phys.*, 17, 347–54.

—— (1970) 'Ultraproducts of m-valued Models and a Generalization of the Löwenheim-Skolem-Gödel-Malcev Theorem for Theories Based on m-Valued Logics', *Bull. Acad. Polon. Sci., Serie Sci. Math. Astr. Phys.*, 18, 415–20.

—— (1972) 'The Craig Interpolation Theorem for m-valued Predicate Calculi', *Bull. Acad. Polon. Sci., Serie Sci. Math. Astr. Phys.*, 20, 141–6.

—— (1973) 'On Generalized Post Algebras of Order ω^+ and ω^+-valued Predicate Calculi', *Bull. Acad. Polon. Sci., Serie Sci. Math. Astr. Phys.*, 21, 209–19.

—— (1974a) *An Algebraic Approach to Non-Classical Logics* (Amsterdam: North-Holland).

—— (1974b) 'Post Algebras as a Semantic Foundation of m-Valued Logic', *The Mathematical Association of America Studies in Mathematics*, 9, 92–142.

—— (1977) 'Many-Valued Algorithmic Logic as a Tool to Investigate Programs', in G. Epstein, and J. M. Dunn (eds) *Modern Uses of Multiple-Valued Logic* (Dordrecht: Reidel), 79–102.

Ratsa, M. (1982) 'On Functional Completeness in the Intuitionistic Propositional Logic', *Problemy Kibernetiki*, 39, 107–150 [in Russian].

Rautenberg, W. (1981) '2-element Matrices', *Studia Logica*, 40, 315–53.

Rescher, N. (1969) *Many-Valued Logic* (New York: McGraw Hill). 2nd edn printed in 1993.

Rose, A. (1952) 'The Degree of Completeness of the m-valued Łukasiewicz Propositional Calculus', *London Mathematical Society*, 27, 92–102.

Rose, A., and Rosser, J. B. (1958) 'Fragments of Many-Valued Statement Calculi', *Transaction of the American Mathematical Society*, 87, 1–53.

Rosenberg, I. G. (1980) 'The Ramifications of Słupecki Criterion in Many-Valued Logic', in *Proceedings of the 24th Conference on the History of Logic, Cracow, April 28–30, 1978* (Cracow: Jagiellonian University), 58–74.

Rosser, J. B., and Turquette, A. M. (1952) *Many-Valued Logics* (Amsterdam: North-Holland). 2nd edn printed in 1958.

Scott, D. (1974) 'Completeness and Axiomatizability in Many-Valued logics' in L. Henkin *et al.* (eds) *Proceedings of the Tarski Symposium* (Providence, RI: AMS), 411–35.

—— (1976) 'Does Many-Valued Logic Have Any Use?', in S. Körner (ed.) *Philosophy of Logic* (Berkeley: University of California Press), 64–74.

Shramko, Y., and Wansing, H. (2010) 'Truth values', in E. N. Zalta (ed.) *Stanford Encyclopedia of Philosophy*, http://plato.stanford.edu/entries/truth-values/, accessed 20 October 2012.

Słupecki, J. (1936) 'Der volle dreiwertige Aussagenkalkül', *Comptes Rendus des Séances de la Société des Sciences et des Lettres de Varsovie*, Cl. III, 29, 9–11. Engl. trans: 'The Full Three-Valued Propositional Calculus', in McCall (1967), 335–7.

—— (1939a) 'Kryterium pełności wielowartościowych systemów logiki zdań', *Comptes Rendus des Séances de la Société des Sciences et des Lettres de Varsovie*, Cl. III, 32, 102–9. Engl. trans: 'A Criterion of Fullness of Many-Valued Systems of Propositional Logic', *Studia Logica*, 30 (1972), 153–7.

—— (1939b) 'Dowód aksiomatyzowalności pełnych systemów wielowartościowych rachunku zdań', *Comptes Rendus des Séances de la Société des Sciences et des Lettres de Varsovie*, Cl. III, 32, 110–128. Engl. trans: 'Proof of Axiomatizability of Full Many-Valued Systems of Calculus of Propositions', *Studia Logica*, 29 (1971), 155–68.

—— (1946) 'Pełny trójwartościowy rachunek zdań', *Annales Universitatis Mariae Curie-Skłodowska*, Sectio F 1, 193–209.

—— (1964) 'Próba intuicyjnej interpretacji logiki trójwartościowej Łukasiewicza', in *Rozprawy logiczne. Księga pamiątkowa ku czci K. Ajdukiewicza* (Warsaw: PWN), 185–91.

Słupecki, J., Bryll, J., and Prucnal, T. (1967) 'Some Remarks on the Three-Valued Logic of J. Łukasiewicz', *Studia Logica*, 21, 45–70.

Sobociński, B. (1936) 'Aksiomatyzacja pewnych wielowartościowych systemów teorji dedukcji', *Roczniki prac naukowych zrzeszenia asystentów Uniwersytetu Józefa Piłsudskiego w Warszawie, 1. Wydział matematyczno-przyrodniczy*, 1, 399–419.

Stachniak, Z. (1988) 'Two Theorems on Many-Valued Logics', *Journal of Philosophical Logic*, 17, 171–9.

Suchoń, W. (1974) 'Definition des foncteurs modaux de Moisil dans le calcul n-valent des propositions de Łukasiewicz avec implication et négation', *Reports on Mathematical Logic*, 2, 43–7.

Surma, S. J. (1971) 'Jaśkowski's Matrix Criterion for the Intuitionistic Propositional Calculus', *Prace z Logiki*, 6, 21–54.

—— (1974a) 'A Method of the Constructions of Łukasiewiczian Algebras and its Application to Gentzen-Style Characterization of Finite Logics', *Reports on Mathematical Logic*, 2, 49–54. Reprinted in (Wójcicki, Malinowski, 1977), 93–9.

—— (1974b) 'An Algorithm for Axiomatizing Every Finite Logic', *Reports on Mathematical Logic*, 3, 57–61. Reprinted in D. C. Rine (ed.) *Computer Science and Multiple-Valued Logic: Theory and Applications* (Amsterdam: North-Holland), revised edition, 143–9.

Surma, S. J. Wroński, A. and Zachorowski, S. (1975) 'On Jaśkowski-Type Semantics for the Intuitionistic Propositional Logic', *Studia Logica*, 34(2), 145–8.

Suszko, R. (1975) 'Remarks on Łukasiewicz's Three-Valued Logic', *Bulletin of the Section of Logic*, 4(3), 87–90.

—— (1977) 'The Fregean Axiom and Polish Mathematical Logic in the 1920s', *Studia Logica*, 36(4), 377–80.

Tarski, A. (1930a) 'Über einige fundamentale Begriffe der Metamathematik', *Comptes Rendus des Séances de la Société des Sciences et des Lettres de Varsovie*, Cl. III, 23, 22–9. Engl. trans.: 'On Some Fundamental Concepts of Metamathematics' in (Tarski, 1956), 30–7.

—— (1930b) 'Fundamentale Begriffe der Methodologie der deduktiven Wissenschaften I', *Monatshefte für Mathematik und Physik*, 37, 361–404. Engl. trans.: 'Fundamental Concepts of the Methodology of the Deductive Sciences', in (Tarski, 1956), 60–109.

—— (1938) 'Der Aussagenkalkül und die Topologie', *Fundamenta Mathematicae*, 31, 103–34. Engl. trans.: 'Sentential calculus and topology', in (Tarski, 1956), 421–54.

—— (1956) *Logic, Semantic, Metamathematics: Papers from 1926 to 1938* (Oxford: Clarendon Press), 2nd edn (Indianapolis: Hackett Publishing Company), 1983.

Tokarz, M. (1974a) 'Invariant Systems of Łukasiewicz Calculi', *Zeitschrift für mathematische Logic und Grundlagen der Mathematik*, 20, 221–8.

—— (1974b) 'A Method of Axiomatization of Łukasiewicz Logics', *Studia Logica*, 33, 333–8. Reprinted in (Wójcicki and Malinowski, 1977), 113–18.

—— (1977) 'Degrees of Completeness of Łukasiewicz Logics', in (Wójcicki and Malinowski, 1977), 127–34.

Trzęsicki, K. (1993) 'Łukasiewicz on Philosophy and Determinism', in F. Coniglione, R. Poli, J. Woleński (eds) *Polish Scientific Philosophy: The Lvov-Warsaw School* (Amsterdam-Atlanta: Rodopi), 251–97.

Tsuji, M. (1998) 'Many-Valued Logics and Suszko's Thesis Revisited', *Studia Logica*, 60(2), 299–309.

Tuziak, R. (1988) 'An Axiomatization of the Finite-Valued Łukasiewicz Calculus', *Studia Logica*, 47(1), 49–55.

Ulrich, D. (1986) 'On the Characterization of Sentential Calculi by Finite Matrices', *Reports on Mathematical Logic*, 20, 63–86.

Urquhart, A. (1973) 'An Interpretation of Many-Valued Logic', *Zeitschrift für mathematische Logic und Grundlagen der Mathematik*, 19, 111–14.

Wajsberg, M. (1931) 'Aksjomatyzacja trójwartościowego rachunku zdań', *Comptes Rendus des Séances de la Société des Sciences et des Lettres de Varsovie*, Cl. III, 23, 126–45. Engl. trans.: 'Axiomatization of the Three-Valued Calculus' in (Wajsberg, 1977), 12–29.

—— (1935) 'Beiträge zum Metaaussgenkalkül I', *Monatshefte für Mathematik und Physik*, 42, 221–42. Engl. trans.: 'Contributions to Meta-Calculus of Propositions I', in (Wajsberg, 1977), 89–106.

—— (1936) 'Untersuchungen über Unabhängigkeitsbeweise nach der Matrizen-methode', *Wiadomości Matematyczne*, 41, 33–70. Engl. trans.: 'On the Matrix Method of Independence Proofs', in (Wajsberg, 1977), 107–31.

—— (1977) *Logical Works*, S. J. Surma (ed.) (Wrocław: Ossolineum).

Wansing, H., and Shramko, Y. (2008) 'Suszko's Thesis, Inferential Many-Valuedness and the Notion of a Logical System', *Studia Logica*, 88, 405–29 [*Erratum* in volume 89, 2008, p. 147].

Wojtylak, P. (1979) 'Strongly Finite Logics: Finite Axiomatizability and the Problem of Supremum', *Bulletin of the Section of Logic*, 8, 99–111.

—— (1984) 'An example of a Finite Though Nonfinitely Axiomatizable Matrix', *Reports on Mathematical Logic*, 17, 39–46.

Woleński, J. (1985) *Filozoficzna Szkoła Lwowsko-Waszawska* (Warsaw: PWN). Engl. trans.: *Logic and Philosophy in the Lvov-Warsaw School* (Dordrecht: Reidel), 1989.

—— (1990) 'Kotarbiński, Many-Valued Logic and Truth', in J. Woleński (ed.) *Kotarbiński: Logic, Semantics and Ontology* (Dordrecht: Kluwer), 191–8.

—— (1995) 'Mathematical Logic in Poland 1900–1939: People, Circles, Institutions, Ideas', *Modern Logic*, 5(4), 363–405; repr. in J. Woleński (ed.) *Essays in the History of Logic and logical Philosophy* (Cracow, 1999), 59–84.

—— (2001) 'The Rise of Many-Valued Logic in Poland', in W. Stelzner, M. Stöckler (eds) *Zwischen traditioneller und moderner Logik. Nichtklassische Ansätze* (Paderborn: Mentis Verlag), 193–204.

—— (2003a) 'The Achievements of the Polish School of Logic', in T. Baldwin (ed.) *The Cambridge History of Philosophy 1870–1945* (Cambridge: Cambridge University Press), 401–16.

—— (2003b) 'Determinism and Logic', *Voprosy Filosofii*, 5, 71–81 [in Russian]. Spanish version in J. Woleński, P. Domínguez (eds) *Lógica y filosofía* (Madrid: Publicaciones de la Faculdad de Teología 'San Dámaso', 2005), 83–102.

—— (2004a) 'Lvov-Warsaw School', in *Philosophy. Encyclopedic Dictionary* (Moscow: Gardariki), 456–7 [in Russian].

—— (2004b) 'Polish Logic', *Logic Journal of the IGPL*, 12(5), 399–428.

—— (2009) 'The Principle of Bivalence and Suszko Thesis', *Bulletin of the Section of Logic*, 38(3/4), 99–100.

—— (2010) 'Lvov-Warsaw School', in E. N. Zalta (ed.) *Stanford Encyclopedia of Philosophy*, http://plato.stanford.edu/entries/lvov-warsaw/, date accessed 20 October 2012.

Woleński, J., and Zygmunt, J. (1989) 'Jerzy Słupecki (1904–1987): Life and Work', *Studia Logica*, 48, 401–11.

Wójcicki, R. (1969) 'Logical Matrices Strongly Adequate for Structural Sentential Calculi', *Bulletin de l'Académie Polonaise des Sciences, Série des Sciences, Mathematiques, Astronomiques et Physiques*, 6, 333–5.

—— (1970) 'Some Remarks on the Consequence Operation in Sentential Logics', *Fundamenta Mathematicae*, 68, 269–79.

—— (1973a) 'Matrix Approach in the Methodology of Sentential Calculi', *Studia Logica*, 32, 7–37.

—— (1973b) 'On Matrix Representations of Consequence Operations of Łukasiewicz's Sentential Calculi', *Zeitschrift für mathematische Logic und Grundlagen der Mathematik*, 19, 239–47. Reprinted in (Wójcicki, Malinowski, 1977), 101–11.

—— (1977) 'Strongly Finite Sentential Calculi', in (Wójcicki and Malinowski, 1977), 53–77.

—— (1984) *Lectures on Propositional Calculi* (Wrocław: Ossolineum).

—— (1988) *Theory of Logical Calculi: Basic Theory of Consequences Operations* (Dordrecht: Reidel).

Wójcicki, R., and Malinowski, G. (eds) (1977) *Selected Papers on Łukasiewicz Sentential Calculi* (Wrocław: Ossolineum).

Wójcicki, R., and Zygmunt, J. (2003) 'Polish logic of the Postwar Period', in V. F. Hendricks, and J. Malinowski (eds) *Trends in Logic: 50 Years of Studia Logica* (Dordrecht: Kluwer), 11–33.

Wroński, A. (1974) 'On the Cardinalities of Matrices Strongly Adequate for the Intuitionistic Propositional Logic', *Reports on Mathematical Logic*, 3, 67–72.

—— (1979) 'A Three Element Matrix whose Consequence Operation is not Finitely Based', *Bulletin of the Section of Logic*, 8, 68–71.

Zawirski, Z. (1934a) 'Znaczenie logiki wielowartościowej i związek jej z rachunkiem prawdopodobieństwa', *Przegląd Filozoficzny*, 37, 393–8.

—— (1934b) *Stosunek logiki wielowartościowej do rachunku prawdopodobieństwa* (Poznań: Poznańskie Towarzystwo Przyjaciół Nauk); German partial translation in Z. Zawirski 'Über das Verhaltnis mehrwertiger logik zur Wahrscheinlichkeitsrechnung', *Studia Philosophica*, I, 407–42.

Zygmunt, J. (1981) 'The Logical Investigations of Jan Kalicki', *History and Philosophy of Logic*, 2, 41–53.

—— (1983) 'An Application of the Lindenbaum Method in the Domain of Strongly Finite Sentential Calculi', *Acta Universitatis Vratislaviensis*, 517; *Logika*, 8, 59–68.

—— (1984) 'An Essay in Matrix Semantics for Consequence Relations', *Acta Universitatis Vratislaviensis*, 741 (Wrocław).

2
The Dependence and Independence of Quantifiers: Truth, Proof and Choice Functions

Gabriel Sandu

2.1 Quantifier dependence and the *epsilon-delta* definition of continuity

When one looks at the *use* of quantifiers in the mathematical vernacular, one very rarely finds quantifiers in isolation. Instead, one is often confronted with a sequence of interdependent quantifiers. The classical example is the mathematical notion of the limit of a function which goes back to Bolzano (1817). It is commonly acknowledged that Bolzano had the basic ingredients of the so-called *epsilon-delta* definition of continuous functions, which, in the form in which it is used today, due to Cauchy, goes roughly like this:

A function f is continuous at a point x_0 if

given any $\varepsilon > 0$
one can choose $\delta > 0$

so that

for all y,
when x_0 is within distance δ from y, then $f(x_0)$ is within distance ε from $f(y)$.

The last statement is expressed in mathematical notation by:

$$|x_0 - y| < \delta \rightarrow |f(x_0) - f(y)| < \varepsilon.$$

After pushing the conditions expressing the restrictions on the quantifiers to the right, and quantifying over the points x_0 we end up with the definition of continuous function which has the form:

$$\forall x_0 \forall \varepsilon \exists \delta \forall y R(x_0, \varepsilon, \delta, y).$$

The order of the quantifiers in this formula makes explicit the fact that the choice of δ depends on ε (and obviously on the point x_0).

It sometimes turns out that one can find a δ which works *no matter* what x_0 is. If this is so then the function f is said to be *uniformly continuous*. In order to represent uniform continuity in mathematical symbolism, one has to be able to represent the fact that the choice of δ depends only on ε but does not depend on x_0. In other words, the choice of δ depends on ε but is independent of x_0. This is accomplished by the formula:

$$\forall x_0 \forall \varepsilon (\exists \delta / \{x_0\}) \forall y R(x, \varepsilon, \delta, y).$$

The resulting logic, that is, the logic with quantifiers

$$(\exists x / W), (\forall y / V)$$

where W and V are finite sets of variables has been called Independence-Friendly Logic (IF logic) and has been introduced by Hintikka and Sandu (1989). When W and V are empty, we recover the ordinary first-order quantifiers.

IF logic will not concern us in this paper. The thing we want to emphasize at this point is that any satisfactory account of quantifiers has to tackle the kind of quantifier dependence (and independence) arising from the epsilon-delta technique.

In a seminal paper, Goldfarb (1979) convincingly argues that it is misleading to portray Frege, as it is often done, as having provided *all* the central notions of quantification. Although he does not mention the epsilon-delta technique, Goldfarb's main point in his paper is that it is precisely the understanding of the phenomenon of quantifier dependence that constituted the source of major progress in the development of logic in the twentieth century, (Goldfarb, 1979, p. 357): 'The connection between quantifiers and choice functions or, more precisely, between quantifier-dependence and choice functions, is the heart of how classical logicians in the twenties viewed the nature of quantification'.

Goldfarb traces the connection between quantifier-dependence and choice functions in the work of Skolem and Hilbert in some detail. In the next section I will highlight some of his points.

2.2   Skolem and Hilbert

Before Skolem and Hilbert it was customary to assimilate quantifiers to (possibly infinite) sums and products whose contributions to the relevant formal system were given through algebraic manipulations. This conception goes back to Peirce (the third volume of his *Lectures on the Algebra of Logic* (1895)), but the main illustration comes from Schröder's calculus of classes. Schröder's work was continued by Löwenheim, whose distinction between quantification over individuals and quantification over relations opened the door to the separation of first-order from second-order logic. Goldfarb emphasizes, however, that 'Löwenheim's interest in the first-order fragment of the relative fragment of the relative calculus seems motivated by purely algebraic, rather than foundational considerations' (1979, p. 355). The problem with the algebraic tradition, and Frege saw this perhaps better than anyone else, is that it continuously mixed up the propositional and the class-theoretical considerations, ignoring the question of predication and the fact that logical assertions have content, that is, the goal of logic is to make assertions about the universe. This is perhaps the main difference between the algebraic school of Schröder and Löwenheim and the logicist school of Frege and Russell.

Frege regarded an existentially quantified sentence $\exists x\varphi$ as saying that the class of individuals satisfying φ denoted by $\| \varphi \|$, is not empty. A universally quantified sentence $\forall x\varphi$ says that $\| \varphi \|$ is the entire universe. In modern model-theoretical terms, given a structure M, the existential quantifier is identified with the collection of all nonempty subsets of M, and the universal quantifier with the collection whose only member is the universe itself:

$$\exists_M = \{X \subseteq M : X \neq \emptyset\}, \forall_M = \{M\}.$$

The model-theoretical versions of truth-conditions for existentially and universally quantified formulas fall out quite naturally from the semantic conventions:

- $\exists x\varphi$ is true in M if and only if $\| \varphi \| \in \exists_M$
- $\forall x\varphi$ is true in M if and only if $\| \varphi \| \in \forall_M$.

Logicians like Skolem and Hilbert broke with both the algebraic tradition which treated quantifiers as (potentially infinite) sums and products, and with Frege's conception of quantifiers as higher-order relations.

Both of them recognized something special about the nature of quantification: the phenomenon of quantifier dependence, that is, the idea of a quantifier depending on others. Both Skolem and Hilbert expressed quantifier-dependence through the use of choice functions, with the purpose, for the former, of giving an alternative proof of Löwenheim's theorem, and with the hope, for the latter of turning formal proofs into some finitistically manageable, quantifier-free manipulations.

In his proof of Löwenheim's theorem, Skolem (1920) makes the connection between the satisfiability of a proposition in a certain universe and the existence of appropriate (Skolem) functions which correspond to the existential quantifiers. Each such function yields an appropriate value when it takes as arguments the values for the universal quantifiers superordinate to the existential quantifier in question. The proof of the Löwenheim's theorem is then straightforward: any countable, satisfiable collection of propositions can involve only countably many Skolem functions.

The idea of quantifier dependence is also at work in Hilbert's formulation of the laws of quantification by means of the ε-terms (Hilbert, 1923). An epsilon term $\varepsilon_x\varphi(x)$ is intended to denote an element x of which φ holds (if any). Thus when φ contains free variables other than x, the epsilon term $\varepsilon_x\varphi(x)$ represents a function which yields an appropriate value for the arguments which are the values of those free variables. The standard quantifiers can then be defined in terms of the ε-symbol:

$$\exists x\varphi(x) = \varphi(\epsilon_x\varphi(x)), \forall x\varphi(x) = \varphi(\varepsilon_x(\neg\varphi(x))).$$

The remarkable thing is that these definitions together with the ε-axiom

$$\varphi(t) \to \varphi(\varepsilon_x\varphi(x))$$

yield all the usual quantificational rules.

Hilbert, unlike Skolem, was concerned with the role of quantifiers in formal proofs, and his use of the choice functions encoded in the ε-terms is subordinated entirely to this purpose:

> I am suggesting that behind Hilbert's interest in proving by finitistic hook or crook, the consistency of formal systems, lies a deeper point: that of using the proxy choice-functions to provide in some measure an explication of the meaning of the quantification used in formal proofs. (Goldfarb, 1979, p. 361)

Given that any formal proof is finite, it contains only a finite number of ε-axioms. The idea is now to assign successively, during the proof, effective values to the ε-terms with the hope of transformation of the whole proof in a manipulation of quantifier-free formulas. To this purpose Hilbert did not need the full power of the choice functions; it sufficed, instead 'to obtain finitely-based functions (functions that are zero everywhere but on a finite number of arguments) that approximate the "real" choice functions'. (Goldfarb, 1979, p. 361).

We know that in the end Hilbert's project did not work but we shall return to the use of choice functions later on.

2.3 Game theory

The idea of treating quantifiers game-theoretically may be found already in Peirce. In his second Cambridge Conference lecture, *Types of Reasoning*, Peirce noticed the following:

> When I say 'every man dies,' I say you may pick out your man for yourself and provided he belongs to *this here* world you will find he will die. The 'some' supposes a selection from 'this here' world to be made by the *deliverer* of the proposition, or made in his interest. The 'every' *transfers* the function of selection to the *interpreter* of the proposition, or to anybody acting in his interest. (Peirce, 1992, pp. 129–30)

Obviously Peirce's way to look at quantifiers stands in deep contrast to Frege's conception of quantifiers as 'ranging over' a class of entities. However, Peirce's idea of the 'deliverer of a proposition' or of somebody acting 'in his interest' is rather obscure, as are his remarks about 'the interpreter of the proposition.' He is nevertheless on the right track: one can spell out the interpretation of the standard quantifiers in terms of picking out individuals from a universe of discourse. And he is also on the right track about the picking out being done by two distinct agents, one for each quantifier. What is missing, however, for a full-blown game-theoretical interpretation is the strategic interaction between quantifiers, that is, quantifier dependence.

Quantifier dependence is the main motivation behind Hintikka's game-theoretical semantics (GTS). It spells out quantifier dependence in terms of the strategic interaction of two players in a semantical game. The connection between the existence of certain (Skolem) functions and satisfiability is a natural by-product of this analysis. More remarkably perhaps, from the game-theoretical analysis of quantification and of quantifier-dependence, one can obtain Tarski's semantical interpretation for first-order languages (assuming the Axiom of Choice). In order for this correlation to work one needed to find a class of games which are 'well-behaved' and to formulate semantical games for first-order languages as a subclass of such games. This is what Hintikka did. Again, I will only sketch the main steps. Contrary to the standard exposition of these matters, I will start with the basic notions from game theory; then I will formulate Hintikka's game-theoretical semantics; finally I will present Tarski's semantics for ordinary first-order languages as a special case. The reader is sent to (Mann, Sandu & Sevenster, 2011) for details and full proofs.

2.4   Extensive, finite games of perfect information

A standard textbook in game theory describes games as '[...] a description of strategic interaction that includes the constraints on the actions that the players *can* take and the players' interests, but do not specify the actions that the players *do* take'. (Osborne & Rubinstein, 1994, p. 2) Classical game theory makes a distinction between *extensive* and *strategic* games. The former games are sequential, representing the order of the choices of the players in a possible play of the game. This is made obvious by the formal definition of an *extensive game of perfect information*, which is represented by a tuple

$$G = (N, H, Z, P, (u_i)_{i \in N})$$

where

1. N is the set of players
2. H is a set of finite sequences called *histories*, or *plays*
3. Z is the set of terminal histories of the game
4. $P : H \setminus Z \to N$ is the *player function* which assigns to every nonterminal history the player whose turn is to move.
5. For each $i \in N$, u_i is the *payoff function* for player i, a function that specifies the payoff of player i at terminal histories.

If h is a history, then any nonempty initial segment of h is also a history. We sometimes call it a subhistory of h. For every nonterminal history $h = (a_1, \ldots, a_m)$ the player $P(h)$ chooses an action a' to continue the play. The action is chosen from the set

$$A(h) = \{a : (a_1, \ldots, a_m, a) \in H$$

and the play continues from

$$(a_1, \ldots, a_m, a').$$

We shall be interested in the class of games with two players, that is, $N = \{I, II\}$. A two-player extensive game is *strictly competitive* if the players have no incentive to cooperate, that is, if for all $h, h' \in Z$:

$$u_I(h) \geq u_I(h') \Leftrightarrow u_{II}(h') \geq u_{II}(h).$$

A *constant-sum* game is one in which the sum of the player payoffs is constant, that is, there exists a $c \in R$ such that for every terminal history h we have $u_I(h) + u_{II}(h) = c$. We shall be particularly interested in the games for which $c = 1$, that is, *one sum* games. We notice that every constant-sum game is strictly competitive but not vice versa. In addition, in a one-sum game, $u_{II}(h) = 1 - u_I(h)$ for every terminal history h.

If the only possible payoffs are 1 and 0, we say that player $p \in \{I, II\}$ *wins* a terminal history h if $u_p(h) = 1$ and *loses* if $u_p(h) = 0$. An extensive game is a win-lose if exactly one player wins each terminal history. In this case we can replace the players' utility functions with the *winner function*:

$$u : Z \rightarrow N$$

which indicates the winner of each terminal history.

Let $H_p = P^{-1}(p)$ denote the set of histories where it is player p's turn to move. A *strategy for player p* is a function σ which for every history $h \in H_p$ it gives the player a choice $\sigma(h) \in A(h)$. A player p *follows a strategy* σ during a history $h' = (a_1 \ldots a_n)$ if, whenever $h = (a_1 \ldots a_m) \in H_p$ is a subhistory (initial segment) of h', the history

$$(h, \sigma(h)) = (a_1, \ldots, a_m, \sigma(h))$$

is either h' or a subhistory of h'.

The following sets are important:

- H_σ, the set of histories where a particular strategy σ is followed.
- $Z_\sigma = H_\sigma \cap Z$, the set of maximal histories in which σ is followed and, in a win-lose game.
- Z_p, the set of maximal histories h which are a win for player p.

A strategy σ is a *winning* one for player p if $Z_\sigma \sqsubseteq Z_p$. In other words, σ is a winning strategy if and only if p wins every maximal play on which he or she follows σ.

Theorem 1 (Gale-Stewart Theorem). *Every two-player win-lose game with finite horizon and a unique initial history is determined.*

Proof. We let H_m denote the set of histories of height m. Suppose the game tree has height n. We extend the winner function to a labelling $\hat{u} : H \rightarrow \{I, II\}$ of every history in a canonical way. If the game has height

n then every history in H_{n-1} is terminal. For each such history h we let $\hat{u}(h) = u(h)$. Suppose $\hat{u}$ has been defined for all histories in H_m. Extend $\hat{u}$ to H_{m-1} in the following way. If $h \in H_{m-1}$ is terminal, then let $\hat{u}(h) = u(h)$. If h is not terminal and $P(h) = p$, then define

$$\hat{u}(h) = \begin{cases} p, & \text{if there is an } a \in A(h) \text{ such that } \hat{u}(h,a) = p \\ \bar{p}, & \text{otherwise} \end{cases}$$

where $\bar{p}$ denotes the opponent of p. Let $h_0 \in H_0$ be the unique initial history, and let $\hat{u}(h_0) = p_0$. We claim that p_0 has a winning strategy σ defined as follows. For all $h \in H_{p_0}$, if $\hat{u}(h) = p_0$ choose $\sigma(h)$ such that $\hat{u}(h, \sigma(h)) = p_0$. If $\hat{u}(h) = \overline{p_0}$ then choose $\sigma(h)$ arbitrarily. We show that σ is a winning strategy by showing that for all $h \in H_\sigma$ we have $\hat{u}(h) = p_0$. First observe that $h_0 \in H_\sigma$ because $\hat{u}(h) = p_0$ by hypothesis. Now suppose $(h,a) \in H_\sigma$. Then $h \in H_\sigma$ and so by inductive hypothesis, $\hat{u}(h) = p_0$. Now either $P(h) = p_0$ or $P(h) = \overline{p_0}$. In the first case we have

$$\hat{u}(h,a) = \hat{u}(h, \sigma(h)) = p_0$$

by the definition of σ. In the second case we must have $\hat{u}(h,a) = p_0$ because otherwise we would have $\hat{u}(h) = \overline{p_0}$ contrary to the inductive hypothesis. We have shown that $h \in H_\sigma$ implies $\hat{u}(h) = p_0$. In particular, $h \in Z_\sigma$ implies $u(h) = \hat{u}(h) = p_0$.

It should be noticed that Gale and Stewart used trees to study the class of two-player, win-lose games in which there is a unique initial history, and every terminal history has infinite length ω. Actually, for the game-theoretical interpretation considered here, the Gale-Stewart Theorem is actually a version of the much older Zermelo theorem which says that in any finite two-person game of perfect information in which the players move alternatively and in which chance does not affect the decision-making process, one of the players must have a winning strategy (assuming that the game cannot end in a draw). Zermelo considered the class of two-person games without chance, where players have strictly opposing interests and where only a finite number of positions are possible. (Zermelo published his article in 1913. It appears in English in Schwalbe & Walker, 2001, pp. 123–37.)

2.5   Game-theoretical semantics

Hintikka's formulation of game-theoretical semantics (Hintikka and Kulas, 1983; 1985) did not strictly follow the game-theoretical format displayed above, but a less formalized version that I will shortly describe. With each formula φ in negation normal form (the negation symbol occurs only in front of atomic formulas), structure M which interprets the non-logical vocabulary of φ and assignment s in M whose domain includes the free variables of φ a semantical game $G(M, s, \varphi)$ is associated. The game rules can be informally described as follows:

- The game has reached the position (θ, r) with θ an atomic formula or its negation and r and assignment: No move takes place. If r satisfies θ in M (that is, $M, s \models \theta$), then Eloise wins. Otherwise Abelard wins.
- The game has reached the position $(\theta \vee \theta', r)$: Eloise chooses one of the two formulas $\chi \in \{\theta, \theta'\}$ and the game continues from the position (χ, r).
- The game has reached the position $(\theta \wedge \theta', r)$: Abelard chooses one of the two formulas $\chi \in \{\theta, \theta'\}$ and the game continues from the position (χ, r).
- The game has reached the position $(\exists x \theta, r)$: Eloise chooses an individual $a \in Dom(M)$ and the game continues from the position $(\theta, r(x/a))$.
- The game has reached the position $(\forall x \theta, r)$: Abelard chooses an individual $a \in Dom(M)$ and the game continues from the position $(\theta, r(x/a))$.

It is straightforward to reformulate this informal description of the game as a win-lose extensive game.

Skipping over many details, we let the set of players $N = \{\text{Eloise}, \text{Abelard}\}$ and the set of histories $H = \bigcup \{H_\psi : \psi \text{ is a subformula of } \varphi\}$, where H_ψ is defined recursively. The initial history is (s, φ). Supposing that H_ψ has been formed:

- if ψ is $\chi_1 \vee \chi_2$, then $H_{\chi_i} = \{(h, \chi_i) : h \in H_{\chi_1 \vee \chi_2}, i = 1, 2\}$.
- if ψ is $\chi_1 \wedge \chi_2$, then $H_{\chi_i} = \{(h, \chi_i) : h \in H_{\chi_1 \wedge \chi_2}, i = 1, 2\}$.
- if ψ is $\exists x \chi$, then $H_\chi = \{(h, (x, a)) : h \in H_{\exists x \chi}, a \in M\}$.
- if ψ is $\forall x \chi$, then $H_\chi = \{(h, (x, a)) : h \in H_{\forall x \chi}, a \in M\}$.

The set of maximal histories Z is simply

$$Z = \bigcup \{H_\chi : \chi \text{ is a literal, subformula of } \varphi\}.$$

As for the player function, disjunctions and existential quantifiers are decision points for Eloise, while conjunctions and universal quantifiers are decision points for Abelard:

$$P(h) = \begin{cases} \exists, \ if \ h \in H_{\chi \vee \chi'} \ or \ h \in H_{\exists x \chi} \\ \forall, \ if \ h \in H_{\chi \wedge \chi'} \ or \ h \in H_{\forall x \chi} \end{cases}$$

As for winning and losing, Eloise wins a maximal history $h \in H_\chi$ if the atomic formula χ is satisfied by the current assignment; Abelard wins if it is not:

$$u(h) = \begin{cases} \exists, \ if \ M, s_h \models \chi \\ \forall, \ if \ M, s_h \not\models \chi \end{cases}$$

An example may help. Consider the semantical game associated with the sentence $\exists x \forall y (y \leq x)$ played on the standard model of arithmetic $\mathbb{N}$. For convenience let φ denote the original sentence and ψ denote the subformula $\forall y (y \leq x)$. Assume the initial assignment is empty so that $H_\varphi = \{(\emptyset, \varphi)\}$. Eloise moves first, choosing a value for x. Then

$$H_\psi = \{(\emptyset, \varphi, (x, a)) : a \in \omega\}.$$

After that Abelard picks up a value for y, and the game ends with

$$Z = \{(\emptyset, \varphi, (x, a), (y, b)) : a, b \in \omega\}.$$

Eloise wins if $b \leq a$; otherwise Abelard wins. We notice that Abelard has a winning strategy σ:

$$\sigma(\emptyset, \varphi, (x, a)) = (y, a + 1).$$

Everything is now set for the definition of truth and falsity:

Definition   *Let φ be a first-order formula, M a suitable structure and s an assignment in M whose domain contains the free variables of φ. Then*

$M, s \models^+ \varphi$ *iff Eloise has a winning strategy in $G(M, s, \varphi)$.*

$M, s \models^- \varphi$ *iff Abelard has a winning strategy in $G(M, s, \varphi)$.*

When φ is a sentence, and we let $\emptyset$ denote the empty assignment, we have as a special case:

φ is true in M ($M \models^+ \varphi$) iff Eloise has a winning strategy in $G(M, \varphi)$.

φ is false in M ($M \models^- \varphi$) iff Abelard has a winning strategy in $G(M, \varphi)$.

This definition together with the Gale-Stewart theorem yields the principle of bivalence:

For every first-order sentence φ and adequate structure M, either φ is true or φ is false in M.

2.6   Negation

We now relieve the restriction that the negation symbol occurs only in front of an atomic formula. Hintikka took every occurrence of negation to indicate the *role reversal* of the players. At the beginning of the game Eloise is the Verifier and Abelard is the Falsifier, but each time they encounter an occurrence of the negation symbol, they switch roles.

Obviously, we can always tell which player is the verifier in a history by counting the number of role reversals of the form $\neg\chi$ to χ. If there is an even number of such transitions, then Eloise is the Verifier; if there is an odd number, then Abelard is the Verifier.

Disjunctions and existential quantifiers are decision points for the verifier; conjunctions and universal quantifiers are decision points for the falsifier. Also the rules of winning and losing are redefined accordingly: The player who is the Verifier wins a maximal play if the atomic formula with which the play ended up is satisfied by the current assignment. The falsifier wins if it is not. One of the advantages of treating negation in this way is that we still remain within the class of two-player, win-lose games of perfect information. The definition of truth and falsity in a structure (with respect to an assignment) also remains intact. Therefore we can still apply the Gale-Stewart theorem and obtain the principle of bivalence as a special case.

Here is an example which illustrates the so-called *copy cat* strategy.

Example 1   We compare the semantical games $G(\mathbb{N}, \emptyset, \varphi)$ and $G(\mathbb{N}, \emptyset, \neg\varphi)$ where φ is the sentence $\exists x \forall y (y \leq x)$ from our earlier example. As observed, Abelard has a winning strategy σ in the first game, given by

$$\tau(\emptyset, \varphi, (x, a)) = (y, a + 1)$$

for any natural number a.

In the second game Eloise has a winning strategy σ that she borrows from Abelard's strategy in the first game

$$\sigma(\emptyset, \neg\varphi, \varphi, (x, a)) = \tau(\emptyset, \varphi, (x, a)) = (y, a + 1).$$

The last example should make clear that the following holds:

Proposition 1   For any first-order formula φ, structure M and assignment s:

- Eloise has a winning strategy in $G(M, s, \neg\varphi)$ iff Abelard has a winning strategy in $G(M, s, \varphi)$.

- Eloise has a winning strategy in $G(M, s, \varphi)$ iff Abelard has a winning strategy in $G(M, s, \neg\varphi)$.

This fact and the Gale-Stewart theorem yield:

Proposition 2 Let φ be a first-order formula, M a suitable structure, and s an assignment whose domain includes the free variables of φ. Then

$$M, s \models^+ \neg\varphi \Leftrightarrow M, s \not\models^+ \varphi.$$

Proof. The extensive game $G(M, s, \neg\varphi)$ is a two-player win-lose game which is finite and has a unique initial history. If Eloise has a winning strategy in $G(M, s, \neg\varphi)$, then by the previous fact Abelard has a winning strategy in $G(M, s, \varphi)$. Because the game is a win-lose game, Eloise cannot have a winning strategy in $G(M, s, \varphi)$.

Conversely, if Eloise does not have a winning strategy in $G(M, s, \varphi)$, then by the Gale-Stewart theorem, Abelard must have one. By the previous fact, Eloise has a winning strategy in $G(M, s, \neg\varphi)$.

2.7   Tarskian semantics as a special case

We are now ready to prove the Tarski-type semantics as a special case.

Theorem 2 (Assuming the Axiom of Choice). *The following holds for all first-order formulas φ, ψ, structure M, and s an assignment whose domain includes the free variables of φ, ψ.*

$$M, s \models^+_{GTS} \neg\varphi \Leftrightarrow M, s \models GTS^+ \varphi$$

$$M, s \models^+_{GTS} \varphi \vee \psi \Leftrightarrow M, s \models^+_{GTS} \varphi \text{ or } M, s \models^+_{GTS} \psi$$

$$M, s \models^+_{GTS} \varphi \wedge \psi \Leftrightarrow M, s \models^+_{GTS} \varphi \text{ and } M, s \models^+_{GTS} \psi$$

$$M, s \models^+_{GTS} \exists x\varphi \Leftrightarrow M, s(x/a) \models^+_{GTS} \varphi \text{ for some } a \in M$$

$$M, s \models^+ \forall x\varphi \Leftrightarrow M, s(x/a) \models^+_{GTS} \varphi \text{ for every } a \in M.$$

Proof. We have already established the case for negation. We consider two other cases.

Suppose Eloise has a winning strategy σ for $G(M, s, \varphi \vee \psi)$. Then $\sigma(s, \varphi \vee \psi) = \theta$ where θ is either φ or ψ. But then the strategy σ'

$$\sigma'(s, \theta, \dots) = \sigma(s, \varphi \vee \psi, \theta, \dots)$$

which mimics σ after the choice of θ is a winning strategy in $G(M, s, \theta)$.

Conversely, suppose that Eloise has a winning strategy in $G(M, s, \theta)$, for θ either φ or ψ. Define a winning strategy for Eloise in $G(M, s, \varphi \vee \psi)$ by

$$\sigma(s, \varphi \vee \psi) = \theta$$

$$\sigma(s, \varphi \vee \psi, \theta, \ldots) = \sigma'(s, \theta, \ldots).$$

Suppose now that Eloise has a winning strategy σ in $G(M, s, \forall x\varphi)$. For every $a \in Dom(M)$ define

$$\sigma_a(s(x/a), \varphi, \ldots) = \sigma(s, \forall x\varphi, (x, a) \ldots).$$

That is, σ_a mimics σ after Abelard chooses a. But then σ_a is winning for $G(M, s(x/a), \varphi)$.

Conversely, suppose that for every $a \in M$, Eloise has a winning strategy in $G(M, s(x/a)\varphi)$, for every $a \in M$. Choose one, say σ_a. Define now a winning strategy for $G(M, s, \forall x\varphi)$ by

$$\sigma(s, \forall x\varphi, (x, a), \ldots) = \sigma_a(s(x/a)\varphi \ldots)$$

That is, after the choice of a by Abelard, Eloise will mimic her winning strategy σ_a.

2.8 Strategies in semantical games as Skolem functions

Let us define in a precise way the Skolemization of a first-order formula which is, as already mentioned, the process of eliminating existential quantifiers by appropriate Skolem function symbols.

Let φ be a first-order variable and U a set of variables which includes the free variables in φ.

The *Skolem form* $Sk(\varphi)$ is defined by induction on subformulas ψ:

$Sk_U(\psi) = \psi$ (ψ is a literal)

$$Sk_U(\psi \vee \psi') = Sk_U(\psi) \vee Sk_U(\psi')$$

$$Sk_U(\psi \wedge \psi') = Sk_U(\psi) \wedge Sk_U(\psi')$$

$$Sk_U(\exists x\psi) = Sub(Sk_{U \cup \{x\}}(\psi xf(y_1, \ldots, y_n))$$

$$Sk_U(\forall x\psi) = \forall x Sk_{U \cup \{x\}}(\psi)$$

where $y_1, \ldots, y_n$ are the variables in U. The term $f(y_1 \ldots, y_n)$ is called a *Skolem term*. We abbreviate $Sk_\emptyset(\varphi)$ by $Sk(\varphi)$.

We now define *truth in a structure in the Skolem semantics*:

Definition   *Let φ be a first-order formula, M a suitable structure and s an assignment in M whose domain contains the free variables of φ. Then*

$M, s \models^+_{SK} \varphi$ *iff there are appropriate functions $g_1, \ldots, g_n$ in M to be the interpretation of the function symbols $f_1, \ldots, f_n$ in the Skolemization of φ so that*

$$Mg_1, \ldots, g_n, s \models Sk_{dom(s)}(\varphi).$$

Example 2   We consider the Skolemization of $\forall x \exists y x = y$.

- $Sk_{\{x,y\}} (x = y)$ *is* $x = y$.
- $Sk_{\{x\}} (\exists y x = y) = Subst(Sk_{\{x,y\}} (x = y), y, f(x)) = x = f(x)$.
- $Sk_{\emptyset} (\varphi) = \forall x Sk_{\{x\}} (\exists y x = y) = \forall x x = f(x)$.
- We notice that if in any structure M we interpret the function f by $g(x) = x$ then

$$M, g, \emptyset \models \forall x x = f(x).$$

- We conclude that

$$M, \emptyset \models^+_{SK} \forall x \exists y x = y.$$

Example 3   Let φ be the sentence (Shoenfield, 1967)

$$\exists x \forall y \exists z \forall w B(x, y, z, w)$$

where $B(x, y, z, w)$ is an atomic formula. For convenience we use the following abbreviations:

$A(x, y, z)$ for $\forall w B(x, y, z, w)$

$C(x, y)$ for $\exists z \forall w B(x, y, z, w)$

$D(x)$ for $\forall y \exists z \forall w B(x, y, z, w)$

Then:

- $Sk_{\{x,y,z,w\}} (B(x, y, z, w))$ is $B(x, y, z, w)$
- $Sk_{\{x,y,z\}} (A(x, y, z))$ is $\forall w B(x, y, z, w)$
- $Sk_{\{x,y\}} (C(x, y))$ is $\forall w B(x, y, f(x, y), w)$
- $Sk_{\{x\}} (D(x))$ is $\forall y \forall w B(x, y, f(x, y), w)$
- $Sk_{\emptyset}(\varphi)$ is $\forall y \forall w B(c, y, f(c, y), w)$.

Theorem 3   *Let φ be a first-order formula, M a suitable structure, and s an assignment restricted to the free variables of φ. Then φ is game-theoretically* true in M (relative to s) if and only if φ is true in M (relative to s) in the Skolem semantics. In symbols:

$$M,s \models^+_{GTS} \varphi \text{ iff } M,s \models^+_{Sk} \varphi.$$

For the proof see Theorem 4.13 in Mann, Sandu, and Sevenster (2011). In the proof one decomposes a supposed winning strategy for Eloise in the semantic game $G(M,s,\varphi)$ into appropriate Skolem functions which serve as witnesses for the function symbols of $Sk(\varphi)$; and vice versa. Thus the Skolem functions may be thought of as 'local' strategies for Eloise.

2.9   Two ways of Skolemizing

Actually it is customary to define the arguments of the Skolem functions for ordinary first-order sentences to be only the universally quantified variables, and not all the quantified variables in which the relevant existential quantifier occurs. This way of Skolemizing is based on a top down (and not bottom up) procedure.

We call a formula *universal* if it is in prenex normal form and all the quantifiers in its prefix are universal. The top down procedure for Skolemization starts with an arbitrary sentence φ in prenex normal form and then produces a sequence $\varphi^0, \varphi^{00}, \ldots, \varphi^{000}$ by dropping one by one all existential quantifiers until one reaches a universal sentence A_S:

If A is a universal sentence, then A_S is A.

If A is not a universal sentence, then it must have the form $\forall x_1 \ldots \forall x_n \exists y B$, for some $n \geq 0$. We let A^0 be $\forall x_1 \ldots \forall x_n Sub(B,y,f(x_1,\ldots,x_n))$.

If A^0 is not a universal sentence, then we form the sequence $A^{00}, \ldots, A^{000}$ until we reach a universal formula which is our A_S.

When we apply this procedure to our earlier example φ

$$\forall y \exists z \forall w B(x,y,z,w)$$

the result is φ_S which is $\forall y \forall w B(c,y,g(y),w)$. It is straightforward to show that for ordinary first-order sentences, the two procedures coincide (call them Sk and Sk^*).

Instead of giving a full proof, we work out an example.

Let M be an arbitrary structure. Suppose $\exists x \forall y \exists z \forall w B(x,y,z,w)$ is true in M, that is, there is an individual a and a function $f : M^2 \to M$ so that $M,a,f \models \forall y \forall w B(c,y,f(c,y),w)$. We show that there is an individual b and a function $h : M \to M$ so that, $b,h \models \forall y \forall w B(c,y,g(y),w)$. Just let $a = b$ and define for all $d \in M$: $h(d) = f(a,d)$. The converse is similar.

The two Skolemization procedures for ordinary first-order sentences have a clear game-theoretical counterpart, as our example illustrates:

the bottom-up procedure which results in $\forall y \forall w B(c, y, f(c, y), w)$ amounts to Eloise's local strategies being defined on all possible choices made earlier in the game, including Eloise's own ones. On the other side, the top-down procedure, which in our example results in $\forall y \forall w B(c, y, g(y), w)$, amounts to Eloise's local strategies being defined only on the opponent's earlier choices.

2.10   Skolemization for IF logic

Let me say a few things about Skolemization in IF logic so that the reader understands the difference between the ordinary first-order sentence expressing continuity

$$\forall x_0 \forall \varepsilon \exists \delta \forall y R(x_0, \varepsilon, \delta, y)$$

and the IF sentence expressing uniform continuity

$$\forall x_0 \forall \varepsilon (\exists \delta / \{x_0\}) \forall y R(x, \varepsilon, \delta, y).$$

Skolemization in IF logic goes exactly like in ordinary first-order logic, except for the clause

$$Sk_U((\exists x / W)\psi) = Subst(Sk_{U \cup \{x\}}(\psi), x, f(y_1, \ldots, y_n))$$

where $y_1, \ldots, y_n$ are the variables in $U - W$.

Perhaps it is worth emphasizing that the two ways of Skolemizing are not any longer equivalent (as already observed by Hodges, 1997) as the following example shows.

Example 4 The bottom-up Skolemization of the sentence $\forall x \exists z (\exists y / \{x\}) x = y$ is obtained through the following stages:

- $Sk_{\{x,z,y\}} (x = y)$ is $x = y$
- $Sk_{\{x,z\}} ((\exists y / \{x\}) x = y) = Subst(Sk_{\{x,z,y\}} (x = y), y, f(z)) = (x = f(z))$
- $Sk_{\{x\}} (\exists z (\exists y / \{x\}) x = y) = Subst(Sk_{\{x,z\}} (\exists y / \{x\}) x = y), z, g(x)) = Subst(x = f(z), z, g(x)) = x = f(g(x))$
- $Sk_{\emptyset} (\forall x \exists z (\exists y / \{x\}) x = y) = \forall x (x = f(g(x)))$.

On the other side, it is straightforward that the top-down Skolemization of $\forall x \exists z (\exists y / \{x\}) x = y$ is $\forall x (x = g(x))$.

2.11   Falsity and Kreisel's counter-examples

By analogy with the Skolem form of an ordinary first-order sentence, the Kreisel form $Kr(\varphi)$ is defined by induction on subformulas ψ:

- $Kr_U(\theta) = \neg\theta$ (θ is atomic)
- $Kr_U(\neg\theta) = \theta$ (θ is atomic)
- $Kr_U(\psi \vee \psi') = Kr_U(\psi) \wedge Kr_U(\psi')$
- $Kr_U(\psi \wedge \psi') = Kr_U(\psi) \vee Kr_U(\psi')$
- $Kr_U(\exists x\psi) = \forall x Kr_U(\psi)$
- $Kr_U(\forall x\psi) = Subst(Kr_{U\cup\{x\}}(\psi) \, xf(y_1,\ldots,y_n))$

where $y_1,\ldots,y_n$ are the variables in U.

Example 5 Suppose φ is the sentence from our previous example. We use the same abbreviations as before. The Kreisel form of φ is obtained in the following stages:

- $Kr_{\{x,y,z,w\}}\big(B(x,y,z,w)\big) = \neg B(x,y,z,w)$
- $Kr_{\{x,y,z\}}\big(A(x,y,z)\big) = \neg B\big(x,y,z\big), g(x,y,z)$
- $Kr_{\{x,y\}}\big(C(x,y)\big) = \forall z \neg B(x,y,z,g(x,y,z))$
- $Kr_{\{x\}}\big(D(x)\big) = \forall z \neg B(x,f(x),z,g(x,f(x),z))$
- $Kr_\emptyset\big(D(x)\big) = \forall x \forall z \neg B\big(x,f(x),z,g(x,f(x),z)\big).$

We now define *falsity in a structure in the Kreisel semantics* (for uniformity we use the same subscript as earlier):

Definition Let φ be a *first-order formula, M a suitable structure and s an assignment in M whose domain contains the free variables of φ. Then*

- $M,s \models^-_{SK} \varphi$ *iff there are appropriate functions $h_1,\ldots,h_m$ in M to be the interpretation of the function symbols $f_1,\ldots,f_m$ in the Kreisel form of φ so that*

$$M,h_1,\ldots,h_m,s \models Kr_{dom(s)}(\varphi).$$

We then note an analogous result:

Theorem 4 Let φ *be a first-order formula, M a suitable structure, and s an assignment restricted to the free variables of φ. Then φ is game-theoretically false in M (relative to s) if and only if φ is false in M (relative to s) in the Skolem semantics. In symbols:*

$$M,s \models^-_{GTS} \varphi \text{ iff } M,s \models^-_{Sk} \varphi.$$

As in the case of Skolemization, there is a top-down procedure to define the Kreisel form of an ordinary first-order sentence such that the arguments of the relevant functions are only the existentially quantified variables in the appropriate sentence. Applied to our example, this procedure will yield $\forall x \forall z \neg B(x, h(x), z, k(x, z))$ as the Kreisel form of φ. The two procedures turn out to be equivalent. More exactly, for every suitable structure M for the language of φ:

there are functions f, g such that

$$M, h, k, \emptyset \models \forall x \forall z \neg B(x, h(x), z, k(x, z))$$

iff

there are functions f', g' such that

$$M, f', g', \emptyset \models \forall x \forall z \neg B(x, f(x), z, g(x, f(x), z)).$$

Let us fix a suitable structure M for the language of φ (and ignore the empty assignment). Our definition of falsity together with our previous remarks tell us that, $M, \emptyset \models_{SK}^{-} \varphi$ iff there are functions f, g such that

$$M, f, g \models \forall x \forall z \neg B(x, h(x), z, k(x, z)).$$

By Theorem 4, the left side is equivalent to $M \models_{GTS}^{-} \varphi$. Thus φ is (game-theoretically) false iff there are functions f, g such that

$$M, f, g \models \forall x \forall z \neg B(x, h(x), z, k(x, z)).$$

Shoenfield calls the functions f, g counter-examples to φ and this has motivated our terminology in this section.

On the other side, we recall from Theorem 2 that

$$M \models_{GTS}^{+} \neg\varphi \Leftrightarrow M \not\models_{GTS}^{+} \varphi$$

Equivalently

$$M \models_{GTS}^{-} \varphi \Leftrightarrow M \not\models^{+} \varphi$$

that is,

$$M \models_{GTS}^{-} \varphi \Longleftrightarrow M \models_{GTS}^{+} \varphi.$$

In other words, φ is (game-theoretically) true in M iff φ is not (game-theoretically) false in M. By Theorem 4 and the definition of $M \models_{SK}^{-} \varphi$ we obtain:

$M \models_{GTS}^{+} \varphi$ iff there are no functions f, g such that

$$M, f, g \models \forall x \forall z \neg B(x, h(x), z, k(x, z)).$$

Applying the prenex operations to negation and bringing the second-order quantifiers to the object language we get:

$M \models_{GTS}^{+} \varphi$ iff $M \models \forall f \forall g \exists x \exists z B(x, f(x), z, g(x, z)).$

Thinking of Eloise's finding x and z as deterministic processes, we apply again the Axiom of Choice and introduce functionals $F(f,g)$ and $G(f,g)$, respectively:

(*) $M \models^+_{GTS} \varphi$ iff

$$M \models \exists F \exists G \forall f \forall g B \left(F\left(f,g\right), f\left(F\left(f,g\right)\right), G(f,g), g\left(F\left(f,g\right), G\left(f,g\right)\right)\right).$$

We have reached a characterization of game-theoretical truth in terms of the no-counterexample interpretation.

2.12   Some remarks on Dialectica interpretation

Gödel's *Dialectica* Interpretation (Gödel, 1958/1980) reduces the consistency of elementary number theory to the notion of validity in a system which employs the notion of computable functions of finite type over the natural numbers together with elementary principles for constructing such functions. The basic concept is that of 'computable function of type t' that Gödel explains as follows:

1. The computable functions of type are the natural numbers.
2. If the concepts 'computable function of type t_0', 'computable function of type t_1',..., 'computable function of type t_k' have been defined, then a computable function of type $(t_0 t_1, \ldots, t_k)$ is defined to be an operation which is always effective (and constructively recognizable as such), and which assigns a computable function of type t_0 to each k-tuple of computable functions of types $(t_1, t_2, \ldots, t_k)$.

The system also contains the axioms of identity (including those for functions), Peano's third and fourth axioms, the rule of substitution for free variables and a few axioms allowing us to define functions. Gödel remarks that 'Thus the axioms of the system (call it T) are formally much the same as those of primitive recursive number theory, except that the variables (apart from those on which induction is performed) and the defined constants can be of any finite type over the natural numbers' (Gödel, 1980, p. 135).

After introducing the system T, Gödel correlates every formula F of number theory with a formula F' of the system T having the form $\exists y \forall z A(x, y, z)$, where y and z are finite sequences of variables of any type. The correlation between F and F' proceeds by induction on the number k of logical operators in F. Here are the clauses of the correlation (we think of s and t as number variables):

- For $k = 0$, we let F' be F.
- Suppose now $F' = \exists y \forall z A(y, z, x)$ and $G' = \exists v \forall w B(v, w, u)$ have been defined. Then we stipulate:

1. $(F \wedge G)' = \exists y v \forall z w \left(A\left(y, z, x\right) \wedge B\left(v, w, u\right) \right)$.
2. $(F \vee G)' = \exists y v t \forall z w \left((t = 0 \wedge A\left(y, z, x\right)) \vee (t = 1 \wedge B(v, w, u)) \right)$.
3. $(\forall s F)' = \exists Y \forall s z A\left(Y\left(s\right), z, x\right)$.
4. $(\exists s F)' = \exists s y \forall z A(y, z, x)$.
5. $(F \rightarrow G)' = \exists V Z \forall y v \left(A\left(y, Z\left(yw\right), x\right) \rightarrow B\left(V\left(y\right), w, u\right) \right)$.
6. $(\neg F)' = \exists Z \forall y \neg A\left(yZ\left(y\right), x\right)$.

Hintikka (2002) observes that the rationale behind Gödel's interpretation rules can be given, at least in the basic cases, in game-theoretical terms. In such cases $F' = \exists y \forall z A(y, z, x)$ is a second-order formula which expresses the existence of a winning strategy y for Eloise in the game associated with F which wins the game against any strategy z of Abelard.

An example may help us to understand Hintikka's point. The translation of the formula $\forall s \exists t R(s, t)$ is obtained through the following steps (here $R(s, t)$ stands for a number-theoretical atomic formula):

- $(R(s, t))' = R(s, t)$.
- $(\exists t R(s, t))' = \exists t R(s, t)$ (Rule 4).
- $(\forall s \exists t R(s, t))' = \exists f \forall s R(s, f(s))$ (Rule 3).

We notice right away that $\exists f \forall s R(s, f(s))$ is nothing else but the existential quantification over the function symbols in the Skolem form of $\forall s \exists t R(s, t)$ (see Example 2.) The game-theoretical interpretation in this case is straightforward: the translation asserts the existence of a winning strategy f in the game associated with $\forall s \exists t R(s, t)$ (and the relevant structure.) Hintikka then entertains the hope that the above interpretation remains within the second-order existential ($\sum_{1}^{1}-$) fragment of second-order logic for the whole language of arithmetic. But he quickly observes that some of the rules, that is, rules (5) and (6) go beyond this fragment. For instance, the application of rule (6) to $\neg \forall s \exists t R(s, t)$ yields:

$$(\neg \forall s \exists t R(s, t))' = \exists Z \forall f \neg R\left(Z\left(f\right), f\left(Z\left(f\right)\right)\right).$$

In order to avoid the type-theoretical ascent, Hintikka proposes new interpretation rules for negation and conditionals. For him, the problem with rule (6) is that it is motivated by a conception of negation as classical, contradictory negation. This can be easily seen by way of our example. We first obtain, as above, the interpretation $\exists f \forall s R(s, f(s))$ of $\forall s \exists t R(s, t)$. Then when applying rule (6), we take negation to be the operation on $\exists f \forall s R(s, f(s))$ which asserts that there is no winning strategy in the game $\forall s \exists t R(s, t)$. We then apply the prenex operations to obtain $\forall f \exists s \neg R(s, f(s))$. Finally we apply the Axiom of Choice and introduce the functional Z.

Hintikka proposes replacing rule (6) with a new rule which better reflects the dual, switching role of the game-theoretical negation. By Proposition 1, Eloise has a winning strategy in the game associated with $\neg F$ if and only if Abelard has a winning strategy in the game associated with F. Whence rule (6) should be replaced with:

$(6^*)\,(\neg F)' = \exists z \forall y \neg A(y, z, x).$

When (6^*) is applied to $\neg \forall s \exists t R(s, t)$, the result is $\exists s \forall f \neg R(s, f(x))$, which gives Hintikka's intended interpretation: there is a strategy s by Abelard which wins the game $\forall s \exists t R(s, t)$ against any strategy f of Eloise.

Let me say two things about this proposal. First, it is true that it reduces the type-theoretical assent induced by rule (6). Nevertheless we still need a further reduction if we are to remain within the $\sum_{1}^{1}$ – fragment of second-order logic: $\exists s \forall f \neg R(s, f(x))$ is not yet a $\sum_{1}^{1}$ – formula. I guess Hintikka's reply would be that we further replace the universal second-order quantifier $\forall f$ by $\forall x$ or even by $\forall x / \{s\}$. I do not think that this will do the job but I shall not pursue this issue in this paper.

Second, and more importantly, I fail to see how the proposed interpretation would be able to do the job intended by Gödel, although I do not have an impossibility result. We recall Goldfarb's remarks about the use of choice functions by Skolem in the context of satisfiability (the proof of Löwenheim's theorem), and their use by Hilbert in the context of proof. I think that similar remarks apply to the present case.

The purpose of Gödel's interpretation is to show what kind of resources one needs for the proof of the consistency of arithmetic. He uses the translation to obtain the following result:

If a number-theoretical formula F is provable in Heyting arithmetics, then appropriate functions in T can be defined to serve as values of the existential quantifiers of F' so that the instantiation of F' with these functions is provable in T.

Gödel's result highlights the difference between the use of choice functions in the context of truth and satisfiability, on one side, and their use in the context of proof. In the case of our example $\forall s \exists t R(s, t)$, game-theoretical semantics shows how to pass from the truth in a structure M of the sentence $\forall s \exists t R(s, t)$ to the existence in M of a (Skolem) function f such that $\forall s R(s, f(s))$ is true in M. If in addition, $\forall s \exists t R(s, t)$ is provable in Heyting arithmetics, then Gödel's result shows us how we can define a function f in the system T so that $R(a, f(a))$ is provable in T for every choice of a. We clearly obtain additional information on f.

The IF-alternative to Gödel's *Dialectica* interpretation proposed by Hintikka restricts the translation to the $\sum_1^1 -$ fragment of second-order logic. The problem, as I see it, is that this fragment can only express the existential claims associated with the truth-conditions of number-theoretical sentences. It won't give us any additional information about the strategies of the players when these sentences are provable.

2.13   Kreisel's no counter-example interpretation

The concluding remarks of this last section are not new. Shoenfield (1967) emphasizes the distinction between the existence of appropriate functions and functionals associated with the truth of a sentence, on the one hand, and their being type-recursive (definable in the system T) when the formula is provable, on the other. To this purpose he chooses a correlation between number-theoretical formulas and formulas of a calculus Y of computable functions of finite type (essentially the systems T and Y are the same) different from Gödel's (Gödel points out at the end of his short paper that an analogous result can be achieved for classical number theory.) Whereas Gödel translates each number-theoretical formula F into a formula $F' = \exists x \forall y A(x, y, z)$, Shoenfield translates F as $F' = \forall x \exists y A(x, y)$. Here are the clauses of the translation (the underlying logic of the object language is classical):

1. If A is atomic then $A^* = A$.
2. If A is $\neg B$ and $B^* = \forall x \exists y B'(x', y', z)$ then $A^* = \forall y' \exists x \neg B'(x, y'(x), z)$ where y' is a sequence of new variables of the appropriate type.
3. If $A = B \vee C$, $B^* = \forall x \exists y B'(x, y, z)$ and $C^* = \forall x' \exists y' C'(x', y', z)$, then $A^* = \forall x x' \exists y y' (B'(x, y, z) \vee C'(x', y', z))$.
4. If $A = \forall w B$ and $B^* = \forall x \exists y B'(x, y, z)$, then $A^* = \forall w \forall x \exists y B'(x, y, z)$.

As in the case of Gödel's *Dialectica* interpretation, one can think of the translation as having a game-theoretical meaning in the basic cases. However, given the form of the translation, this meaning is not any longer 'there exists a winning strategy for Eloise' but 'there exists no counterexample for Abelard'.

Shoenfield (1967) has the following argument.

We call a formula *existential* if it is in prenex normal form and all the quantifiers in its prefix are existential. There is a top-down procedure which associates with every arbitrary formula φ in prenex normal form an existential formula. The procedure is analogous to the one which associates with every formula in prenex normal form a universal formula. That is, we form a sequence $\varphi^*, \varphi^{**}, \ldots$ by dropping all universal quantifiers one by one until one ends up with an existential formula A_H. Shoenfield (1967, p. 223) then proves the following theorem known as *Kreisel's no counterexample interpretation*:

Theorem 1 *Let φ be a sentence in prenex normal form which is a theorem of P. Let φ_H be $\exists x_1 \ldots \exists x_m B$ with B a quantifier-free formula, and let $f_1, \ldots, f_n$ be the new function symbols of φ_H. Then there are type-recursive functionals $F_1, \ldots, F_m$ such that*

$$B(F_1(f_1, \ldots, f_n), \ldots, F_m(f_1, \ldots, f_n))$$

is true for every choice of $f_1, \ldots, f_n$.

Shoenfield gives an illustration of the significance of this result using the sentence $\varphi = \exists x \forall y \exists z \forall w B(x, y, z, w)$. In this case $\varphi_H = \exists x \exists z B(x, h(x), z, g(x, z))$. Now the above theorem tells us there are type-recursive functionals F and G such that

$$B\left(F(h, k), f\left(F(h, k)\right), G(h, k), g\left(F(h, k), G(h, k)\right)\right)$$

is true for every choice of F and G.

The connection with the game-theoretical interpretation is now obvious. Recall our earlier equivalence (*):

(*) $M \vDash_{GTS}^{+} \varphi$ iff

$$M \vDash \exists F \exists G \forall f \forall g B\left(F(f, g), f\left(F(f, g)\right), G(fg), g\left(F(f, g)\, G(f, g)\right)\right).$$

Once again, the game-theoretical interpretation tells us that if φ is (game-theoretically) true in M, then there are functionals F and G in M. But if φ is provable in Peano arithmetic, then from the no counterexamples interpretation theorem, we get some additional information: the functionals F and G may be chosen-type recursive. All this is due to Shoenfield. The new thing in this paper is the connection between the counter-examples f and g and the winning strategy of Abelard in semantical games.

References

Bolzano, B. (1817). *Rein analytischer Beweis des Lehrsatzes, dass zwischen je zwey Werthen, die ein entgegengesetzes Resultat gewähren, wenigstens eine reelle Wurzel der Gleichung liege*. Translated by W. Engelmann as *Purely analytic proof of the* theorem that between any two values which give results of opposite sign, there lies at least one real root of the equation. Ewald (1996), 225–48.

Buld, B. *et al.* (eds) (2002) *Kurt Gödel: Wahrheit und Beweisbarkeit* (Wien: Hölder-Pichler-Tempsky).

Gödel, K. (1958/1980) 'Über eine bisher noch nicht benützte Erweiterung des finiten Standpunkt', *Dialectica*, 12, 280–7. Translated as 'On a Hitherto Unexploited Extension of the Finitary Standpoint', *Journal of Philosophical Logic*, 9, 133–42.

Goldfarb, D. W. (1979) 'Logic in the Twenties: The Nature of the Quantifier', *The Journal of Symbolic Logic*, 44, 351–68.

Hilbert, D. (1923) 'Die logischen Grundlagen der Mathematik', *Mathematische Annalen*, 88, 151–65.

Hintikka J., Kulas J. (1983) *The Game of Language: Studies in Game-Theoretical Semantics and Its Applications* (Dordrecht: D. Reidel).

Hintikka J., Kulas J. (1985) *Anaphora and Definite Descriptions: Two Applications of Game-Theoretical Semantics* (Dordrecht: D. Reidel).

Hintikka J., Sandu, G. (1989) 'Informational Independence as a Semantical Phenomenon' in Fenstad J. E. *et al.* (eds) *Logic, Methodology and Philosophy of Science* (Amsterdam: Elsevier), vol. VIII, 571–89.

Hintikka, J. (2002) 'Die Dialektik in Gödels Dialectica-Interpretation', in Buld *et al.* (2002), 67–90.

Hodges W. (1997) 'Compositional Semantics for a Language of Imperfect Information' in *Logic Journal of the IPGL*, 5, 539–63.

Mann, A., Sandu, G., and Sevenster, M. (2011) *Independence-Friendly Logic: A Game-theoretical Approach* (Cambridge: Cambridge University Press).

Osborne, M. J., and Rubinstein A. (1994) *A Course in Game Theory* (Cambridge, MA: The MIT Press).

Peirce, C. S. (1992) *Reasoning and the Nature of Things* (Cambridge, MA: Harvard University Press), 129–30.

Schwalbe, U., and Walker, P. (2001) 'Zermelo and the Early History of Game Theory', *Games and Economic Behavior*, 34, 123–37.

Shoenfield, J. (1967) *Mathematical Logic* (Reading, MA: Addison-Wesley).

Skolem, T. (1920) 'Logisch-kombinatorische Untersuchungen über die Erfüllbarkeit oder Beweisbarkeit mathematischer Sätze nebst einem Theorem über dischte Mengen', *Videnskapsselskapet Skrifter, I. Matematisk-naturvidenskabelig Klasse*, 6, 1–36.

3
Sixty Years of Stable Models

David Pearce

3.1 Introduction

This paper relates some episodes in the history of logic from the mid-twentieth Century to a highly influential line of research in logic programming and artificial intelligence that developed independently from around the end of the 1980s.[1] In 1988 Michael Gelfond and Vladimir Lifschitz published a celebrated paper (1988) on the stable model semantics of logic programs. Today, having built on and enlarged those key ideas of 24 years ago, answer set programming (often abbreviated as ASP) has emerged over the last decade as a flourishing paradigm of declarative programming, rich in theoretical advances and maturing applications.[2] This is one aspect of the legacy of stable models, and a very important one. Another aspect, equally important, but somewhat farther from the limelight today, lies in the ability of stable models to provide us with a valuable method of reasoning – to give it a name let us call it *stable reasoning*. In this essay I examine some of the foundational concepts underlying the approach of stable models. I try to answer the question: "What is a stable model?" by searching for a purely logical grasp of the stability concept. In so doing, I shall discuss some concepts and results in logic from more than 60 years ago. In particular, I look at questions such as:

- How does a notion of stability presented in a work on intuitionistic mathematics in 1947 relate to the Gelfond-Lifschitz concept of 1988?
- How does the notion of constructible falsity published in 1949 help to explain properties of negation arising in the language of ASP?
- Why is a seminal paper by McKinsey and Tarski (1948) important for understanding the relations between answer sets and modal and

epistemic logics, including recent results about modal embeddings of ontologies and rules?

Relating stable models and answer sets to logical concepts and results from the 1940s and even earlier sets the stage for our second line of discussion. I shall consider two different techniques for studying the mathematical foundations of stable reasoning and ASP. One of these is based on classical, propositional and predicate logic. Its main advantage is its familiarity, its wealth of results and its suitability for rapid prototyping. Its drawback is that it lies, in a sense that can be made precise, one level removed from the action. We first have to translate, manipulate and modify before we obtain relevant representations in classical logic that we could have obtained in simpler fashion using a non-classical logic. This need not but sometimes can add an *ad hoc* flavour to the modelling unless it is carefully spelt out why certain features are attached to the formalism at hand. Moreover in some cases it can add an unnecessary layer of complexity that also increases the difficulty of establishing properties and theorems.

The second approach to understanding stable models is more direct and immediate. It uses a non-classical logic in which stable models can be represented as minimal models, and formulas, programs and theories that are in a robust sense equivalent under stable reasoning can be shown to be logically equivalent. By making use of logical and metalogical results, this approach to the foundations of stable reasoning has considerable explanatory power, as I hope to show in the remainder of the paper. I shall use it to make some recommendations on not only how stable reasoning and the foundations of ASP can best be grasped and further studied, but also on some specific topics for future research.

This paper is more about concepts and methods than establishing new research results. It is also an essay about some episodes from the history of logic and the persons who have shaped them. However it does not aim to present a cultural or scientific history. It offers rather a patchwork of different ideas sewn together rather haphazardly in the hope of explicating a beautifully simple yet extraordinarily rich theory of reasoning.

3.2   A short history of stability

3.2.1   Stable part + minimal model = stable model

There is a strong Russian thread to the story of stable models. If we look back to the history of constructive mathematics and logic, we find that alongside the names of Brouwer and his student Arend Heyting, the name of Andrei Kolmogorov looms large.[3]

While Brouwer developed his ideas on intuitionism already in the early years of the twentieth century, it was only in 1930 that Heyting published the now standard axiomatization of intuitionistic logic. However almost all of Heyting's calculus was anticipated by Kolmogorov in 1925. In fact, Kolmogorov provided the axioms for a system now known as minimal logic, that is, intuitionistic propositional logic minus the 'explosive' axiom $p \to (\neg p \to q)$. Actually, this axiom was a controversial matter in constructivism and even Heyting himself regarded it as suspect. Kolmogorov (1925) also provided the first embedding of classical into intuitionistic mathematics, using the *double negation* translation, a variant of which was later rediscovered and made famous in (Gödel, 1933). Around the same time the key idea emerged that, in contrast to the classical interpretation of logic, one should understand the basic logical connectives in terms of (operations on) constructive proofs. This cornerstone of constructivism has since become known as the Brouwer-Heyting-Kolmogorov interpretation, which in Kolmogorov's case was expressed in terms of a logic of problems (Kolmogorov, 1932).

As is well known, intuitionism rejects the double-negation law in the form $(\neg\neg p \to p)$, and so in intuitionistic theories such as arithmetic we cannot in general infer a proposition p from the falsity of $\neg p$ (that is, infer p from $\neg\neg p$), nor conclude in general that either a proposition p or its negation $\neg p$ must be true. However within an intuitionistic mathematical theory some sentences, typically sentences about finite, constructively presented objects, may have a definite truth value. Moreover there are propositions p such that $\neg\neg p \to p$ is a valid statement. In 1947 the Dutch mathematician David van Dantzig proposed to call the set of such statements the *stable part* of an intuitionistic theory.[4] van Dantzig was a follower and correspondent of Gerrit Mannoury, a founding member of the Dutch Significs movement that included also his student Brouwer, as well as Evert Beth who was to make major contributions to intuitionistic logic.[5] van Dantzig's stable mathematics (van Dantzig, 1947; 1951) can be regarded as an attempt to extend the double negation embedding of the Kolmogorov-Gödel kind in order to develop an elementary part of classical analysis within intuitionistic mathematics.[6]

Within a non-classical, intuitionistic theory, therefore, one may identify a stable part that in a logical sense behaves almost classically. At the time van Dantzig was writing in the 1940s, intuitionistic logic and mathematics had of course a proof theory, but no established semantics or a counterpart to the classical theory of models being developed by Tarski, Robinson and others. The beginnings of an intuitionistic semantics emerged only in the 1950s with Beth's work on trees and later the semantics of possible worlds.

If the stable part comprises sentences of an intuitionistic theory that satisfy double negation and in this sense behave classically, it is also tempting to call *stable* a model in the intuitionistic sense that corresponds to a classical model.[7] Recall that in the semantics of possible worlds we can model intuitionistic logic by considering a partially order sets, or *posets*, where each element or node w is associated with a set of atomic sentences that are *verified* at that point. Without loss of generality we can restrict attention to finite posets containing a single initial node or *root* as in Figure 1.

Formally they correspond to the idea of developing stages of knowledge, where every later stage (higher up the poset) includes all knowledge verified at its predecessor nodes. Looking at such finite, rooted frames, therefore, the classical models are simply those that comprise a single node or world, as in the example above on the right of the picture. Generally speaking, a theory whose semantics is described by such relational models may have among them several or even many such "classical" models. This idea of a single-world model is the first ingredient of the concept of a stable model.

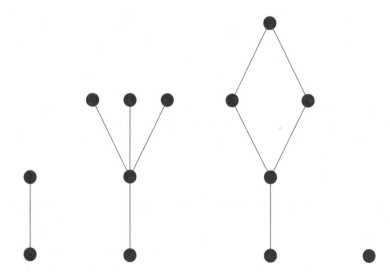

Figure 1:   Rooted posets

The second component of the stability concept arises from the fact that the stable model semantics yields a non-monotonic form of reasoning. Virtually all such forms of reasoning can be represented by means of minimal or preferred models (even if they are initially presented in some purely syntactical fashion). Interestingly, of the two founders of stable models, it was Vladimir Lifschitz who was most closely associated with minimal model reasoning through his work on circumscription, while Michael Gelfond had noticed an interesting connection between logic programs and autoepistemic reasoning, where models are defined via a fixpoint construction.

Let us then try to add minimality to our previous idea of classicality. In other words, among those intuitionistic possible world models that correspond to classical models, which if any could be said to be *minimal*? Intuitively we want to reproduce the idea that there is no other model 'smaller' than it. Take such a total, single node model (Figure 2).

Evidently we can compare it directly with some non-classical relational models, in fact all those models that terminate in a single node, for example Figure 3.

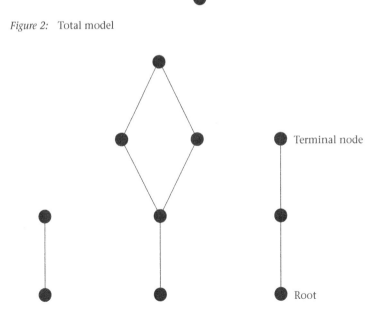

Figure 2: Total model

Figure 3: Terminating models

Let us call them *terminating* models. If such a possible world model agrees with our classical model at its terminal node and has a 'smaller' initial node or root (in the sense of verifying fewer sentences), then, since it is the root node that determines what is true in the model as a whole, we can legitimately claim this model to be comparable with and smaller than our initial classical model.[8] On the other hand, if our initial model has no such smaller relational model, among all the models of the theory, then indeed we can regard it as minimal. This, I suggest, is the concept of stable model: the combination of stable part, or classicality when transferred to the setting of models, together with minimality.

Evidently we have simplified and left out plenty of possible world structures. We have ignored say all those posets that fork and possess more than one maximal node. Why? One reason is this. When we identify our initial, classical model with a terminating model, we are essentially saying that these two models precisely agree on what is false. In both models therefore falsity is behaving classically. First, they both agree on the set of all $\neg p$ that are true. And more particularly our non-classical model has the feature that a proposition is false in the model *as a whole* just in case it is false at that single, terminal, 'classical' node. Evidently the same would not be true of any 'fork' model: thus Figure 4.

Even if one of its end nodes agreed with our initial model, refutation in such models is highly non-classical, in the example below we refute $a \wedge b$ but we do not refute either a or b individually.

Another way to view our simplification is by interpreting possible world models as representing the development of knowledge over time, or perhaps in the Kolmogorov spirit, as the solving of problems over time. Although at each node our state of knowledge is only partial,

Figure 4:   Fork model

and the development of knowledge may branch over time, *in the end* everything becomes decided and all problems solved.

Here then is our first characterization of the semantics of Gelfond and Lifschitz:

- a stable model is a total model that is minimal over the terminating intuitionistic frames

These frames characterise the logic KC of weak excluded middle, that is, intuitionistic logic plus the axiom $\neg\neg p \vee \neg p$. It is interesting to note that in this logic stability implies decidability, and it is actually the weakest super-intuitionistic logic for which this property holds. We thereby verify the equation

- stable part + minimal model = stable model

Now it turns out that when considering intuitionistic frames we can actually simplify matters considerably. Let $\mathcal{M}$ be a total model for some theory T that verifies the set of atoms X. Suppose it is not stable in the above sense, so there is a terminating model $\mathcal{M}'$ of T, such that $\mathcal{M}' < \mathcal{M}$. The picture looks like this.

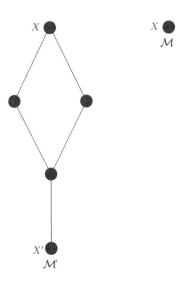

Figure 5: Comparing models

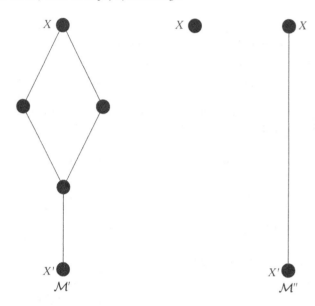

Figure 6:   A smaller linear model

where $X' \subset X$. Then there is a linear, two-node model, say $\mathcal{M}''$ of T such that $\mathcal{M}'' < \mathcal{M}$, Figure 6.

In fact we can choose $\mathcal{M}''$ to consist precisely of the root node and the terminal node $\mathcal{M}'$. An easy induction argument shows that if φ is a formula true in $\mathcal{M}'$, then φ is also true in $\mathcal{M}''$.

Linear, rooted frames with two nodes like $\mathcal{M}'$, are also called *here-and-there* frames. They characterize a super-intuitionistic logic called here-and-there, in symbols *HT*, and sometimes referred to as Gödel's 3-valued logic or as the Smetanich logic, after another Russian mathematician (Smetanich, 1960).[9] Now a total model that is minimal over the here-and-there frames has been called an *equilibrium* model, (Pearce, 1997; 2006). Evidently, if $\mathcal{M}$ is not an equilibrium model, there is a smaller *HT*-model which is clearly also a *KC*-model; hence $\mathcal{M}$ is not stable. While if $\mathcal{M}$ is not stable our argument shows there is a smaller *HT*-model, and so $\mathcal{M}$ is not an equilibrium model. So we can conclude that stable models and equilibrium models coincide. Thus we are led to our second characterization of the semantics of Gelfond and Lifschitz:

- a stable model is a total model that is minimal over the here-and-there frames.

In other words we can equate stable reasoning with minimal model reasoning in the super-intuitionistic logic *HT*.

The original definition of stable model given in 1988 looks quite different to this, but is fully equivalent to it.[10] Moreover our definition has the advantage of being fully logical and thus applicable to arbitrary formulas, while the original was given for special formulas, so-called rules of logic programs, with a restricted syntax.[11]

There are other advantages to a logical approach. One clear benefit is that we can use logical deduction, in this case deduction in *HT*, to infer properties of stable reasoning. For example, formulas or theories that are equivalent in *HT*, that is, have the same models, must also have the same stable models, so they are also equivalent under stable reasoning. Moreover they are equivalent in the following, strong sense: if T_1 and T_2 are *HT* equivalent then for any additional formulas T, also $T_1 \cup T$ and $T_2 \cup T$ are equivalent under stable models. This, strong form of equivalence means that T_1 and T_2 behave similarly whatever the larger context in which they are embedded. In non-monotonic reasoning it is important to distinguish these two concepts of equivalence since only the stronger form guarantees inter-substitutability *salva veritate*.

Some years ago it was shown (Lifschitz, Pearce & Valverde, 2001) that theories are strongly equivalent under stable model semantics *if and only if* they are equivalent in *HT*. So *HT* provides a means to divide formulas, theories and logic programs into equivalence classes of equivalent objects. Moreover, it is a maximal logic with this property. In fact, since it is the strongest super-intuitonistic logic that is properly contained in classical logic (and in turn it properly contains all others), it is easy to see that it is a maximal logic with the property that equivalent theories always have the same stable models.

Just as logicians study and apply valid and admissible inference rules in a logic, so in logic programming it is customary to study program transformations that allow programs to be simplified and optimized. Indeed such transformations are often built into stable reasoners and answer set solvers. To be useful, such transformations must preserve stable models, but many do more than this: they preserve stable models no matter what the context is, in other words they preserve strong equivalence. Now our previous observations provide a welcome link between inference rules and program transformations: namely the transformations preserving strong equivalence are precisely those that result from equivalence-preserving inference rules of *HT*. On the other hand a transformation that preserves stable models but *not* strong equivalence must be nonmonotonic in nature; it cannot be an instance of an equivalence preserving inference rule in *HT*.[12] So there is a natural distinction here that is important in program analysis and transformation, between monotonic rules entirely captured within *HT* and stronger, nonmonotonic rules.

Let me conclude these remarks by noting a few more recent results of logic that are relevant here. Returning to our earlier discussion of the logic *KC* of weak excluded middle, de Jongh and Hendriks (2003) showed that *KC* is the weakest super-intuitionistic logic that precisely captures the strong equivalence of *nested* logic programs.[13] So for these programs strong equivalence is captured by all and only those logics in which van Dantzig stability implies decidability. On the other hand, Cabalar and Ferraris (2007) showed that in *HT* any theory is equivalent and hence strongly equivalent to a special form of disjunctive logic program, where negation is allowed in the heads of rules. Even more recently, Grigori Mints announced a cut-free Gentzen style sequent calculus for *HT* and observed that this logic is conservative over *KC* on formulas without nested occurrences of implication.[14]

These recent results can be used to derive further insights into the nature of stable reasoning. If we are looking at super-intuitionistic logics that can serve as a basis for stable model semantics it seems rational to restrict attention to those logics in which we can capture strong equivalence. de Jongh and Hendriks showed that *KC* is the weakest such logic, even if you are interested in just normal logic programs (not just nested programs). On the other hand, *HT* is the strongest such logic and it also captures strong equivalence for arbitrary theories. Evidently, *HT* is also the weakest and hence the only such logic that works for arbitrary theories. If there were a strictly weaker such logic *L* between *KC* and *HT* there would be sentences that are *HT* equivalent yet not *L*-equivalent, and so their strong equivalence would not be captured in *L*. By the same token it follows that other than *HT* itself there is no logic *L* between *KC* and *HT* that allows arbitrary sentences to be reduced to the form of nested programs. Again by $L \subset HT$ there would be sentences that are *HT*-equivalent but not *L*-equivalent. After conversion into the normal form of nested programs they would continue to be equivalent in *HT*, and hence strongly equivalent, but obviously not in L. But this contradicts that by strong equivalence and the results of de Jong and Hendriks (2003) the sentences would also be KC-equivalent, so they would have to be L-equivalent after all.

In short, while any logic from KC to HT can serve as a basis for reasoning about logic programs under stable model semantics, HT is the only one that works for arbitrary theories and indeed permits replacing such theories by fully equivalent logic programs.

3.2.2   A tale of two negations

Two years after the first presentation of stable model semantics, Michael Gelfond and Vladimir Lifschitz contributed a further paper to the 1990 edition of the International Conference of Logic Programming (ICLP). It was entitled 'Logic Programs with Classical Negation', (Gelfond & Lifschitz, 1990), and introduced the nomenclature *answer set* for the first time. The authors observed that, besides the negation-as-failure or negation-by-default operator typical of logic programming, it is very useful to have an additional form of negation that directly corresponds to falsity. In virtue of its properties they labelled it 'classical' negation. At first glance this terminology might seem appropriate, but on closer inspection it turns out to be quite misleading. Several fundamental properties of classical logic did not hold for this operator, they include *excluded middle, contraposition* and *modus tollens*. A much closer conceptual fit from logic was the notion of *strong* negation.[15]

Strong negation in logic originates from a paper by David Nelson (1949) on constructible falsity. Some sources cite the Russian mathematician and logician Andrei Andreevich Markov as a co-founder of this concept. Indeed, while Nelson's paper appeared in the Spring of 1949, in October of that year Markov was already lecturing on this topic in Leningrad University's Mathematical Institute. Only an extended abstract of that lecture was published as (Markov, 1949), but it clearly refers to Nelson's work and takes it as a point of departure. What is clear is that Markov was already analysing logical properties of strong negation in constructive logic and proposing a strong version of logical equivalence which he called *complete* equivalence. Shortly thereafter it is evident that constructible falsity became a research topic in Markov's group at Leningrad University. It was studied in particular by one of Markov's pupils, Nikolai Nikolaevich Vorob'ev, who in 1952 provided the first axiomatic system for strong negation (Vorob'ev, 1952a; 1952b).[16]

It may seem odd at first to speak of an axiomatic system for strong negation, rather than specifying a particular logic containing strong negation as one of its operators. In fact the Vorob'ev axioms have the remarkable property that one can add them to any super-intuitionistic logic, including classical logic, and one obtains a conservative extension of that logic. Moreover many key properties of a logic are preserved when passing to its least strong negation extension (Kracht, 1998).

Vorob'ev already established an important property of strong negation. Many years later, the same property became a design feature of the way the second negation operator was introduced into answer set semantics: the new negation was only allowed to stand directly in front of an atom. This convention was retained throughout successive extensions of the syntax of programs, from normal programs, through to disjunctive programs and even programs with nested expressions (Lifschitz *et al.*, 1999). In each case therefore the only difference between a program rule with or without strong negation was that in the former case the basic elements were literals rather than atoms.[17] This convention allowed for a very simple extension of the definition of stable model and an easy proof of some important properties of it. However, once the second negation is identified as strong negation in the Nelson and Markov sense it is easy to see that this syntactic restriction is quite harmless. Vorob'ev showed that strong negation can always be driven-in to stand directly in front of atoms; the resulting *negation normal form* is logically equivalent to the original formula.

No sooner has the student of answer set programming encountered the second, strong negation operator than he or she learns a reduction technique that shows how answer sets can be characterized in terms of the stable models of a program without strong negation. This reduction technique involves introducing a new atom or predicate symbol for each strongly negated literal. An earlier version of this idea also has its roots in constructive logic and was also developed by a Russian logician, albeit this time not a student from Leningrad. Yuri Gurevich, a former student of the Ural State University, elaborated in (Gurevich, 1977) a method for reducing the completeness problem for constructive predicate logic (with strong negation) to that of ordinary intuitionistic predicate logic. It involved an application of the Vorob'ev technique to drive strong negation into formulas and then a replacement of negated atoms by new (positive) atoms. Thus one associates with any formula φ of the original language a strong negation-free formula, say φ'. If T is any theory in the language of constructive logic, set $T' = \{\varphi' : \varphi \in T\}$. Let S stand for the collection of all formulas of the form $P'(x) \rightarrow \neg P(x)$. Then Gurevich's technique establishes that

$$T \vdash_N \varphi \text{ iff } T' \cup S \vdash_I \varphi' \tag{1}$$

where $\vdash_N$ (respectively $\vdash_I$) stands for inference in Nelson's logic N (respectively intuitionistic predicate logic I).[18] This forms the basis for reducing the completeness of constructive logic to that of intuitionistic logic. It is easy to see that (1) continues to hold if we replace intuitionistic inference by that of (quantified) here-and-there logic, QHT, and replace N by the least strong negation extension of here-and-there. By Gurevich's method one can quickly derive the main property of Gelfond and Lifschitz's second negation, established in Gelfond and Lifschitz (1990): that the answer sets of T can be recovered from the answer sets or stable models of T', by translating the primed atoms back into strongly negated literals. The pleasing aspect of the Gurevich technique is its generality: since it works for arbitrary theories it covers the case of answer sets for any syntactical class of logic program.

3.2.2.1   *Negation and contraposition*

The logical approach to understanding the two negations of stable reasoning helps to clarify their interrelations. For example, while the above method may be seen as a way to eliminate strong negation when computing answer sets, we also know from logic that '~' is genuinely stronger that '¬', since $\sim p \to \neg p$ is a theorem of N. Moreover, it is also well known than '¬' is definable in N and hence eliminable from it; so in principle we can do logic programming with just strong negation providing we accept nested implications: we set $\neg p := p \to \sim p$.

The logical approach also helps to dispel some common misunderstandings about properties of negation in logic programming. The subject of contrapositive reasoning in logic programming is one where errors and misunderstandings have been frequent. For a long time it was accepted wisdom in logic programming to regard program formulas or rules as having a unique direction and not to obey any form of contraposition. The clearest evidence came from a simple formula such as

$$a \leftarrow \neg b \tag{2}$$

(in logic programming notation) that behaves differently from its 'contrapositive'

$$b \leftarrow \neg a. \tag{3}$$

The first has {a} as a stable model, the second has {b}. So it seems that '←' is directional, more like an inference rule, and such rules do not entail their contrapositives. The case of strong negation is apparently similar. While

$$a \leftarrow \sim b \tag{4}$$

and

$$b \leftarrow \sim a \tag{5}$$

have the same (empty) answer set, as soon as we add say $\sim b$ to each, their behaviour is quite different: a becomes derivable in the first case and not in the second. One might even think that the failure of contraposition is a result of the special properties of '←' making it different from ordinary logical implication. But nothing could be further from the truth. First there is nothing special about '←' as an implication connective, providing one observes that its underlying logic is nonclassical. Second, the behaviours of '¬' and '$\sim$' are actually quite different. While the former obeys a form of contraposition, the latter does not. The source of the confusion and common misunderstanding stems from wrongly interpreting (3) as the contrapositive of (2) by supposing that the underlying logic is classical. Providing weak negation '¬' is understood as intuitionistic, it is clear that the contrapositive form of (2) is actually

$$\neg\neg b \leftarrow \neg a \tag{6}$$

which is neither intuitionistically nor *HT*-equivalent to (3).[19] In fact, (6) is derivable from (2), though not equivalent to it (indeed (6) has no stable model). So in stable reasoning, formulas containing weak negation entail their contrapositives but (3) does not follow from (2) because '¬' does not obey the law of double-negation. In the case of strong negation exactly the converse holds: '$\sim$' obeys the double-negation law but fails contraposition. (5) is a contrapositive of (4) but not entailed by it. This has to do with the strong, constructive nature of the negation. The constructive meaning of (4) is that any refutation of b can be effectively converted into a proof of a. This is quite different from (5) which states that a refutation of a can be converted into a proof of b.

3.2.3   Modal negation

When the Polish logician, Alfred Tarski, set sail for America on the *Piłsudski* in August 1939 he had no tenured university appointment in the offing; merely an invitation to the *Unity of Science Congress* and a promise from Quine to arrange him some guest lectures at US universities. It was to take three years before Tarski obtained a tenured professorship, at the University of California, Berkeley, in 1942. In the meantime he had formed a close friendship with the logician J. C. C. 'Chen' McKinsey who was one of his first visitors at Berkeley, on a Guggenheim Fellowship in 1942–43. They began to work on algebras of topologies for propositional logics, taking up a theme from Tarski's pre-war publications. Later, in 1944, Tarski visited McKinsey in Montana where he was then working, and this led to their first joint paper, on closure algebras, published in 1946. Two years later their collaboration resulted in a further joint paper (McKinsey & Tarski, 1948). Using algebraic methods, they proved there a conjecture that Gödel (1933) had formulated in a short paper. Gödel had observed that Heyting's intuitionistic calculus *I* could be translated into the modal logic now known as *S4*, by interpreting an intuitionistic proposition p as 'p is provable', or as $\Box p$ in the modal language. Gödel's extended his translation, call it *G*, to the intuitionistic connectives by setting

$$(p)^G = \Box p \tag{7}$$

$$(\varphi \wedge \psi)^G = \varphi^G \wedge \psi^G \tag{8}$$

$$(\varphi \vee \psi)^G = \Box \varphi^G \vee \Box \psi^G \tag{9}$$

$$(\varphi \rightarrow \psi)^G = \Box \varphi^G \supset \Box \psi^G \tag{10}$$

$$(\neg \varphi)^G = \neg \Box \varphi^G \tag{11}$$

Gödel (1933) stated without proof that the translation of any theorem of *I* is a theorem of *S4*, and he conjectured that the converse would also be true, so that *G* yields an embedding of *I* into *S4*. He also mentioned an alternative translation, obtained by placing a $\Box$ before each subformula of φ.[20]

Using algebraic methods, McKinsey and Tarski proved Gödel's conjecture using still further variants of the translation. One of these, that has come to be known as the Tarski interpretation (Rautenberg, 1979), is like the second Gödel translation except that the translation of a conjunction remains as in (8).[21] The McKinsey-Tarski embedding was studied intensively in the ensuing years. By the 1970s and 80s a systematic study had emerged dealing with the general problem of how super-intuitionistic logics are embedded into modal logics, in particular how the lattice of extensions of *I* is related to the lattice of normal

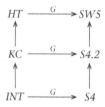

Figure 7: Some model embeddings

extensions of *S4*; an important reference here is the work of the Russian logicians, Maksimova and Rybakov (1974). Following another seminal work by the Georgian logician Leo Esakia (1979), modal counterparts of super-intuitionistic logics became known as *modal companions*.[22] Each super-intuitionistic logic has an infinite number of modal companions, for instance of the logics we met earlier *KC* has *S4.2* as a modal companion, while *HT* has the companions *S4F* and *SW5*.[23] These are both normal extensions of S4 contained in S5. So the picture looks like this (Figure 7) where up arrows represent inclusion and right arrows indicate an embedding under the Gödel-McKinsey-Tarski translation, *G*.

What happens now if we pass to the nonmonotonic versions of these logics? We have seen that, based on minimal models in *HT*, we can regard stable reasoning as a nonmonotonic extension of *HT*, elsewhere called *equilibrium logic* (Pearce, 2006). In the case of a normal modal logic *S*, one can define what are called the S-*expansions* of a set of modal sentences *T* by

$$E = Cn_S(T \cup \{\neg\Box\varphi : \varphi \notin E\}) \tag{12}$$

where Cn_S is the inference operation of S. Then the nonmonotonic variant of S is determined by truth in each such S-expansion. In each case, let us denote these nonmonotonic variants by adding a star '*', so we obtain *HT**, *S4F**, *SW5** and so on. And now the following question springs naturally to mind: Can we lift the embedding G from the monotonic to the nonmonotonic case? In other words can we, for instance, make this diagram commute? (Figure 8)

The answer is, *Yes, we can!* This can be shown in several ways but let me mention one approach in particular that has not yet been fully exploited in the literature. When we use the logical representation of stable reasoning it is not hard to discover a counterpart to equation (12), in other words a logical, fixpoint condition for stability. The condition

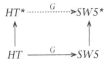

Figure 8:

is this (see Pearce, 1999a,b):

$$\mathcal{M} \text{ is a stable model of } T \text{ iff } Th(\mathcal{M}) = Cn_{HT}(T \cup \{\neg\varphi : \varphi \notin Th(\mathcal{M})\}) \quad (13)$$

where T is a theory, $Th(\mathcal{M})$ is the set of sentences true in $\mathcal{M}$ and Cn_{HT} is HT-consequence.

The right-hand side of (13) bears an uncanny resemblance to condition (12) defining the modal expansions. (13) can also be fine-tuned. For example, if T is a logic program it suffices to build this equation from negated atoms. And in general, thanks to an observation from Mauricio Osorio and his group (Osorio, Navarro Perez & Arrazola, 2005), it is easy to check that instead of adding the negation of all sentences fulfilling the condition, it suffices to add negated literals (that is, atoms or their negations). Now, using (12) and (13) as our starting point, we can begin to apply the monotonic embedding G which will relate an HT-theory T to its modal counterpart, T^G. Eventually, after some calculations we shall reach our target, a correspondence between HT and $SW5^*$, as follows.

Theorem 1   *Let T be a theory and $\mathcal{M}$ a total model in HT whose true atoms are M. Then $\mathcal{M}$ is a stable model of T if and only if ST(M) is an SW5-expansion of T^G.*

To understand this, we need to explain one more piece of notation. Following Stalnaker (1980) a set of modal or epistemic sentences E is said to form a stable belief set if E is closed under classical propositional consequence and

- $\varphi \in E$ implies $\Box\varphi \in E$;
- $\varphi \notin E$ implies $\neg\Box\varphi \in E$.

It is easy to see that for normal modal logics S every S-expansion of a theory is a stable belief set containing that theory. In fact it turns out that, given a set X of non-modal propositions, there is a unique stable belief set, in symbols $ST(X)$, such that the non-modal propositions in $ST(X)$ are exactly the classical consequences of X, (Marek &

Truszczyński, 1993a). This explains the notation of Theorem 1 and the correspondence between HT^* and $SW5^*$.

Given the close connections between provability and modal logic it is not surprising that efforts were made quite early in the history of logic programming to provide a modal interpretation for Prolog's concept of negation as failure-to-prove. One such attempt was made by Michael Gelfond in 1987, which aimed to show how logic programming could be used as a deductive mechanism for a special form of nonmonotonic reasoning, known as *autoepistemic* logic. Gelfond used the epistemic or belief interpretation under which a formula

$$A_1 \wedge \ldots \wedge A_m, \neg A_{m+1} \wedge \ldots \neg A_n \rightarrow A_0 \tag{14}$$

of a normal logic program was translated as

$$A_1 \wedge \ldots \wedge A_m, \neg \Box A_{m+1} \wedge \ldots \neg \Box A_n \rightarrow A_0 \tag{15}$$

under the belief reading of '$\Box$'.

For the class of so-called 'stratified' logic programs, that possess unique least Herbrand models, Gelfond's translation established an exact correspondence between query answering under SLDNF-resolution and provability in the nonmonotonic system of autoepistemic logic. More precisely, a query φ is derivable from a (stratified) program $\prod$ if and only if the translation of φ belongs to every autoepistemic expansion of the above translation of $\prod$.[24]

As we now know, Gelfond's discovery was to have a more profound effect on the foundations of logic programming than on the development of autoepistemic logic. It was to take him and Vladimir Lifschitz along the path to developing stable models.[25] Applying his modal interpretation of negation-as-failure offered up a new possibility, to provide a foundation for logic programming based not on minimal (classical) reasoning but on reasoning with stable belief sets (initially autoepistemic but, as we have seen, this is not an essential feature of it). Since the stable, S-expansions or autoepistemic expansions are defined for theories in general, and not only for stratified programs, one could apply this idea for arbitrary programs $\prod$, not only those possessing a least Herbrand model, namely: let the intended models of $\prod$ be those that correspond to the stable, autoepistemic expansions of the modal translation of $\prod$. Thus, the stable model semantics for logic programming was soon born (Gelfond & Lifschitz, 1988). The great virtue of their 1988 concept was that Gelfond and Lifschitz were able to define it without reference to modalities; but they did prove that stable models have indeed the property just indicated.

Notice that the Gödel translation of a program rule of form (14) would be

$$\Box(\Box A_1 \wedge \ldots \wedge \Box A_m, \neg\Box\Box A_{m+1} \wedge \ldots \neg\Box\Box A_n) \to \Box\Box A_0 \qquad (16)$$

Applying some simple equivalences valid in *SW5* or *S4F*, this expression can be reduced to

$$\Box A_1 \wedge \ldots \wedge \Box A_m, \Box\neg\Box A_{m+1} \wedge \ldots \Box\neg\Box A_n \to \Box A_0 \qquad (17)$$

Following Gelfond's initial effort and the subsequent work with Lifschitz (Gelfond and Lifschitz, 1988), there were several further attempts to relate the stable semantics of programs with nonmonotonic modal reasoning. In Lifschitz and Schwartz (1993) and Marek and Truszczyński (1993; 1993a), for example, precisely translation (17) is applied to establish a correspondence with *SW5**, along the lines of Theorem 1. However, (17) and other equivalent translations were proposed directly, rather than being derived from the Gödel-McKinsey-Tarski embedding.[26]

What became of Gelfond's initial quest to relate logic programming semantics to autoepistemic logic? For more complex formulas than (14), the simple translation (15) no longer works. Since autoepistemic logic is based on *KD45*, its models are not reflexive, unlike the extensions of *S4*. However, there is a standard technique in modal logic for interpreting reflexive in non-reflexive logics. Let ^{0} be the translation defined by setting φ^0 to be the result of replacing each subformula of φ of the form $\Box\psi$ by $\varphi \wedge \Box\psi$. Using this translation for example, the modal logic *S4Grz* can be embedded into the provability logic *GL*. This technique was first used by the Russian logicians, Kuznetsov and Muravitsky (1977; 1980), and independently discovered in Goldblatt (1978) and Boolos (1980).

By composing *G* and ^{0}, super-intuitionistic logics can be embedded into non-reflexive modal logics. For example, since *S4Grz* is a modal companion of intuitionistic logic, *I*, the latter can be embedded in *GL* via the translation $G^0 : \varphi^{G^0} = (\varphi^G)^0$. Since autoepistemic logic is *KD45**, its frames are exactly like those of *SW5* except for containing a single non-reflexive world. Consequently, it is easy to see that *SW5* can be embedded in *KD45* by the ^{0} translation (Marek & Truszczyński, 1993a) and *KD45* is a modal companion of *HT* via the G^0 translation. From this we can derive the embeddings of stable model semantics for programs into autoepistemic logic first presented in Lifschitz and Schwartz (1993) and Marek and Truszczyński (1993).

3.3   Where do we go from here?

It is now time to end our reflections on the past and draw some conclusions for the future. Where does stable reasoning now stand? What are the open questions and the promising lines of research? At this point let us also bring answer set programming, as a computational paradigm, closer into the picture. Indeed, it is the computational model of ASP that has developed so rapidly over the past ten years. How can our account of the foundations of stable reasoning contribute to the computational model and its application to real-world problem-solving?

So far I have concentrated mainly on propositional logic because this is a simpler story to tell and moreover ASP has developed largely as a propositional form of reasoning. Even today, although answer set solvers take first-order formulas as input, they generally work by first implementing a complex grounding or instantiation process before stable models are computed. Until recently, most of the effort spent on computational foundations went on reasoning at the level of propositions or generalizations such as QBFs (quantified Boolean formulas). At this level, the logical model I have been describing has made and continues to make contributions to ASP. For example, work on different kinds of program correspondence and equivalence continues to evolve and contribute to issues such as program optimization and modularity. Only recently a complete, logical characterization of the important concept of uniform equivalence was found by examining countermodels in the logic *HT* (Fink, 2008). So, for reasoning about programs, our logical model still has life left in it and may still contribute to software engineering issues, even at the propositional level. Another area of active research is that of language extensions, where the logical model can suggest ways to extend and apply the ASP language to enhance its knowledge representation and problem-solving capabilities. Recent examples include a modal version of the *HT* logic, developed in Wang and Zang (2005), and a temporal extension of *HT* presented in Aguado *et al.* (2008), which promises to bring interesting new tools to the planning domain.

On the other hand, at the basic level of computing stable models for standard languages, the propositional version of our logical foundations is unlikely to provide further improvements. So much development work has gone into existing systems that, in one form or another, they already incorporate all the deductive techniques that propositional logic can provide. It is to first-order logic that we must now look for inspiration and computational gains.

First-order (and even higher-order) logic is where much work on ASP foundations is currently focused. Some of this work looks at the traditional language of disjunctive programs, but now from a more 'first-order' viewpoint. Other efforts aim at new or extended languages. Examples are open and guarded open answer set programming in Heymans, van Nieuwenborgh and Vermeir (2005; 2008) and the new language RASPL-1 in Lee, Lifschitz and Palla (2008) based on the generalized definition of stable model for first-order formulas given in Ferraris, Lee and Lifschitz (2007). How does our logical account using *HT* fare in this company? First-order versions of *HT* have been around for some time and their properties studied (for example, Ono, 1983). Recent work has focused on finding a first-order version of *HT*, now called *quantified here-and-there*, *QHT*, suitable as a foundation for ASP, and finding a compact and complete set of axioms for it (Pearce & Valverde, 2004, 2005, 2006; Lifschitz, Pearce & Valverde, 2007). In these works it is shown how one can define in *QHT* a concept of minimal model analogous to the propositional case and that this captures an appropriate concept of stable reasoning for first-order theories. *QHT* also characterizes the strong equivalence of programs and theories in the first-order case.

So most of our story from Section 2 can be retold in first-order logic, but now with more promise for providing new tools and techniques with a potential impact on the practice of ASP. Areas to be targeted include:

- Automated deduction in *QHT*: for aiding the computation of stable models and checking theory equivalence.
- Discovering and validating inference rules in *QHT* for program transformation and optimization prior to grounding.
- Decidable fragments and new conditions of safety for first-order formulas to guarantee efficient computation.
- New kinds of splitting theorems and other techniques to achieve greater modularity.
- Study of interpolation and other metalogical properties that facilitate logical characterizations of relations between theories and knowledge modules.
- New ways of combining nonmonotonic rules with ontologies, description logics and other formalisms.
- Language extensions that support epistemic, temporal, causal and fuzzy or approximate reasoning.

Our other topics discussed in Section 2 can also be studied at the first-order level. The case of strong negation is routine, since the techniques of Gurevich (1977) for establishing reductions to intuitionistic logic and proving completeness were already presented in first-order guise. Extending QHT with strong negation is therefore a simple exercise.

The Gödel-McKinsey-Tarski embeddings can also be extended to the first-order case though this territory is less well charted than for propositional logic. Still, we can expect to find a close correspondence between the nonmonotonic version of QHT and first-order modal logics. This topic is now of special interest in the area of hybrid knowledge bases that combine first-order classical theories and nonmonotonic rules, for example, for reasoning about ontologies. Variants of first-order autoepistemic logic (AEL) are being applied in this context and recently embeddings from formalisms such as dl-programs and HEX programs into first-order AEL have been studied (de Bruijn, Eiter & Tompits, 2008). It remains to be seen whether these can also be derived as extensions of the translation G.

3.3.1   Which logical methodologies?

Despite the optimistic title of a recent paper by Vladimir Lifschitz: 'Twelve Definitions of a Stable Model' (2008), in my view there are just two main logical methodologies now guiding foundational research on stable models and answer set programming.[27] The first and most natural one is that based on the logics we have encountered here: primarily HT and its first-order extension QHT. These capture stable reasoning directly as a simple form of minimal model reasoning. They capture useful metatheoretic properties, they help us pinpoint open questions and research lines, and they guide us towards potential solutions. In addition, I believe the methodology based on these two logics has considerable explanatory power. We have seen an example of this in Section 3.2.3. Once we identify stable reasoning as a nonmonotonic form of reasoning in a super-intuitionsitic logic, we have at our disposal a whole theory with a rich, 60 year history about how these systems relate to modal logics and other formalisms. Strong negation provides another example on a smaller scale.

But for first-order programs and general theories, another approach has also emerged as a viable methodology, based on classical first and higher order logic, as exemplified by the second-order definition of a stable model given by Ferraris *et al.* in (2007) and subsequent developments such as Lee *et al.* (2008) and Lee, Lifschitz and Palla (2008a). A further approach is that of open answer set programming (Heymans *et al.*, 2008), though I would consider this as falling within the classical logic tradition.[28] The paradigm of classical logic has much to commend it. It is well-known and well accepted and it is suitable for rapid prototyping, having many computational systems to support its application. Should we prefer it to the non-classical approach? How can we compare the two?

First, it is important to be clear about one fact that can easily be overlooked: in many cases the classical and non-classical approaches deal with exactly the same objects and properties, albeit characterized in different ways. The new concepts of stable model and answer set defined in Ferraris *et al.* (2007) and Lee *et al.* (2008) coincide with the first-order versions of (stable) equilibrium models described in Pearce and Valverde (2005; 2006) and Lifschitz *et al.* (2007) and based on the logic *QHT*. In other words, while they may employ different techniques and heuristics, these two approaches are often in agreement about the basic objects of study and even the main research results. The reason is that one can reformulate many of those non-classical concepts and results in terms of classical logic. So for example to define stability, instead of referring to a special kind of minimal possible world model in *QHT*, one may refer to a classical model that satisfies a certain second-order formula re-expressing that minimality condition. Or, instead of reducing the strong equivalence of programs to logical equivalence in *QHT*, we can re-formulate the reduction in entirely classical terms. And there will be many more examples of this kind. The reason is simple: there is a general embedding of *QHT* into classical first-order logic that allows us to translate statements of *QHT* and properties of its models into equivalent classical ones. This embedding is described in detail in Pearce and Valverde (2008) and used to derive some classical encodings of ASP properties. The upshot of this is that, alongside the non-classical methodology of *QHT*, there is an almost equivalent, shadow methodology expressible in classical logic.

It seems there are still two important features that speak for the non-classical view of stable reasoning and ASP foundations. The first is simplicity. Being more familiar does not mean that classical logic is simpler to use and it certainly doesn't mean that classical representations are simpler to formulate and understand. When we use the shadow methodology we very often have to transform and manipulate simple expressions that then acquire an extra level of complexity and seem to capture their subject matter less directly. Sometimes one may find the shadow methodology being used with little or no explanation about how or why the classical representations perform as they do. For example the embedding of *QHT* into classical logic involves expanding the language to add new predicates. If a classical representation of some property now suddenly introduces new predicates not present in the original language, most likely it is a result of that *QHT* reduction and can easily be explained as such. But if an author simply omits to mention this property, then the representation may look quite awkward or even incomprehensible.

The second feature supporting the non-classical view is that it too can draw on a wealth of history, tradition, theory and practice. Besides giving a very simple, direct and natural representation, it is backed by its own set of tools and techniques that may not be so easily available to the classical logician. In Pearce and Valverde (2008) for instance it is shown how the proof theory of *QHT* can be directly used to shed light on problems of Θ-subsumption in ASP. More generally, it is hard to imagine that the shadow, classical methodology would readily identify the second negation of ASP as Nelson's strong negation or translations into epistemic logic as being instances of the Gödel-McKinsey-Tarski embedding. This means it is poorer as a result, since it is unable to take advantage of the insights and theorems that accompany these concepts. Besides, the shadow methodology hardly makes sense at all unless one understands it in light of the *QHT* reduction.

The non-classical, logical methodology has been slowly gaining adherents and winning acceptance. I think we can look forward to many more years of stable reasoning, many more logical theories being built upon it and many more developments in the computational model of ASP. I venture to hope that the non-classical methodology can play an important role in these future developments. By all means let us use the shadow methodology too, and, where needed, bring also techniques from classical logic to bear on computational and software engineering issues.

Notes

1. It is a great pleasure to dedicate this paper to Jan Woleński who has done so much to raise awareness and increase our understanding of Polish logic and philosophy in the twentieth century.
2. Vladimir Lifschitz is Professor in Computer Sciences at the University of Texas at Austin. He obtained a BSc in Mathematics from Leningrad University at the age of 19 and one year later received his PhD in Mathematics from the Steklov Mathematical Institute in Leningrad with a thesis on constructive counterparts of Gödel's completeness theorem, supervised by Nikolai Shanin. Lifschitz emigrated to the US in 1976 and later began his pioneering work on nonmonotonic reasoning and on logic programming, leading to the development of stable model semantics in 1988. Subsequently his major

achievements in artificial intelligence include seminal work on action theory, planning, nonmonotonic reasoning and answer set programming.

3. Andrei Nikolaevich Kolmogorov (1903–87) is regarded as one of the greatest mathematicians of the twentieth century. He is most famous for his work on probability theory and stochastic processes. He also solved Hilbert's thirteenth problem and developed the theory of complexity that carries his name in algorithmic information theory. In 1925 he published a pioneering paper on intuitionistic logic: 'On the principle of the Excluded Middle'. Kolmogorov was professor at Moscow State University. He received numerous awards including the Lenin and Lobachevsky Prizes as well as the Wolf Prize, mathematics' equivalent to the Nobel.

4. Note that in general, in intuitionistic logic stability ($\neg\neg p \to p$) is not the same as decidability ($p \vee \neg p$), since, for example, all negated propositions are stable in virtue of $\neg\neg\neg p \vdash \neg p$, but need not be decidable.

5. David van Dantzig (1900–59) was also a co-founder of the famous National Resarch Institute for Mathematics and Computer Science (CWI) in Amsterdam. Like Kolmogorov, van Dantzig is best known for his work on probability and statistics and his efforts to establish this field in the Netherlands.

6. It may be of interest to note that alternative embeddings of classical arithmetic into intuitionistic arithmetic were provided in the 1950s by Nikolai Shanin who later supervised the doctoral thesis of Vladimir Lifschitz. This topic is also close to the one studied by Michael Gelfond in his doctoral thesis of 1975. Shanin in turn was a pupil of A. A. Markov (Jr.) whom we shall encounter in Section 3.2.2.

7. Although in intuitionistic logic stability does not always imply decidability, if an intuitionistic possible worlds *model* verifies $\neg\neg p \to p$ for each p then it is equivalent to a classical model and verifies $p \vee \neg p$. Let us call such a model *total*.

8. Let us denote this relation by $\leq$, or $<$ when it is strictly smaller.

9. The logic *HT* was first presented in Heyting's paper (1930) formalizing intuitionistic logic but it was the Polish logician Jan Łukasiewicz who first axiomatized the logic in a paper (Łukasiewicz, 1941) presented in 1938.

10. See Gelfond and Lifschitz (1988) or any introduction to answer set programming.

11. Initially they were formulas of form (14) mentioned in Section 3.2.3. This restriction was gradually removed in successive extensions of answer set semantics, but it was only fully eliminated in Ferraris (2005).

12. Note that from our observations about maximality, it cannot be the result of applying something monotonic but stronger than *HT*. In that case it would have to be a rule valid in something at least as strong as classical logic; but classically equivalent theories need not have the same stable models.

13. For our purposes, by a curious twist of language, a nested logic program is a set of formulas *without* nested occurrences of implication. The original definition can be found in Lifschitz, Tang and Turner (1999).

14. I am grateful to Grisha Mints for showing me a draft manuscript of his paper (Mints, 2010) .

15. This was independently introduced in the context of logic programming in Pearce and Wagner (1990) presented at a conference in Helsinki in May 1989.

16. Andrei Andreevich Markov (1903–79) was the son of the famous mathematician A. A. Markov. He founded the Leningrad school of constructive mathematics but also made important contributions to topology, dynamical systems and the theory of algorithms. Markov can surely be regarded as a close intellectual forefather of Vladimir Lifschitz, since he was a teacher and mentor of Lifschitz's doctoral advisor, Shanin.

17. Here a *literal* is understood to be an atom or its strong negation. This restriction is somewhat surprising in the case of nested programs where all other connectives, except for implication, are allowed to be nested.

18. Gurevich (1977) deals with the case where T is empty. The above version follows by compactness.

19. This author also became infected by the contraposition bug, wrongly claiming in Pearce (2006) that (2) is equivalent to (6). I should have said that (2) is equivalent to (2) plus (6); in other words that (6) is derivable from (2).

20. Not quite. While most logic texts define this to be Gödel's translation it is not exactly what appears in Gödel's paper. Gödel's second translation is like the first except that $(\neg\varphi)^G$ becomes $\Box\neg\Box(\varphi)^G$, and $(\varphi \wedge \psi)^G$ becomes $\Box(\varphi)^G \wedge \Box(\psi)^G$.

21. There are several variants of the Gödel-McKinsey-Tarski translation; they lead to similar results as they are provably equivalent in *S4*. For example, Chagrov and Zakharyaschev (1997) use the translation T defined over the connectives $(\bot, \rightarrow, \wedge, \vee)$ by $T(\bot) = \Box\bot$; $T(\varphi \rightarrow \psi) = \Box(T(\varphi) \rightarrow (\psi))$; $T(\varphi \star \psi) = T(\varphi) \star T(\psi)$, for $\star = \wedge, \vee$.

22. Leo Esakia (1934–2010) created a vibrant school of modal logic in the Georgian capital, Tbilisi. By extending the algebraic and topological methods of McKinsey and Tarski, he helped to establish the systematic study of the relations between modal and super-intuitionistic logics.

23. Characterizations of the logics *S4F* and *SW5* can be found in Marek and Truszczyński (1993a). For a detailed overview of results on modal companions of super-intuitionistic logics see (Chagrov & Zakharyaschev, 1992).

24. Although originally defined in a different way, autoepistemic expansions are precisely the *S*-expansions as in (12) where *S* is the normal logic *KD*45.

25. Michael Gelfond is Professor in the Department of Computer Science at Texas Tech University. He received his PhD in Mathematics from the Steklov Mathematical Institute in Leningrad in 1975 with a thesis on classes of formulae of classical analysis compatible with the constructive interpretation. Gelfond emigrated to the US in 1977 and joined the University of Texas at El Paso in 1980. His subsequent research on nonmonotonic reasoning and logic programming led to the development of stable model semantics in 1988. He has since made major contributions to artificial intelligence, with important work on action theory, knowledge representation, nonmonotonic reasoning and answer set programming.

26. These works also treat disjunction, again following the pattern of G, as well as strong negation. However it is well known that super-intuitionistic logics with strong negation also have modal companions. The translation extends G by setting $(\sim\varphi)^G = \Box\neg\varphi^G$.

27. I stress logical methodologies here, because there are other foundational methods that are not directly based on logic, for example, from

argumentation theory (Dung, 1993) or algebra (Denecker, Marek & Truszczyński, 1999).

28. Though broadly speaking classical, these approaches focus however on different techniques. That of Ferraris *et al.* (2007) is closer to the methods of circumscription, while Heymans *et al.* (2008) draws inspiration from fixpoint logic. Another classical approach, combining first-order logic and algebraic methods is that of ID-logic (Mariën, Gilis & Denecker, 2004; Wittocx *et al.*, 2006).

References

Aguado, F. *et al.* (2008) 'Strongly Equivalent Temporal Logic Programs', in Hölldobler, S. *et al.* (eds) *Lecture Notes in Artificial Intelligence*, 5293: *11th European Conference on Logics in Artificial Intelligence (JELIA'08)* (Berlin: Springer), 8–20.

Banda, M. G., and Pontelli, E. (eds) (2008) *Logic Programming: 24th International Conference, ICLP 2008 Udine, Italy, December 9–13 2008 Proceedings (LNCS/Programming and Software Engineering)*, LNCS, 5366 (Berlin: Springer).

Baral, C. *et al.* (eds) (2005) *Logic Programming and Nonmonotonic Reasoning, 8th International Conference, LPNMR 2005, Diamante, Italy, September 5–8, 2005*, LNCS, 3662 (Berlin: Springer).

Boolos, G. (1980) 'On systems of modal logic with provability interpretations', *Theoria*, 46, 7–18.

de Bruijn, J., Eiter, T., and Tompits, H. (2008) 'Embedding Approaches to Combining Rules and Ontologies into Autoepistemic Logic', in G. Brewka, and J. Lang (eds) *Principles of Knowledge Representation and Reasoning: Proceedings of the Eleventh International Conference, KR 2008, Sydney, Australia, September 16–19, 2008* (Chicago: AAAI Press).

Cabalar, P., and Ferraris, P. (2007) 'Propositional Theories are Strongly Equivalent to Logic Programs', *Theory and Practice of Logic Programming*, 7(6), 745–59.

Chagrov, A., and Zakharyaschev, M. (1992) 'Modal Companions of Intermediate Propositional Logics', *Studia Logica*, 51, 49–82.

—— (1997) *Modal Logic* (Oxford: Oxford University Press).

van Dantzig, D. (1947) 'On the Principles of Intuitionistic and Affirmative Mathematics', *Indagationes Mathematicae*, 9, 429–40, 506–17.

—— (1951) 'Mathématique Stable et Mathématique Affirmative' in *Congrès International de Philosophie des Sciences, Paris 1949*, Vol. II, *Logique, Actualités Scientifiques et Industrielles* (ASI) (Paris: Hermann), 1951, 123–35.

Denecker, M., Marek, V., and Truszczyński, M. (1999) 'Approximating Operators, Stable Operators, Well-Founded Fixpoints and Applications in Nonmonotonic Reasoning', in J. Minker (ed.) *NFS-Workshop on Logic-Based Artificial Intelligence* (Dordrecht: Kluwer), 1–26.

Dung, P. M. (1993) 'An Argumentation Semantics for Logic Programming with Explicit Negation', in D. Warren (ed.) *Proceedings ICLP 1993: Proceedings of the Tenth International Conference on Logic Programming, June 21–24, 1993, Budapest, Hungary* (Cambridge, MA: The MIT Press), 616–30.

Esakia, L. (1979) 'To the Theory of Modal and Superintuitionistic Systems', (Russian), in V. A. Smirnov (ed.) *Logical Deduction* (Moscow: Nauka), 147–72.

Ferraris, P. (2005) 'Answer Sets for Propositional Theories', in Baral *et al.* (2005), 119–31.

Ferraris, P., Lee, J., and Lifschitz, V. (2007) 'A New Perspective on Stable Models', in *Proceedings of International Joint Conference on Artificial Intelligence (IJCAI)*, 372–9.

Fink, M. (2008) 'Equivalences in Answer Set Programming by Countermodels in the Logic of Here-and-there', in Banda and Pontelli (2008), 99–113.

Gelfond, M. (1987) 'On Stratified Autoepistemic Theories', in K. Forbus, and H. E. Shrobe (eds) *AAAI-87: Proceedings of the 6th National Conference on Artificial Intelligence, Seattle, WA, July 1987* (San Francisco: Morgan Kaufmann), 207–11.

Gelfond, M., and Lifschitz, V. (1988) 'The Stable Model Semantics for Logic Programs', in K. Bowen, and R. Kowalski (eds) *Logic Programming, Proceedings of the Fifth International Conference and Symposium, Seattle, Washington, August 15–19* (Cambridge, MA: The MIT Press), 1070–80.

—— (1990) 'Logic Programs with Classical Negation', in D. Warren, and D. Szeredi (eds) *Logic Programming, Proceedings of the Seventh International Conference, Jerusalem, Israel, June 18–20, 1990* (Cambridge, MA: The MIT Press), 579–97.

Gödel, K. (1933) 'Eine Interpretation des intuitionistischen Aussagenkalkuls', *Ergebnisse eines mathematischen Kolloquiums*, 4, 39–40; reprinted in Gödel (1986).

—— (1986) *Collected Works*, Vol. I, K. Feferman *et al.* (eds) (Oxford: Oxford University Press).

Goldblatt, R. (1978) 'Arithmetical Necessity, Provability and Intuitionistic Logic', *Theoria*, 44, 36–46.

Gurevich, Y. (1977) 'Intuitionistic Logic with Strong Negation', *Studia Logica*, 36, 49–59.

Heymans, S., van Nieuwenborgh, D., and Vermeir, D. (2005) 'Guarded Open Answer Set Programming', in Baral *et al.* (2005), 92–104.

—— (2008) 'Open Answer Set Programming with Guarded Programs', in *ACM Transactions on Computational Logic (TOCL)* 9(4), 1–53.

de Jongh, D., and Hendriks, L. (2003) 'Characterization of Strongly Equivalent Logic Programs in Intermediate Logics', *Theory and Practice of Logic Programming*, 3(3), 259–70.

Kolmogorov, A. N. (1925) 'On the Principle of the Excluded Middle', *Mat. Sb.* 646–67 [in Russian]. Engl. trans. in van Heijenoort J. (1967) *From Frege to Gödel: A Source Book in Mathematical Logic* (Cambridge: Harvard University Press), 414–37.

—— (1932) 'Zur Deutung der intuitionistische Logik', *Mathematische Zeitschrift*, 35, 58–65.

Kracht, M. (1998) 'On Extensions of Intermediate Logics by Strong Negation', *Journal of Philosophical Logic*, 27(1), 49–73.

Kuznetsov, A. V., and Muravitsky, A. J. (1977) 'Magari's Algebras', in *Proceedings of the 14th USSR Algebraic Conference*, Part 2 (Novosibirsk), 105–6 [in Russian].

—— (1980) 'Provability as Modality' in *Actual Problems of Logic and Methodology of Science* (Kiev: Naukova Dumka), 193–230 [in Russian].

Lee, J., Lifschitz, V., and Palla, R. (2008a) 'A Reductive Semantics for Counting and Choice in Answer Set Programming', in *Proceedings of the Twenty-Third*

National Conference on Artificial Intelligence, 2008 (AAAI-08) (Chicago: AAAI Press), 472–9.

—— (2008b) 'Safe Formulas in the General Theory of Stable Models', [Preliminary report] in Banda and Pontelli (2008), 672–6.

Lifschitz, V. (2008) 'Twelve Definitions of a Stable Model', in Banda and Pontelli (2008), 37–51.

Lifschitz, V., and Schwarz, G. (1993) 'Extended Logic Programs as Autoepistemic Theories', in L. M. Pereira, and A. Nerode (eds) *Logic Programming and Non-Monotonic Reasoning* (Cambridge, MA: The MIT Press), 101–14.

Lifschitz, V., Pearce, D., and Valverde, A. (2001) 'Strongly Equivalent Logic Programs', *ACM Transactions on Computational Logic*, 2(4), 526–41.

—— (2007) 'A Characterization of Strong Equivalence for Logic Programs with Variables', in C. Baral *et al.* (eds) *Logic Programming and Nonmonotonic Reasoning, 9th International Conference, LPNMR 2007, Tempe, AZ, USA, May 15–17, 2007, Proceedings.* LNCS 4483 (Berlin: Springer), 188–200.

Lifschitz, V., Tang, L., and Turner, H. (1999) 'Nested Expressions in Logic Programs', *Annals of Mathematics and Artificial Intelligence*, 25 (3–4), 369–89.

Łukasiewicz, J. (1941) 'Die logic und das Grundlagenproblem', in Les Entretiens de Zürich sur les fondements et la méthode de sciences mathématiques, 12, 6–9 (1938), Zürich, 82–100.

McKinsey, J. C., and Tarski, A. (1948) 'Some Theorems about the Sentential Calculus of Lewis and Heyting, *Journal of Symbolic Logic*, 13, 1–15.

Maksimova, L., and Rybakov, V. (1974) 'On the Lattice of Normal Modal Logics', *Algebra and Logic*, 13, 188–216 [in Russian].

Marek, V., and Truszczyński, M. (1993a) 'Reflexive Autoepistemic Logic and Logic Programming', in L. M. Pereira, and A. Nerode (eds) *Logic Programming and Non-monotonic Reasoning* (Cambridge, MA: The MIT Press), 115–31.

—— (1993b) *Nonmonotonic Logic* (Berlin: Springer).

Mariën, M., Gilis, D., and Denecker, M. (2004) 'On the Relation between ID-Logic and Answer Set Programming', in J. J. Alferes, and J. A. Leite (eds) *Logics in Artificial Intelligence, 9th European Conference, JELIA 2004*, LNCS 3229 (Berlin: Springer), 108–20.

Markov, A. A. (1949) 'Constructible Falsity', *Mathematical Seminar of the LOMI*, 6 Oct [in Russian].

Mints, G. (2010) 'Cut-Free Formulations for a Quantified Logic of Here and There', *Annals of Pure and Applied Logic*, 162 (3), 237–43.

Nelson, D. (1949) 'Constructible Falsity', *Journal of Symbolic Logic*, 14, 16–26.

Ono, H. (1983) 'Model Extension Theorem and Craig's Interpolation Theorem for Intermediate Predicate Logics', *Reports on Mathematical Logic*, 15, 41–58.

Osorio, M., Navarro Perez, J. A., and Arrazola, J. (2005) 'Safe Beliefs for Propositional Theories', *Annals of Pure and Applied Logic*, 134, 63–82.

Pearce, D. (1997) 'A New Logical Characterisation of Stable Models and Answer Sets', in J. Dix *et al.* (eds) *Nonmonotonic Extensions of Logic Programming. Proceedings NMELP 96*, LNAI 1216 (Berlin: Springer), 57–70.

—— (1999a) 'From Here to There: Stable Negation in Logic Programming', in D. Gabbay, and H. Wansing (eds) *What is Negation?* (Dordrecht: Kluwer).

—— (1999b) 'Stable Inference as Intuitionistic Validity', *Journal of Logic Programming*, 38, 79–91.

—— 'Equilibrium Logic', *Annals of Mathematics and Artificial Intelligence*, 47, 3–41.

Pearce, D., and Valverde, A. (2004) 'Towards a First Order Equilibrium Logic for Nonmonotonic Reasoning', in J. J. Alferes, and J. Leite (eds) *Logics in Artificial Intelligence, Proceedings JELIA 2004*, LNAI 3229 (Berlin: Springer), 147–60.

—— (2005) 'A First-Order Nonmonotonic Extension of Constructive Logic', *Studia Logica*, 80(2–3), 321–46.

—— (2006) 'Quantified Equilibrium Logic' [Tech. report, Univ. Rey Juan Carlos]. Available at http://www.satd.uma.es/matap/investig/tr/ma06_02.pdf [Accessed 18 October 2012].

—— (2008) 'Quantified Equilibrium Logic and Foundations for Answer Set Programs', in Banda and Pontelli (2008), 546–60.

Pearce, D., and Wagner, G. (1990) 'Reasoning with Negative Information I: Strong Negation in Logic programs', in L. Haaparanta *et al.* (eds) *Language, Knowledge and Intentionality: Perspectives on the Philosophy of Jaakko Hintikka* (Helsinki: Acta Philosophica Fennica), 49, 430–53.

Rautenberg, W. (1979) *Klassische und nichtklassische Aussagenlogik* (Braunschwieg and Wiesbaden: Vieweg Verlag).

Smetanich, Y. S. (1960) 'On the Completeness of a Propositional Calculus with an Additional Unary Operation', in *Proceeding of Moscow Mathematical Society*, 9, 357–71 [in Russian].

Stalnaker, R. (1980) *A Note on Non-Monotonic Modal Logic* [Manuscript] (Cornell University). Reprinted in *Artificial Intelligence*, 64 (1993), 183–96.

Vorob'ev, N. N. (1952a) 'A Constructive Propositional Calculus with Strong Negation', in *Doklady Akademii Nauk SSSR*, 85 465–8 [in Russian].

—— (1952b) 'The Problem of Deducibility in Constructive Propositional Calculus with Strong Negation', in *Doklady Akademii Nauk SSSR*, 85, 689–92 [in Russian].

Wang, K., and Zang, Y. (2005) 'Nested Epistemic Logic Programs', in Baral *et al.* (2005), 279–90.

Wittocx, J. *et al.* (2006) 'Predicate Introduction under Stable and Well-Founded Semantics', in S. Etalle, and M. Truszczyński (eds) *Logic Programming, 22nd International Conference, ICLP 2006, Seattle, WA, USA, August 17–20, 2006, Proceedings.* LNCS 4079 (Berlin: Springer), 242–56.

4
Wooden Horses and False Friends: On the Logic of Adjectives

Maria van der Schaar

4.1 Introduction

Kazimierz Twardowski wrote a small paper 'On the logic of adjectives', published in Polish in 1927, in which he proposed a classification for adjectives like 'counterfeit', 'purported', 'true' and 'actual'. Twardowski's classification has been a direct inspiration for the classification of non-attributive adjectives I present here. I thus hope to show that the ideas of philosophers from the Lvov-Warsaw school are still of value today. Without Jan Woleński's work, these ideas would not be well known to us, and we would thus have missed that special Polish variant of analytic philosophy. Modern analytic philosophy needs to have an open mind towards alternative philosophical traditions that are equally analytic in their methods. Philosophers from the Lvov-Warsaw school are not only of historical interest, they are still of value for philosophy today.

Terms like 'fake', 'pure', 'mere', 'true', 'real', 'actual', 'authentic', 'apparent', 'proper', 'genuine' have peculiar logical properties, and they are of interest to the linguist and logician alike. These terms behave in a systematic way, and are therefore not on a par with irony, or other forms of non-literal meaning. It is perfectly clear what is meant by 'stone lion' and the meaning is literal, which cannot be said of the phrase 'iron lady'. There is not much work done on these terms, but one can take Twardowski's paper on the logic of adjectives and Terry Parsons' paper on the logic of grammatical modifiers as a starting-point.[1]

When Timothy Williamson (2000, p. 47) says that mere belief is a kind of botched knowing, it should be made clear how the meaning of the term 'mere belief' relates to that of 'belief', and how the meaning of 'botched knowing' is related to that of 'knowing'. Philosophers from all traditions have used these terms. They say such things as that the 'things that are really real last a very short time' (Russell about sense data in his 1918, p. 274). Or, that symbolic judgements are substitutes for real judgements (*die wirkliche Urteile*, Husserl, 1890, p. 361).[2] Or, that the bed you are sleeping in is not a real bed, because the really real bed is the Form bed. When Plato says that the Form is more real than its sensible instances, which fall between the purely real and the wholly unreal (*Republic* 477a), 'real' is not to be understood in its existential sense. Plato is not asserting that there are different degrees of reality (cf. Vlastos, 1965, p. 219); he is using the term 'real' or 'really real' ('ὄντως οὐσης' *Rep.* 597D) rather in a non-existential sense. According to Göran Sundholm, the normative notion of rightness, or 'truth of things', is at issue when we deny that someone is a true friend (Sundholm, 2004, p. 439). We use these terms in ordinary language, too: 'Were you really listening (or were you just pretending)?' 'This is real gold (not a substitute metal)'. 'Is it a real Rembrandt (or is it painted by one of his pupils)?' 'These flowers aren't real (they're artificial, but you can't see the difference from real flowers)'.

There is something enigmatic about adjectives like 'fake', 'false', 'mere', 'true' and 'real'. Some of these terms behave strangely in inferences: a fake Rembrandt is not a Rembrandt. Others seem to be redundant: there is no difference between gold and true gold. Furthermore, it seems impossible to give a conceptual analysis of complex concepts like *false friend* or *fake gun*. None of our methods of analysis seems to be able to cope with terms like 'real' and 'fake'. The method of conceptual analysis as we know it from Moore's early writings has an atomistic flavour. The complex concept *bachelor* is analysed into two concepts: *unmarried* and *man*. Conceptual analysis may also be conceived along Fregean lines, as analysis of the content of a judgement. The content of the judgement *John is married to Sue* may be analysed into *John* and *married to Sue*, into *Sue* and *John is married to*, or into *John* and *Sue*, in that order, and *being married to*, depending on the kind of inference one needs to draw. On Frege's account, we obtain three different concepts as a result of the three possibilities to analyse this judgement. Finally, there is the form of logical analysis of which Russell's analysis of definite descriptions is a prototype example. The linguistic sentence of which the definite description forms a part is reconstructed with a logical, and epistemological aim, resulting in a proposition that is either true or false. None of these methods of analysis is able to capture the relation between the concepts *gun* and *fake gun*, between *wine* and *pure wine*, or between *gold* and *real gold*. The concept *real* is not simply a part of the concept *real gold*, nor is the concept *friend* part of the concept *false friend* in any straightforward sense of the term 'part'. And on a Fregean account, one has to treat the concept *fake gun* as an unanalysable unity. It is also unclear how we can give a logical construction of a sentence in which these terms occur. Neither on the Moorean, Fregean or Russellian accounts of analysis can we make sense of the relation between *gun* and *fake gun*. What is it that terms like 'false', 'mere', 'real', 'true', 'alleged' have in common, when we compare them with terms like 'red' and 'round'?

4.2   The distinction between attributive and non-attributive adjectives

Attributive or determining adjectives like 'blue' are standardly used to attribute a property to the object denoted by the noun phrase to which the adjective belongs. Non-attributive terms are not used in this way; they rather qualify the meaning of the noun-phrase in use. Terms like 'fake' and 'apparent' operate on concepts to create new concepts. Formally, they might be understood as functions from classes to classes. But, because many of these terms create intensional contexts, it is preferable to understand them as functions from properties to properties.

In order to understand the meaning of a non-attributive adjective, one needs to know to what noun phrase it belongs. The meaning of 'false' can only be understood together with its noun-phrase, for something may be a false Rembrandt, but a real seventeenth-century painting. And a false Rembrandt is false in a different sense than a false friend is. In this sense, non-attributive terms are to be classified as syncategorematic terms, terms that have a meaning only together with a categorematic term or noun phrase.

In contrast to attributive terms, non-attributive terms are not able to split from the nouns to which they belong, or so it seems. The term 'red' in 'He is wearing a red jacket' may split from the term 'jacket', as in 'He is wearing a jacket that is red'. In contrast, we cannot say 'This is gold that is real' or 'This is knowing that is botched'. Barbara Partee has recently argued that splitting is possible for terms like 'fake' and 'false', at least in some languages, and that these terms are therefore to be classified as attributive. As I am going to argue that these terms are not attributive, and if it is true that these terms do sometimes split, I should have a different criterion to distinguish attributive from non-attributive terms.

At first sight, non-attributive terms do not seem to form a homogenous class, for the criterion to distinguish them from attributive terms appears to be disjunctive. As a first suggestion, one might say that a term is non-attributive if either of the following two inferences is *invalid*, where *A* is an adjective:

I.   *x* is (an) *AN*.

⊢ *x* is (an) *N*.

II.   *x* is (an) *AN*.

⊢ *x* is A*.[3]

Terms for which inference I is invalid are generally called 'non-subsective'. The invalidity of I suffices to classify modifying terms like 'fake' and 'false' as non-attributive, whereas the invalidity of II would make it possible to classify terms like 'real', 'mere' and 'good' as non-attributive. This criterion needs improvement, though, for the second inference is often not simply invalid; for quite some adjectives, it is rather that '*x* is *A*', 'this is pure', does not make sense because these adjectives are non-predicative. It is possible to formulate a positive, simple criterion to distinguish attributive terms from non-attributive terms.

A term *A* is *attributive* precisely if the following inference is valid, otherwise, it is non-attributive:

x is (an) *AN*.

⊢ *x* is A* & *x* is (an) *N*.

4.3   A classification for non-attributive terms

4.3.1   The distinction between relative terms and other non-attributive terms

Relative terms are a special case of non-attributive terms. They are not used to attribute a property to an object, but rather qualify the property attributed. A term like 'good', as in 'good pianist', qualifies the property or concept denoted by the noun phrase, *being a pianist*, thus creating a new concept, *being a good pianist*. In order to understand the meaning of a relative term, one needs to know to what noun phrase it belongs. For a good shot is 'good' in a different sense than a good pianist. Someone may be a good pianist, but a bad leader of the orchestra. Relative terms are *qua* terms: someone who is a good pianist is good as a (*qua*) pianist. Someone may be a good pianist and a man, but not a good man. For relative terms, the following inference is *invalid*, where *R* is a relative term, x is the subject term and N and N' are nouns or noun phrases:

x is an RN & x is an N'.

$\vdash x$ is an RN'.

And the following inference is also invalid for relative terms:

x is an RN.

$\vdash x$ is R^*.[4]

There is an agreement between relative terms and terms like 'false' and 'fake', for something may be false money, but real silver. Both inferences just mentioned are invalid for terms like 'false' and 'fake'.

There is also an important difference (1) between relative terms and terms like 'false' and 'fake'. For relative terms the following inference is valid, whereas it is invalid for 'false' and 'fake', as we will see below:

x is an RN.

$\vdash x$ is an N.

For, a good pianist is a pianist, just as a big mouse is a mouse, but false gold is not gold. This means that relative terms, in contrast to terms like 'false', are subsective.

There is another difference (2) between the group consisting of terms like 'fake', 'real' and 'pure', on the one hand, and terms like 'good' and 'great', on the other hand. 'He is a good pianist' gives an answer to the question: 'to which group of pianists does he belong, the good or the bad ones?'. The good pianists, if there are any, form part of the extension of pianists. Just as small mice form part of the extension of mice, and we may thus speak of two kinds of mice, the small and the big ones. But, there are not two kinds of guns, fake guns and other guns, nor two kinds of gold, gold and real gold. This is precisely because terms like 'fake' and 'real' neither simply qualify the property denoted by the noun phrase, nor are they attributes of the object denoted by the subject term. The question is sometimes raised what kind of adjective the term 'dead' is. Because there are not two kinds of man, the dead and the living, 'dead' seems not to be a relative term. As Aristotle remarks, it is false to call a dead man a man (*De Interpretatione*, chapter xi, especially 21a18–33).

There is another way (3) to distinguish the relative terms from other non-attributive terms. It makes sense to say 'he is good', 'she is skilful', or 'that is a big one' (speaking about a mouse). This is caused by the fact that these terms qualify the subject *qua* so and so. Thus, I can say of him *qua* pianist that he is good. So, if the context is one in which young pianists are contesting for a prize, one simply says 'he is good (but *she* deserves the prize)'. In contrast, it is difficult to give a meaning to the phrases 'this is real' and 'this is pure', and these phrases might even be called ungrammatical. And, although there may a context in which one can say 'this is fake', one cannot say 'it is false' *qua* gold.

4.3.2   The distinction between modifying and non-modifying terms

Within the group of non-attributive terms, the modifying terms like 'fake', 'botched', 'potential' and 'false' may be distinguished from the non-modifying terms like 'real', 'true', 'mere', 'pure' and the relative terms.

For *non-modifying terms A*, inference (I) is valid:

I.   *x* is (an) *AN*.

⊢ *x* is (an) *N*.

It does not follow that *x* is *A*, for '*x* is *A*' does not make sense, and is not well-formed, at least, in the case of non-relative non-modifying terms. 'He is real', 'he is mere' and 'she is true' do not make sense.

A is a *modifying term*, precisely when inference (I) is *in*valid:

I.   *x* is (an) *AN*.

⊬ *x* is (an) *N*.

4.3.3   Restrictive and restorative terms

Among the non-modifying terms one may distinguish between the terms 'mere' and 'pure', on the one hand, and terms like 'real', 'proper' and 'authentic', on the other hand. The latter terms have a counterpart: 'real' and 'fake'; 'true' and 'false'; 'proper' and 'improper'; 'authentic' and 'inauthentic'. When it is suggested that someone is a false friend, the answer might be 'No, he is a real friend'. To understand the meaning of 'real friend', one needs to understand the meaning of 'not being a real friend', that is, of 'being a false friend'. It is the negative use that wears the trousers, as Austin says (Austin, 1962, p. 70). We first have to understand what it is for something to not be real gold, that is, to be a material that looks like gold but does not have the chemical properties that would make it into a piece of gold. Only against this background does the phrase 'real gold' obtain meaning. Saying that this isn't a real Rembrandt, is not denying that it is a painting, nor is one generally claiming that it is a forgery. The question is rather whether it was painted by Rembrandt, or by one of his pupils. Only against the background of this question is the assertion that it is a real Rembrandt given sense. The question might also have been whether it is a forgery or not, and then asserting that it is a real Rembrandt has a different meaning. According to Twardowski, terms like 'real', 'true' and 'actual' may restore the change in meaning that was caused by such terms as 'fake', 'false' and 'former' (Twardowski, 1927, p. 142). It is for this reason that these terms come in pairs: 'true' and 'false'; 'actual' and 'former'. Borrowing Twardowski's terminology, we may call them *restorative* terms.

The terms 'mere' and 'pure' do not have the same complexity as the restorative terms. They seem to contain a negation: mere belief is belief and nothing more, just as a mere child is a child and can therefore not be treated as an adult. 'Pure wine' is wine that is not mixed with anything of lesser value; 'pure wine' is wine and nothing less. The difference between 'mere' and 'pure' is not very strict, for the German translation of 'mere nonsense' is 'reiner Unsinn', and we also can say 'pure nonsense'. These terms function as an operator upon the term that follows, but not precisely in the way negation does. Unlike negation, 'mere' and 'pure' cannot be iterated, and what is negated is not that x is an N. For, pure wine is certainly wine, just as mere nonsense is definitely nonsense. A term like 'pure' and 'mere' *restricts* the meaning of the noun N that follows, in the sense that it denies a purported or expected aspect of things that are N. For this reason I call 'mere' and 'pure' *restrictive terms*. It seems that languages are in need of only two restrictive terms: one corresponding to 'mere', and another corresponding to 'pure'. The term 'pure' can sometimes be classified as a restorative term. If there is a suggestion that the water is contaminated and impure, we may speak about 'pure water', in which case the term 'pure' is used in a restorative way.

Because it does not make sense to say that something is real, actual, pure, or mere, the inference that distinguishes attributive from non-attributive terms is not valid in the case of restorative and restrictive terms.

4.3.4 There are two kinds of modifying terms: privative and modal terms

Modal terms like 'potential' in 'potential candidate' or 'alleged' in 'alleged murderer' are clearly modifying in the sense that an alleged murderer need not be a murderer. But from the fact that he is the alleged murderer, we cannot simply draw the conclusion that he is not a murderer at all. The intensional context makes it impossible to make either inference here. The terms 'putative', 'questionable' and 'disputed' are also *modal modifying terms*. This is an important group of modifiers, and need to be treated in a separate paper.

Privative terms are distinguished from modal terms by the *validity* of the following inference. If *A* is a privative term,

   x is (an) *AN*.

   ⊢ *x* is *not* (an) *N*.

The term *A* may be an adjective, like 'botched' or 'false', an adjectivally used noun, like 'toy', or a prefix, like 'non' or 'in'. The validity of the inference implies that calling someone a false friend means that one is committed to assert that he is not a friend, although one has not asserted yet that he is not a friend.

4.3.5   There are two kinds of privative terms: the pure and non-pure ones

There are a number of prefixes that are purely privative, think of 'non', 'in', and 'un', or the Greek *alpha privative*, and, perhaps, 'ex'. These terms do not simply negate of *x* the quality denoted by *N*; they rather indicate that *x* has the positive contrary of that quality. We call something 'nonsense', not simply because it is lacking sense, but because it is expected to make sense, while not fulfilling this expectation. The purely privative terms should not be understood as negation. Like negation, these are syncategorematic terms. But, whereas negation is a propositional connective, a function from truth-values to truth-values, or from propositions to propositions, the purely privative term denotes a function from properties to properties, or from concepts to concepts.[5] In the case of *purely privative terms*, the following inference is valid:

   x is (an) *AN*.

   ⊢ *x* is not (an) *N* & *x* has the contrary of the quality denoted by *N*.

The Aristotelian notion of privation is the ontological counterpart to the semantic notion of purely privative terms. When a man is called 'blind', he is asserted to lack a capacity he is to have according to his nature. The man is said to be deprived of a certain *habitus*, and the predicate can be constructed by means of an α-privativum, a purely privative term. Perhaps, 'ex' is not purely privative, but works rather like 'former', a non-purely privative term.

The non-purely privative terms form an important and typical group of modifying terms, and may therefore be called *modifying in the strict sense*. The term 'false' modifies the meaning of the noun it precedes by deleting a crucial part of its meaning. At the same time, it substitutes a related meaning. If I call someone a false friend, I imply that he is not a friend, although he has the appearance of a friend. A fake pistol has the appearance of a pistol, but misses the essential ingredient: it cannot be used to kill.

In the case of modifying terms in the strict sense, the following inference is valid:

> x is (an) *AN*.
>
> $\vdash$ x is not (an) *N* & x has the appearance of (an) *N*.

Is it essential that the appearance is deceptive, as in the case of mock-Georgian windows, a false Vermeer, or a fake pistol? If one were to give an affirmative answer, one would exclude terms that are naturally classified as modifying in the strict sense, such as 'toy', and one would exclude the uses of 'mock' or 'dummy' that do not have the deceptive aspect, as in 'mock examination', or 'dummy shells for the cannon'. If we do not demand that the appearance is deceptive, the terms 'presented', 'imagined', 'fictional', 'painted', 'stone' and 'dead' can be considered as modifying terms in the strict sense, too, at least, in one of their meanings. In contrast to the term 'fake', these words can be used both as modifying and as attributive terms. The phrase 'a painted landscape' is ambiguous, as Twardowski has shown.[6] It may be used to talk about a landscape that is painted, in which case 'painted' is used attributively. We use the term attributively to talk about a landscape with a mill and a river near Amsterdam that hasn't changed since Rembrandt painted it. Or, it may be used to talk about a painting that has a landscape as subject. In the latter sense, 'painted' is used as a strictly modifying term: a painted landscape is not a landscape, just as a stone lion is not a lion, although a stone bridge is a bridge. Constitutive material modifiers, as used in 'stone lion' and 'wooden horse', modify the meaning of the noun phrase. 'Rocking' in 'rocking-horse' is not a modifying term, because there are no similar constructions with 'rocking'. The logician has to treat words like 'rocking-horse' and 'teddy-bear' as one word, for there are no other constructions in which 'rocking' or 'teddy' precedes a noun that result in a similar change of meaning. According to Brentano and Twardowski, 'former' is also a modifying term, whose restorative counterpart is 'actual'.[7] I am not sure, though, whether it is a modal or a privative term. If it is always true that a former senator is not a senator, I have to classify 'former' as privative, but it seems to be a modal term, too.

Sometimes the modifying term follows the noun, as in 'president-elect'. And some more complex phrases have characteristics of modifying terms, too. An act that is an assertion on stage is not a real assertion, just as murder on stage is not real murder. The phrase 'on stage' thus functions as a modifying term. The verb phrase 'having in mind' may function as a modifier, too: one may have a million dollars in mind, and be completely broke at the same time.

4.3.6   Questions of method

The method of logical analysis should be broadened so that the relation between concepts such as *friend, false friend,* and *true friend* becomes clear. One can start with a classification that can be given for different kinds of adjectives by considering their logical properties, that is, by considering the way they behave in inferences. Second, one may raise the question as to how a concept such as *wooden horse* is formed. Whereas standard concepts like *red jacket* is obtained by adding to the concept *jacket* the concept *red*, the concept *wooden horse* is obtained by starting with the concept *horse*, of which a part is deleted, namely *living being*; the next step is to substitute for the deleted part, a new concept, here *made of wood*. A concept like *nonsense* is formed by the operator *non* working upon the concept *sense*: by this operation one gets the opposite of the concept *sense*. The concept *false gold* is formed by the operator *false* working on the concept *gold*, thus obtaining the concept *not gold*, and adding the concept *having the appearance of being gold*, and, perhaps, *pretending to be gold*. The concept of *true friend* turns out to be a more complicated one. It is formed by the operator *non* working upon the concept *false friend*. The concept *true friend* is thus identical with the concept *non-false friend*. And the concept *false friend* is obtained by the operator *false* working on the concept *friend*, obtaining the concept *not being a friend*, and adding the concept *appearing to be a friend*, and, perhaps, *pretending to be a friend*. Although the concept *false friend* does not have that of *friend* as its part, it cannot be understood without understanding what a friend is.

4.4   Partee's proposal and answer to Partee

Recently, Barbara Partee has proposed the thesis that there are no privative or modifying terms, apart from the modal adjectives (cf. Partee, 2010). Like standard attributive terms, and like relative terms, privative terms are subsective, that is, inference I is valid for privative terms. She gives three arguments for this original thesis.

(1) Because one can sensibly say 'Is that gun real or fake?', the term 'gun' must include both real and fake guns among its extension.

(2) Unlike modal adjectives, and like attributive adjectives, modifying terms can split in some languages (This is a gun that is fake), and in some languages they can even precede the noun, as in Polish 'Fałszywe znaleźliśmy banknoty' (false we-found banknotes).

Finally, (3) this gives us a possibility to explain in what sense terms like 'real' and 'true' have meaning, and to show that they are not merely redundant or tautologous. For, real guns form now a sub-class of all guns in the appropriate context.

What can we answer to Partee? It is true that there is an important agreement between relative terms like 'good' and 'skilful', and modifying terms, but for a different reason. Their agreement consists in the fact that in both cases the inference

x is (an) AN.

$+ x$ is A^*, and x is (an) N

is invalid. Relative terms may therefore be classified as non-attributive. Furthermore, there is an agreement between relative and privative terms: in the appropriate context, we *can* say 'this is fake', 'this is wooden', and 'this one is false' (speaking about teeth), just as we can say 'he is good' and 'she is skilful'. There are also three differences between terms like 'good' and 'skilful', on the one hand, and modifying terms, on the other hand, as we have seen in the former section (in the first sub-section on the distinction between relative terms and other non-attributive terms).

With respect to Partee's first point (1), one can answer that the fact that 'This Rembrandt is false' makes sense, does not imply that 'false' is subsective. It is not the case that the term '(being a) Rembrandt' is used here as a general term including both true and false Rembrandts. The general term is used in a deviant way for what people *call* a 'Rembrandt'. This makes it possible that we can say both 'That is a false Rembrandt' and 'That Rembrandt is false'. But, there are not two kinds of Rembrandt, the true and the false ones, just as there are not two kinds of gun, the real and the fake ones. In this sense there is a crucial difference between privative terms and subsective ones.

Regarding argument (2), although we can sensibly say 'Zeus is a god who is fictitious' instead of 'Zeus is a fictitious god', this does not mean that there are two kinds of god: fictitious and real ones. Possibility of splitting is apparently not a reliable indication of attributive terms. Instead, one should ask oneself whether it makes sense to say that the extension of the term can be divided in two sets by means of the adjective. For all non-attribute adjectives A that are non-relative, one can say that AN do not form a special kind of N. Just as a false Rembrandt is not a special kind of Rembrandt, a mere child is not a special kind of child, a potential terrorist is not a special kind of terrorist, and real gold is not a special kind of gold.

Regarding (3), I agree with Partee that 'real' in 'real Rembrandt' is not redundant, but not for the reason that real N may form a subclass of N in the appropriate context. The phrase is given a meaning dependent upon the suggestion that the painting is *not* a real Rembrandt. It is the kind of word where the negation wears the trousers.

My conclusion is that for privative terms the following inference is valid, which means that they are not subsective:

x is (an) AN.

$\vdash x$ is not (an) N.

4.5   Conclusion

One may put the different kinds of non-attributive terms in Schema 1

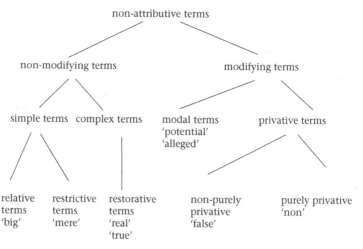

Schema 1

Notes

1. Twardowski (1927) and Parsons (1970), especially the sections on adjectives and prepositions 2.2 and 2.3.
2. Cf. Husserl (1890, p. 368), 'Jede künstliche Operation mit Zeichen dient in gewisser Weise Zwecken der Erkenntnis; aber nicht führt eine jede wirklich zu Erkenntnissen, in dem wahren und echten Sinn ('in the true and real sense') logischer Einsichten'.
3. *A* and *A** are not of the same syntactic and semantic category, for *A** does not, like *A*, stand for a function from properties to properties, but simply for a property, at least, if *x is A** is grammatical and makes sense.
4. See footnote 3. This means that the relative term 'big' is a different word than the absolute term 'big'.
5. Aristotle clearly saw the point in chapter x of the *Categories*: 'Names and verbs that are indefinite (and thereby opposite), such as "not-man" and "not-just", might be thought to be negations without a name and a verb. But they are not. For a negation must always be true or false; but one who says "not-man" – without adding anything else – has no more said something true or false (indeed rather less so) than one who says "man".' (*Cat.* 20a 31–6). Roberto Poli's remark that 'the modifying expression is of a syncategorematic type, that is, that it behaves like a connective' (Poli, 1993, p. 44) should be qualified, for the modifying expression does not behave like the logical connectives of propositional logic.
6. Twardowski (1894, §4, p. 12ff.) uses the distinction between the modifying and the attributive sense of the term 'presented' to explain the distinction between the content and the object of an act. There is a distinction between *presented object as object*, where 'presented' is used in its attributive sense – *x* is presented and *x* is an object – and *presented object as content*, where 'presented' is taken in its modifying sense, because the term modifies the meaning of the term 'object'. If 'painted' is used as modifying term in 'painted landscape', the landscape is a painted one, that is, not a true landscape ('sie ist keine wahrhafte Landschaft', Twardowski, 1894, p. 13).
7. Twardowski (1927, p. 141). Cf. Brentano (1914, p. 46): a former king is no more a king than a beggar is, for he might at this moment be a beggar himself.

References

Ackril, J. L. (1979) Aristotle *Categories and De Interpretatione*. (translation Oxford: Oxford University Press).

Austin, J. L. (1962) *Sense and Sensibilia* (Oxford: Oxford University Press, 1964).

Brentano, F. (1914) *Psychologie vom empirischen Standpunkt*, in O. Kraus, and F. Mayer-Hillebrand (eds) (Hamburg: Felix Meiner), 1968, book III, part I, ch. 5, 37–52.

Husserl, E. (1890) 'Zur Logik der Zeichen (Semiotik)', in L. Eley (ed.) *Philosophie der Arithmetik mit ergänzenden Texten* (Den Haag: Martinus Nijhoff), 1970, 340–73.

Parsons, T. (1970) 'Some Problems Concerning the Logic of Grammatical Modifiers', *Synthese*, 21, 320–34.

Partee, B. (2010) 'Privative Adjectives: Subsective plus Coercion', in R. Bauerle, and T. E. Zimmerman (eds) *Presuppositions and Discourse; Essays Offered to Hans Kamp* (Bradford: Emerald), 273–85.

Poli, R. (1993) 'Twardowski's Theory of Modification against the Background of Traditional Logic', *Axiomathes*, 4, 41–57.

Russell, B. (1918) 'The Philosophy of Logical Atomism', in R. C. Marsh (ed.) *Logic and Knowledge* (London: Allen and Unwin), 1956, 177–281.

Sundholm, G. (2004) 'Antirealism and the Roles of Truth', in I. Niniluoto, M. Sintonen, and J. Woleński (eds) *Handbook of Epistemology* (Dordrecht: Kluwer), 437–66.

Twardowski, K. (1894) *Zur Lehre vom Inhalt und Gegenstand der Vorstellungen; eine psychologische Untersuchung* (Wien: Hölder). Reprinted (München, Wien: Philosophia Verlag), 1982.

—— (1927) 'On the Logic of Adjectives', in Twardowski (1999), 141–3.

—— (1999) *On Actions, Products and Other Topics in Philosophy*, J. Brandl, and J. Woleński (eds) (Amsterdam, Atlanta: Rodopi).

Vlastos, G. (1965) 'Degrees of Reality in Plato', in G. Vlastos, *Platonic Studies* (Princeton: Princeton University Press) 1973, 58–75.

Williamson, T. (2000) *Knowledge and Its Limits* (Oxford: Clarendon Press).

5

Proof, Probability or Plausibility

Joseph Agassi

> The whole of science is nothing more than
> a refinement of everyday thinking.
>
> Albert Einstein

5.1 Proof

The philosophical literature is full of discussions of proof, inductive and deductive plus a multiple of variations on each. What is proof good for?

The standard answer is, we need it to combat the sceptic. The sceptic is the demon that philosophers repeatedly exorcise. Some join David Stove and say, we must try to refute the sceptic even if our effort is doomed to failure, because otherwise the sceptic will win and that is the end of rationalism (Stove, 1981). Moreover, they add, the steady progress of science is a sufficient assurance that the sceptic is in error; that should encourage us in our search for a proper proof that it is indeed erroneous.

It is not clear how serious this argument is. If the steady progress of science refutes scepticism, why should we continue trying to refute it? Those who find a valuable kernel in this argument may try to reword it and then submit it in its improved wording to critical examination. The most valuable option here is this: since the root of this objection to scepticism is the progress of science, we should look at science in action in order to explain what version of scepticism is erroneous, and what version of it may be reasonable. This, however, is not so simple. To make this obvious, let us compare success in science and in other fields, say art, or, to make things easier, narrative art, or, to make things still easier, verbal communication (including writing). How do we achieve success in improving verbal communication? We do not know. We propose the hypothesis that proper speech follows some rules. These rules are not sufficient or else computers will be able to converse with proper speech. But we will agree that grammar is necessary for understanding. Admittedly, there are counter-examples to this, from Gertrude Stein and James Joyce to Salman Rushdie, not to mention William Faulkner's babbling *Sound and Fury*. Let us ignore them for now, since, we agree, for those who wish to improve their ability to communicate the advisable course is to learn some grammar and to try to follow it before thinking of how to violate it with impunity. We learn grammar by listening to good speakers and by studying good texts about grammar. The reference to good speech and good texts makes this characterization a vicious circle; yet a vicious circle may become beneficial as a means of ascent, as pulling oneself up by one's bootstraps. The circle is vicious regarding the wish to justify conduct and it is beneficial regarding the wish for a never-ending course of improvement.

The philosophical literature often justifies scientific theories and is thus circular; it seldom comprises helpful manuals for bootstrapping. Carl Hempel's study of the whiteness of swans comprises justification, not help for students of ornithology (Agassi, 1988, p. 33). Commentators questioned his example as well as his way of justification: it is a vicious circle. Circularity is at times objectionable; not always. The greatest difference between modern, extensional logic and classical essentialist logic is exactly here: classical logic forbids circularity as vicious yet modern logic proudly demands it. Thus, philosophers who complain about the circularity of Alfred Tarski's definition of truth are essentialists who seek what he did not offer (Agassi, 1988, p. 235). Is the whiteness of Hempel's swans the same as the whiteness of Tarski's snow, then? How do we judge this? The answer is, Hempel did and Tarski did not wish to block the sceptic. As Carnap put it in his autobiography, what he had learned from Tarski is the distinction between truth and confirmation (Schilpp, 1963). Taken literally, this is hardly credible. It means, first he demanded to speak of truth only after securing permission to do so (Coffa, 1991, ch. 16). Fortunately, often enough researchers who want to speak up are not afraid to do so without asking for permission. Good researchers do not worry about permissions (Coffa, 1991, p. 414, n. 12), but they are always somewhat sceptical; usually, they both ignore the extreme sceptic and they seldom seek assurances.

Why then worry about the sceptic? Because we need a theory of rationality for the choice of opinions to uphold: we do not want them arbitrary or random. (Even changing one's opinions, said sagacious

Robert Boyle, wants good reasons.) The idea that scepticism imposes arbitrariness sounds very reasonable, let me say by way of an empirical observation. This is but an example of a refuted idea that sounds reasonable; it sounds reasonable until we notice that we are rational yet have scarcely any proof, or even of knowledge of what comprises proof. Moreover, most people hold some opinions accidentally, such as the religion they happen to have been born into, yet they are not quite irrational – not even about their religion. More important, the effort at avoidance of the arbitrary or the accidental has no more than partial success. Still more important, efforts to implement this avoidance of arbitrariness are possible with no theory of rationality as evidence shows abundantly. So the fear that sceptics are right about rationality because they are right about the proof of rationality is erroneous and rests on the confusion of one's failure to act rationally with one's failure to prove that one acts rationally. It is advisable to learn from Tarski to avoid this error, as Carnap says he did.

The difference between the reasonable and the arbitrary or the accidental is obvious. We may want to articulate this as we may want to explain it. What makes this difference? The traditional answer is, proof does. This answer is obviously false. Worse: we do not know what proof is. The requirement for proof is millennia old and proof theory is but a century old! For millennia thinkers had to make do with the paradigm of proof, with Euclidean geometry. In it proof means, deduction from axioms; these are deemed self-evident. Now this already refutes the idea that the rational is the provable, since it allows that self-evident axioms need no proof. The claim that all of Euclid's axioms are self-evident was rejected already in antiquity, with the expression of the idea that the parallel axiom (the fifth postulate) wants proof. Today we view proof as deduction from any set of axioms. The choice of axioms is not arbitrary and not accidental, however: we do offer reasons for it, not proofs. Hence, again, we hold that rationality is no proof. An important aspect of the rationality of the choice of axioms is the ability to pose alternative axiom systems and compare them as fitting some given *desiderata*. (Non-Euclidean geometries should entail some version of Pythagoras' theorem, for example.) At times, admittedly rarely, a new set appears that answers a variant of the *desiderata*. (Thus, differential geometry, geometry of differentiable manifolds, takes axiomatically a generalized version of Pythagoras' theorem.) And then the discussion reopens. More exciting, we may, admittedly much less often, discover new *desiderata*. (A good example is the theory of topological manifolds or the axioms of measurement that demote Pythagoras' theorem.) And then we can compare *desiderata*: we describe some conjectures and try to prove them. How? This is controversial. Proof theory is still in the making.

5.2 Probability

Traditionally, the proof of axioms such as Newton's laws begins with their consequences (*Principia*, Book III). This kind of proof is inductive. What are the rules of inductive logic? Discussions of inductive proofs always referred to its rules (Bacon, 1620; Newton, [1687] 1995). Wording of these rules appeared late (Mill, 1843, III.viii.1). David Hume had argued earlier (Hume, 1739–40) that these cannot be: inductive proof is impossible. This was hardly news. Since Antiquity references to proof-surrogate of sorts appeared: induction, probability, plausibility, or verisimilitude. It was never clear what all this is, let alone what its rules are.

A simple rule is that of eliminative induction, so-called, often ascribed to Francis Bacon or to Sherlock Holmes: 'when you have excluded the impossible, whatever remains, however improbable, must be the truth'. Refutations of all available options but one affirms the remaining one – no matter how improbable it is. To be precise, however low the initial probability of the single remaining option is, its posterior probability is high or even highest in the light of the evidence that refutes all its alternatives. Thus, eliminative induction is deductive: if our list of options is complete, then the sum of prior probabilities of all the given options is one, and then so is posterior probability of the remaining options; the single remaining option is then certain. The Holmes rule is not inductive but deductive.

Even when some options are eliminated, the remaining options become (weakly) more probable. This is the basis of a theory called Bayesian probability, one that rests on Bayes' theorem concerning inverse probabilities (the probability of the causes of a given effect) so that it looks inductive. As Bayes' theorem is a part of a mathematical theory, it renders inductive inference deductively valid, of course. What then is the significance of a higher probability value? It seems to help overcome Hume's scepticism that in truth it simply ignores. Moreover, it is in fundamental error (of Bruno de Finetti, 1936) that is too simple to overlook: Bayes' theorem holds only when all alternative options are known. This is hardly ever the case. The list of suspects rises with the rise of the possible routes of escape and its speed, with the growth of the list of available poisons, etc. Detectives may ignore this, and on some occasions intelligently so; to ignore this indiscriminately, however, is definitely unwise. And as applying Bayesian probability is the claim for knowledge of all possible alternative options, using it regularly is indeed definitely unwise.

Eliminative induction is a serious impediment at crucial junctions – where new options should enter and are kept out by the very use of the assumptions of the eliminative induction that all possible alternative options are known. Nevertheless, eliminative induction is operative in science. The classical place for it is analytic chemistry. The list of options there is the finite list of elements – to the *ad hoc* exclusion of additional options, such as the assumption of the stability of atoms that excludes radioactivity, and such as the assumption of the stability of molecules that excludes the many significant cases where molecules are not clear-cut entities. Analytic chemistry is then a useful but uninteresting tool. And, indeed, eliminative induction makes research a bore.

By the view of science as comprising small and large revolutions, research deviates from Bayesian probability to small and large extents. Analytic chemistry signifies when it violates induction by elimination. Its identification of sugar and alcohol, for example, invites stereo-chemistry to help. Other deviations from analytic chemistry are more interesting.

This is perhaps the place for an aside here on refutation. Most objections to Popper that came my way rest on the reluctance to ascribe falsity to a good theory. Colleagues who hear me say of a good theory of a revered researcher that it is false often respond with, why do you hate him so? They consider refutation disdainful and expressions of disdain improper. Perhaps Thomas Kuhn was right when he declared the variance between his philosophy and Popper's hardly more than terminological (Lakatos & Musgrave, 1970, p. 256). His aim was obviously to evade saying that Newton's theory was false yet without endorsing relativism. He failed. The theory of partial truth of Jan Łukasiewicz may have succeeded here (Kijania-Placek, 2002). This signifies ontologically and is debatable (Zahar, 1980). Epistemologically and methodologically it should not matter as in science proper empirical refutation is the discovery of new facts which is the peak of success; and so scientific criticism depicts respect.

The identification of the truth-surrogate with probability is old; the identification of probability with functions that follow the calculus of probability crystallized in the works of Dorothy Wrinch and Harold Jeffreys (1919) as well as that of John Maynard Keynes (1921). The idea seems reasonable when the rationale of the choice of an option rests on statistical estimates: even though not all swans are white, two assumptions make reasonable the expectation that the next swan that we will encounter is white: that most swans around are white and that our next encounter of one will be random. (Fake randomness appears in many a humorous folk-tale.) Is the assumption of randomness always reasonable? Some inductivists say yes (Williams, 1934, p. 435). This is an implausible ontological hypothesis (Russell, 1948, p. 388) and even in principle so (Landé, 1958, p. 176).

The proposal to follow Bayesian probability is famously faulty. Betting on all of one's resources is irrational regardless of the odds. Similarly, when lives are influenced by a decision, the possibility that it will cause death may not be ignored in all cases of low probability. We do that in some cases of war and of mass immunization, for example, but only reluctantly and in reliance on severely tested theories and in failure of efforts to find better alternative courses of action. Bayesian probability has no room for searches of new options or tests or criticism, only for mute evidence. Efforts to apply Bayesian probability thus share the defect of induction by elimination: the very idea of putting it to use (though vain) suffices to discourage the search for new options. Applicable or not, the very idea of probability as truth-surrogate has a great appeal: it should serve as a mathematical tool for forcing evidence to dictate the choice of a theory and thus of action: it is a dream of having science release us from the heavy burden of responsibility that any tough problem of choice of action imposes.

Karl Popper said, it is not the evidence but the calculus that dictates the tough choice, and it may even conflict with the evidence. He proved that. He proposed three statements, two serving as hypotheses and one as evidence; one hypothesis has low prior probability that the evidence raises and the other has high prior probability that the same evidence reduces, yet the one that the evidence favours is still less probable than the one it undermines. (See Appendix.) In this possible case probability conflicts with evidence. For decades Popper was harshly denounced for this argument (Chattopadhyaya, 1991, ch. 4, §5). This is a remarkable story, especially as it concerns a debate about rationalism.

Why did they ignore Popper's argument? Why do their heirs still not admit that their response to his arguments was a scandal? Because they still cling to Euclid's idea that axioms are self-evident. They use their claim that probability is self-evident to render it appear as quasi-proof-procedure of sorts. Popper suggested that empirical support is probability-increase rather than probability, and that probability-increase deviates from the axioms of probability proper; this raises the fear that probability is no adequate answer to the sceptics. And fear imposes silence. This is unphilosophical: it is neither brave nor wise.

Erroneous as the probability theory of rationality is, some cases seemingly abide by it, as we have noted, namely, some – not all – of the cases of reliance on statistical evidence. As experience shows, such cases are at times encumbered by serious popular errors (Tversky & Kahneman, 1971). Moreover, statistical evidence may mislead. Admittedly, following it repeatedly may be felicitous in the long run. But this is true only in limited cases. Frank Knight and then Keynes stressed this as they distinguished between cases of risk and of uncertainty, where risk is statistical and uncertainty is due to the unknown (LeRoy & Singell, 1987). It is difficult to see how Keynes the economist who limited the applicability of probability could inhabit the same skin as Keynes the philosopher of science who did not (and who devised his principle of limited variety to justify the use of this blinker, thus smuggling ontology into methodology).

5.3   Plausibility

Observations are scientific if and only if they are repeatedly reported and claimed to be repeatable. They merit tests and explanations, and if refuted they merit modification. Empirical studies of rationality cannot justify any theory of justification without begging the question (David Hume, 1739–40) but they can offer results that agree or disagree with any theory of rationality, the theory of rationality as justification included. So let me list some of the extant observation reports that deserve tests and explanations, and possibly modification.

(1) We judge some choices as more reasonable than others. Reference to circumstances is unavoidable as we often judge the reasonableness of choices as depending on the circumstances under which they took place. These are at times social institutions, especially the law of the land, and at times they pertain to public knowledge.

(2) We compare choices for their reasonableness, notably in inquests (*ad hoc* instituted investigations into disasters). We compare alternative options and their choice under given circumstances to decide if a specific disaster is due to negligence.

(3) We deem choices irrational that opt for alternatives that came first for no other reason: this is what we deem prejudice: the inability to improve our choices. Roger Joseph Boscovich (1758) modified Bacon's theory of prejudice: not all preconceived views are prejudices but only those we cling to.

(4) We often deem the reasonable or the plausible publicly permissible, not necessarily obligatory except in specific cases, such as the requirement to administer certain medications to minors in our care regardless of our beliefs.

Choices that rest on tradition may be rational, of course, but they may also be irrational. The preference of tradition on specific occasions for specific reasons thus invites specific discussions. Rational choice theories should approve of some yielding to common prejudice just in order to accommodate neighbours; advocates of these theories conceal this fact. Still, traditionalism may be rationally defended as taking tradition generally as the surviving default option. (This makes the endorsement of the religion of one's parents possibly less irrational than the traditional theories of rationality suggest.) And so the controversy between traditionalists and radicals is still deemed rational. Rightly, I suppose.

The repeatedly observed facts described here deserve empirical study. To the extent that empirical studies of these facts are already published, let me report, such as those of Milton Rokeach (1960) or of Tversky and Kahneman (1971), they criticize the claim that all views of ordinary people are utterly rational. They are thus more concerned with the exposure of the irrationality that is rampant in our midst than to characterize and explain the reasonable and the plausible.

The extant empirical studies that shed some light on the matter are mainly from the psychology of learning, where active learning is repeatedly judged superior to the passive sort still widespread in schools. The active form usually studied is the lowest form of active study: trial and error. Any additional technique is heuristic, which comes next. To a lesser extent also artificial intelligence studies are helpful. These are computer programs that should emulate good learning. Let me not go further into these two fields as the relevant material is easily accessible.

Any guide to trial-and-error studies is a heuristic. It comprises mere suggestions with the expressed denial of any guarantee of fruitfulness. The word was invented by the great William Whewell, who was the first to say, success in research is never guaranteed, and who was dismissed just because people expected guarantees as a matter of course (*Dictionary of National Biography*, Art. Whewell). Karl Popper had better luck. His first book stressed the no-guarantee clause. It also emphatically avoids all discussion of heuristic. Later on Popper added heuristic discussions – while stressing that it is not obligatory in any way. Imre Lakatos declared heuristic obligatory. That silly idea won him the great fame that his philosophy of mathematics deserved and achieved only posthumously. Kuhn, by contradistinction, said, whereas small-time research is guaranteed of success of one sort or another, big-time research is different. He did not specify. Whatever we say of Kuhn's philosophy, we should admit that on this he was reasonable: be his claim true or false, it is plausible. Here we see again that we use the idea of plausibility as a criterion just because the plausible need not be true and it even need not remain plausible with the growth of knowledge: whereas truth is eternal, plausibility is context-dependent. It is therefore useful for historians of science in addition to truth, especially since any idea, no matter how plausible, ceases to be plausible once it is refuted until it is modified in accord with the new information.

Heuristic is an extension of trial and error. It promises nothing but it may cost much, especially when it comes to global politics (not to be confused with international relations). Global politics is concerned with problems of the whole human race that began only in the last century with the appearance of means for its destruction (Agassi, 1985). When offering solutions to global problems, we cannot afford to cause a disaster as this might be our last error. Here critical examinations of solutions in search for plausible ones is urgent, the tools for this may be computer simulations and public debates and more. All this is as rational as at all possible but has nothing at all to do with proof or probability.

To conclude let me point out that both utter rationality and utter irrationality are extremely rare. Hence, it is mostly a matter of degree as well as of kind (Jarvie & Agassi, 1967), not to mention it being a matter of scope. This renders rationality advisable rather than obligatory and eminently open to empirical study. That its minimum characterization is openmindedness is usually recognized. This suffices to prevent circularity and develop helpful bootstrapping. Hopefully.

Nimrod Bar-Am, Ian Jarvie, David Miller, Sheldon Richmond and John Wettersten read earlier versions and suggested improvements. My gratitude to them all.

Appendix: Popper's Anti-Bayesian model

The objects in the model are one throw of a precision die and six possible outcomes of it, all equi-probable.

There are two hypotheses and one piece of evidence such that

(1) $P(h_1) \geq P(h_1, e)$: the evidence undermines h_1

and

(2) $P(h_2) \le P(h_2,e)$: the evidence supports h_2

and yet

(3) $P(h_1,e) \ge P(h_2,e)$: the undermined hypothesis is the more probable.

Thus, support should not be taken as probability.

Popper's detailed example that proves the consistency of his model is this.

h_1 = the next throw is not a 1

h_2 = the next throw is a 2

e = the next throw is 1 or 2 or 3 or 4.

This example satisfies all the three formulas:

$P(h_1) = 5/6$ $P(h_1,e) = 1/2$: the evidence undermines h_1.

$P(h_2) = 1/6$ $P(h_2,e) = 1/4$: the evidence supports h_2.

and $1/4 \le 1/2$: the supported option is the less probable.

References

Agassi, J. (1985) *Technology: Philosophical and Social Aspects. Episteme* (Dordrecht: Kluwer).

—— (1988) *The Gentle Art of Philosophical Polemics: Selected Reviews and Comments* (LaSalle, IL: Open Court).

Agassi, J., and Meidan, A. (2008) *Philosophy from a Sceptical Perspective* (New York and Cambridge: Cambridge University Press).

Agassi, J., and Wettersten, J. R. (1987) 'The Philosophy of Commonsense', *Philosophia*, 17, 421–37.

Bacon, F. (1620) *Novum Organum* (London).

Boscovich, R. J. (1758) *Philosophiae Naturalis Theoria: A Theory of Natural Philosophy.* Translated by J. M. Child (Cambridge, MA: The MIT Press), 1966.

Chattopadhyaya, D. P. (1991) *Induction, Probability and Scepticism* (Albany: SUNY).

Coffa, J. A., and Wessels, L. (1991) *The Semantic Tradition from Kant to Carnap: To the Vienna Station* (Cambridge: Cambridge University Press).

Finetti, B. de (1936) 'Les probabilites nulles', Bulletin de Sciences Mathematiques, 60, 275–88.

Hume, D. (1739–40) *A Treatise of Human Nature* (London: Penguin Books), 1969.

Jarvie, I. C., and Agassi, J. (1967) 'The Rationality of Magic', *Brit. Journal of Sociology* 18, 55–74. Reprinted in B. Wilson (ed.) *Rationality* (Oxford: Blackwell), 1970, 172–93; in S. Bruce (ed.) *The Sociology of Religion*, Vol. 1 (Cheltenham,

UK: Elgar), 148–67 and in J. Agassi, and I. C. Jarvie, *Rationality: The Critical View*, (Dordrecht: Kluwer), 1987.

Keynes, J. M. (1921) *A Treatise on Probability* (London: Macmillan).

Kijania-Placek, K. (2002) 'What Difference Does It Make: Three Truth-Values or Two Plus Gaps?' *Erkenntnis*, 56, 83–98.

Lakatos, I., and Musgrave, A. (1970) *Criticism and the Growth of Knowledge* (Cambridge: Cambridge University Press).

Landé, A. (1958) 'Determinism Versus Continuity in Modern Physics', *Mind*, 67, 174–81.

LeRoy, S. F., and Singell, L. D. Jr. (1987) 'Knight on Risk and Uncertainty', *Journal of Political Economy*, 95, 394–406.

Mill, J. S. (1843) 'A System of Logic, Ratiocinative and Inductive: Being a Connected View of the Principles of Evidence and the Methods of Scientific Investigation', in J. M. Robson (ed.) *Collected Works of John Stuart Mill*, Vols. 7, 8 (Toronto: University of Toronto Press), 1974.

Newton, I. [1687] (1995) *The Mathematical Principles of Natural Philosophy*. Translated by A. Motte (Amherst, NY: Prometheus).

Rokeach, M. (1960) *The Open and Closed Mind* (New York: Basic Books).

Russell, B. (1948) *Human Knowledge: Its Scope and Limits* (London: George Allen and Unwin).

Schilpp, P.A. (ed.) (1963) *The Philosophy of Rudolf Carnap* (LaSalle IL: Open Court).

Stove, D. (1981) *Popper and After: Four Modern Irrationalists* (New York: Pergamon Press).

Tversky, A., and Kahneman, D. (1971) 'Belief in the Law of Small Numbers', *Psychological Bulletin*, 76, 105–10.

Williams, D. C. (1934) 'Truth, Error, and the Location of the Datum', *The Journal of Philosophy*, 31, 428–38.

Wrinch, D., and Jeffreys, H. (1919) 'On Some Aspects of the Theory of Probability', *Philosophical Magazine*, 38, 715–31.

Zahar, E. (1980) 'Einstein, Meyerson and the Role of Mathematics in Physical Discovery', British Journal for the Philosophy of Science, 31(1), 1–43.

Part II

Truth and Concepts

6
Truth Defined and Undefined

Jaakko Hintikka

6.1   What did Tarski prove?

Many recent discussions of truth and its (in)definability can serve to illustrate a well-known fallacy. This fallacy is the fallacy of authority. The authority in question is Alfred Tarski. He is of course a genuine authority when it comes to the definability of truth (see Tarski, [1935] 1983). But then, the fallacy of authority does not mean an authority's mistake, but a misinterpretation of what the authority is supposed to have said.

In Tarski's case the misinterpreted result concerns the indefinability of truth. What Tarski actually proved about the definability and indefinability of truth concerns mainly first-order languages using the traditional ('Frege–Russell') first-order logic (TFO logic). For such a language Tarski showed how to define truth, but not in the same language, only in a richer metalanguage. Needless to say, these are perfectly sound and highly intriguing results. However, Tarski's results do not imply, contrary to what seems to have been often assumed, that truth could not be defined in a first-order language for the same language. What we can try to do is simply to use some other logic. These impossibility results do not imply, either, that truth is not definable and otherwise correctly employable in our ordinary language, which Tarski called 'colloquial language' and which presumably is our own working language. Tarski did claim that it is not, but this conclusion does not follow from his indefinability results. Tarski's stated reasons for his claim about natural language include our inability to specify the structure of natural language sufficiently clearly to apply semantical concepts to it. This impossibility is supposedly manifested in semantical paradoxes, like the Liar Paradox. Yet this supposed obstacle is what generative grammarians have spent half a century to remove, and the implications of the paradoxes depend actually on the logic that the colloquial language is supposed to rely on. For instance, in independence-friendly logic the Liar Paradox does not arise. (Cf. Section 6.6 below)

6.2   T-schema

Even apart from their wider implications, Tarski's results leave us with a conundrum: why the indefinability? This question is especially sharp since Tarski puts forward other ideas that take us tantalizingly close to a definition of truth. The most prominent is what is called Tarski's T-schema. It is formulated by him as

x is a true sentence iff p (1)

Here 'p' is a placeholder for a sentence and 'x' is a placeholder for a name or some other term referring to the same sentence. But how are the schema (1) supposed to be used? Tarski is one of the logicians who have emphasized the need of keeping object language and metalanguage separate. Instances of the schema (1) are formulated in a metalanguage. This metalanguage can be our colloquial language, but it can be something different, for instance an arithmetical language. In any case it must be an antecedently interpreted language, for otherwise the T-schema has no explanatory force. In any case, a substitution instance of 'p' is a sentence in that metalanguage.

But what is the object language of whose sentences truth is predicated? One possibility is that the object language is different from the metalanguage. This means that the object language is to be considered purely formally, abstracted from the meanings of its expressions. Hence any use of the schema (1) involves a formal operation, viz. the construction of the sentence of the metalanguage that replaces 'p' starting from the symbol or symbol combination that replaces 'x.' The function codifying this formal operation will be expressed by G, and $G[x]$ is an expression of the metalanguage that replaces 'p.' Interpretationally, this operation can be viewed as a translation, but it is a true Chinese room translation, that is, one whose input is an uninterpreted string of symbols.

This can be expressed by the schema

x is true iff $G[x]$ (2)

6.3   T-schema in a self-applied language

The use of the T-schema as applied to a foreign object language can help us to cope with that language, but it does not throw any light on the notion of truth as used in our own colloquial metalanguage. Hence, the main case is an application of the T-schema to a language used as its own metalanguage.

This does not involve anything paradoxical or circular. For instance, by means of Gödel numbering, an arithmetical language that can be made to serve as its own metalanguage. It allows the formulation of its own syntax in the same arithmetical language. What happens is that a numeral becomes a double agent: it refers to a number *qua* number, but it can also be interpreted as a codification of an expression of its own language.

Now in this case, too, the language used must be considered purely formally in so far as it figures as the object language. Accordingly, an application of the schema (1) still involves the same kind of formal generation $G[x]$. It is an inverse of the operation of Gödel number formation $g[S]$ in the sense that $G[g(S)] = S$.

The role and the complexity of operation are not always appreciated. In the usual application, x stands for the known Gödel number $g(S)$ of a given sentence S. The straightforward character of the formation of $g(S)$ tends to hide the procedure of actually constructing S from a merely given number $g(S)$, considered in its own right as merely a given number among others. Admittedly, the fundamental operation $G[x]$ does not involve any basic problem anymore than $g(S)$. If the metalanguage used is colloquial language, one way of forming a 'name' n of a sentence S is to put it in quotes. Then the reverse operation $G[n]$ is simply disquotation, which seems to be trivially simple.

However, in principle the situation is analogous to the Gödel numbering case. If we seriously try to use quotation and disquotation in grammatically generated semantics of natural language, we must quote the surface form, SF, of a sentence in discussing our language as an object language. But the back translation must yield a semantically interpretable form of the same sentence, or in Chomsky's occasional jargon, its LF. Now the rules for the purely syntactical procedure of going from SF to LF are far from obvious. (Cf. here Chomsky, 1972; 1981; 2000.)

Hence we can use an arithmetical language as a test case. And when we do so, all sorts of doors seem to open. For instance the following will then be an expression in the language in question:

$$T(x) \leftrightarrow G[x] \tag{3}$$

This formula can apparently be looked upon as a version of the T-schema that can yield an actual truth predicate. The resulting truth condition will be (for a given sentence S)

$$(\forall x)((x = g(S)) \supset (T(x) \leftrightarrow G[x])) \tag{4}$$

This avoids all quantification over propositions, quantification that is extremely tricky to interpret and use.

Definition along these lines seems to work for any S. For any S, a substitution-instance of (4) can be considered as being implied by a suitable instance of the T-schema. But here the initial mystery thickens. According to Tarski's result, (3) should not work. Moreover (4) can be thought of as being obtained from the T-schema, which has been held to be impossible for a genuine truth definition.

6.4   From TFO logic to IF logic

The explanation for the failure of (4) as a truth-definition – which is in effect also an explanation for the indefinability Tarski proved – is implicit in what has been said. In the context of (4), the expression $G[x]$ has the same form as an independent formula whose Gödel number it is. But its semantical interpretation is not the same in the two cases. The semantical force of a quantifier depends on which quantifiers it depends and does not depend. Dependence here means of course that actual (material) dependence of the respective variables. It is in the traditional 'Frege–Russell' logic and in all of its current versions, indicated the nesting of scopes. This aspect of the semantics of quantifiers was not understood by Frege and it has been largely neglected by most subsequent logicians. This neglect gave rise to a flaw in the traditional first-order logic. It is notationally impossible to express all possible patterns of dependence and independence by its means (see Hintikka, 2012a; 2012b).

This defect is partly amended in the so-called independence-friendly (IF) logic. (For IF logic (see Hintikka, 1996; and Mann, Sandu & Sevenster, 2011). Assuming that a formula is in a negation normal form, and existential quantifier $(\exists x)$ can be exempted from its dependence on a universal quantifier $(\forall x)$ within whose formal scope it occurs by writing it $(\exists y / \forall x)$. By means of game-theoretical semantics (GTS) the same idea (and the same notation) can be extended to other quantifier pairs.

In GTS, the independence of $(Q_2 y)$ of $(Q_1 x)$ means that the move prompted by the former is made in ignorance of the move prompted by the latter. This can be extended to the independence of a quantifier of a constant c. It means that the move prompted by it is made in ignorance of what c is. The independence can be expressed by writing the quantifier as $(\exists x/c)$. We have here an example of the significance of dependence relations. In (5), the quantifiers that there are in the complex predicate expression $G[s]$ depend on $(\forall x)$, whereas in a freestanding occurrence of the same expression they do not so depend.

In a definitory formula like (4) the definiens $G[x]$ must not depend (not in the logical sense of dependence just employed, but in the generic sense of being affected by) on the *definiendum*. In (5), the definiendum is an attribution of truth predicate to $g(S) = x$. Hence among other things the quantifiers in the definiens $G[x]$ ought to be independent of $(\forall x)$ and of $g(S)$. But not only are they not; we cannot even make them independent of $(\forall x)$ by means of the traditional first-order logic, for they unavoidably lie in the formal scope of $(\forall x)$. By the same token, in (4) $(\forall x)$ must be independent of $g(S)$ (cf. here Hintikka, 2012a).

This then is the reason for the failure of (4) as a truth definition. It is also the underlying reason for the indefinability of truth in a conventional first-order language for the same language. The deeper reason is a neglect of dependence relations between quantifiers, and the ultimate reason is the neglect of the dependence-indicating function of quantifiers by Frege and by subsequent logicians.

Hence instead of (4) we should write

$$(\forall x/g(S))((x = g(S)) \supset (T(x) \leftrightarrow G^*[x]) \tag{5}$$

where $G^*[x]$ is like $G[x]$ except that all its quantifiers are independent of $(\forall x)$ and of $g(S)$.

Here $G^*[x]$ may seem to be inappropriately defined in that it depends formally on its context. This, however, is a feature of the particular notation used. If we use a notation in which dependence rather than independence has to be explicitly indicated, there would not be any context-dependency.

6.5 Does T-schema help?

We can see that instances of the T-schema are essentially substitution-instances of the definition (4), in other words of the form

$$s = g(S) \supset (T(s) \leftrightarrow G[s]) \tag{6}$$

Since in any actual application it is assumed that $s = g(S)$, this reduces to a substitution instance of (3):

$$T(s) \leftrightarrow G[s] \tag{7}$$

Because (6) is an independent formula not governed by the quantifier $(\forall x)$, some of the sources of problems disappear. We do not need to require that $(\forall x)$ is independent of s or that the quantifiers or Skolem function terms in $G(s)$ are independent of $(\forall x)$. This is the reason why the T-schema works in the way Tarski suggested that it can work even though it cannot according to him be converted into a truth definition by means of the received first-order logic.

Yet the underlying conceptual situation cannot be seen from the T-schema, which therefore is not a good guide to understanding the notion of truth. Some much deeper insights are needed than what can be read from the T-schema. It is a tribute to Tarski's logical acumen that he realized the difference in logical status of a truth definition and the T-schema, even though he did not fully spell out the reasons.

One example of the need of further explanations is that the quantifiers in $G(s)$ must be independent of s. But again, this presupposition is built into the applications of T-schema, as it is normally used. When the technique of Gödel numbering is used, we have to separate two uses of quantifiers, viz. the uses in which we are considering numbers qua numbers and those in which we are considering them as codifications of formulas. The separation does not mean a different range of values, but informational independence. Hence the independence of quantifiers in $G(S)$ of S is tacitly assumed in applications of the T-schema.

Again, Tarski's sharp logical sense is illustrated by the fact that he made the same point indirectly, by emphasizing the purely formal character of the objects to which the truth predicate is to be applied. However, the right way of spelling this out is not in terms of a language–metalanguage distinction, but in terms of informational independence.

6.6   Liar Paradox

But have we avoided the Liar Paradox? Well, do we have to consider it, since it is not clear how it can be formulated in an arithmetical language that does not have first-person pronouns? An answer is given in (Hintikka, 2012a) and it shows that we do have to take seriously the possibility of an apparently paradoxical sentence L ('the Liar sentence') that is equivalent to $\sim T(\ell)$ or perhaps $\neg T(\ell)$ where $\ell = g(L)$. Here $\sim$ is the strong (dual) negation and ¬ is the contradictory negation. What does our proposed truth definition say about them?

Take the former first. What we obtain is

$$(\forall x/\ell)((x = \ell) \supset (T(\ell) \leftrightarrow G^*[\ell])) \tag{8}$$

Arguably here

$$G^*[\ell] \leftrightarrow G[\ell] \tag{9}$$

For the independence is taken automatically into account by the intended interpretation of the technique of Gödel numbering. (See the preceding section.) Thus if we had $(\forall x)$ instead of $(\forall x/\ell)$, we would have, by substituting ℓ for x,

$$T[\ell] \Longleftrightarrow L \tag{10}$$

But L was assumed to equal $\sim T(\ell)$, and we seem to have an outright contradiction. $T(\ell)$ and $\sim T(\ell)$ can have the same truth value, vis. "neither true nor false". Hence (8) is neither true nor false, and no contradiction occurs. Accordingly, $T(\ell)$ is neither true nor false, which is interpretationally as eminently reasonable as one can hope. Hence the Liar Paradox does not arise.

The other case is impossible for ℓ and $\neg T[\ell]$ cannot have the same truth value.

6.7 Definitions of truth on the first-order level

We have thus reached a diagnosis of the problems that have frustrated attempts to define truth for a language in the same language. We have also seen that there is no mystery about the ways in which they can be overcome. The problems are all due to a neglect of the dependence-indicating role of quantifiers; a neglect which goes back to Frege. It has led to the use of an inadequate notation which does not express what it was intended to express. In the case examined here, this is exemplified by the use of $G[x]$ rather than $G^*[x]$.

The other side of the same coin is that once this mistake is corrected, self-applicable truth-definitions of a certain kind become *ipso facto* possible. In particular, it can now seem that there is no reason whatsoever why we could not define truth in fragments of the colloquial language for the same fragment and otherwise use an objective concept of truth in our own working language.

The contrary belief probably ensued from a confusion about the relations of deductive power and expressive power. Since a first-order truth theory seems to lead deductively to paradoxes like the Liar paradox, the logic it used was deemed too strong deductively. Since the colloquial language is presumably stronger still, a consistent use of the concept of truth in it was deemed impossible.

In reality, the problems have been seen to be due to the expressive poverty of traditional FO logic, not to its deductive strength. Since ordinary language is certainly even richer in its expressive power, it poses no obstacles to a self-applicable truth-definition.

How precisely truth could and should be defined in ordinary language or even in IF languages is a separate question that will not be fully investigated here. One particular definition that can be formulated in IF languages, is nevertheless worth mentioning. It can be achieved by appealing to our pretheoretical ideas of truth. When is a quantificational sentence S true? One natural answer is: whenever the 'witness individuals' exist that can correctly show us the truth of S. If S is $(\exists x)F(x)$, then a witness individual a is one that satisfies $F(x)$. If S is $(\forall x)(\exists y)F(x,y)$, the witness individuals must satisfy $F(a,y)$ for the different a's. This example shows that witness individuals may depend on other individuals. The dependence is expressed by the Skolem functions of S. The existence of witness individuals is therefore tantamount to the existence of Skolem functions. (Individuals can be considered as constant Skolem functions.)

The resulting truth definitions say that S is true iff there exists a full set of Skolem functions for S. This definition can be extended to IF languages by choosing the right arguments for the Skolem functions, viz. variables bound to quantifiers on which the quantifiers in question depend.

The existence of Skolem functions is expressed by means of second-order quantifiers. The *definiens* of a resulting definition is therefore a Σ_1^1 formula. This fragment of second-order logic is translatable into the corresponding IF logic. For this first-order level logic, we can thus formulate an explicit truth predicate in the same language.

6.8   Truth definition and the axiom of choice

For the received ('Frege–Russell') first-order logic the Skolem function truth definition agrees with the usual Tarski-type one. Sometimes it is alleged that this argument is not beyond doubt, for it is said to depend on the set-theoretical axiom of choice.

The alleged dependence is nevertheless an illusion. The reason is that in a fundamental sense the principle of reasoning that is allegedly codified in the axiom of choice is itself a first-order logical truth.

The reason why this fact is now obvious is another restrictive feature of the received systems of first-order logic. In them, all the transformations of formulas involved in the rules of logical inferences are triggered by the outmost quantifier or the governing propositional connective. There is no reason for this feature which unnecessarily restricts the manipulation of formulas in logic. Formulating the rules of logical transformation (*salva* equivalence) so as to apply also inside formulas liberates the order of application of rules and hence simplifies proof theory.

For one important instance we can generalize the rule of existential instantiation. In its traditional form it allows us to replace a sentence of the form $(\exists x)F[x]$ by $F[a]$, where a is a new individual constant. When it occurs in the context of a larger sentence

$$S(-(\exists x)F[x]-) \tag{11}$$

we can still eliminate $(\exists x)$ but now we have to replace x by a term $f(y_1, y_2, \ldots, c_1, c_2, \ldots)$. Here f is a new function constant, $(Q_1, y_1)(Q_2, y_2), \ldots$ are all the quantifiers and $c_1, c_2 \ldots$ all the constants on which $(\exists x)$ depends in (11). The result is of the form

$$S(-F[f(y_1, y_2, \ldots, c_1, c_2, \ldots)]-) \tag{12}$$

As can be seen the newly introduced function f is the Skolem function associated with $(\exists x)$.

The result is a reformulation of the old first-order logic, either traditional or IF logic. But if the reformulated first-order logic is used as the underlying logic of set theory or as a part of higher-order logic, all consequences of the axiom of choice are provable.

We will call a truth predicate $T(x)$ valid iff the following equivalence holds unrestrictedly

$$T(x) \leftrightarrow G^*[x] \tag{13}$$

It is now seen that the validity of the Skolem function truth definition is equivalent with the new rule first order inference that does the same job as the axiom of choice. Which side of this equivalence supports the other one more interpretationally is a matter of theoretical taste.

6.9   Theories of truth reformed

The definability of truth puts most of the conventional 'theories' of truth in a sharp new light. Like any theory, they must depart from the definition of their subject, and do justice to it. Now many of these traditional 'theories' have prominently included attempts to show how to live with the alleged indefinability of truth in Tarski's sense. The definability of truth does not refute such theories, but it necessitates a reassessment of these 'theories.' This task promises interesting insights, but it is too large a job to be done in this paper.

Acknowledgment

This paper was written when the author was a Distinguished Visiting Fellow of the Collegium for Advanced Studies of the University of Helsinki. This support is gratefully acknowledged, as is the able assistance of Mr Antti Kylänpää.

References

Chomsky, N. (1972) *Studies on Semantics in Generative Grammar* (The Hague: Mouton).
—— (1981) *Lectures on Government and Binding* (Dordrecht: Foris).
—— (2000) *New Horizons in the Study of Language and Mind* (Cambridge: Cambridge University Press).
Hintikka, J. (1996) *The Principles of Mathematics Revisited* (Cambridge: Cambridge University Press).
—— (2012a) 'IF logic, Definitions and the Vicious Circle Principle', *Journal of Philosophical Logic*, 41(2), 505–17.
—— (2012b) 'Which Mathematical Logic is the Logic of Mathematics?, *Logica Universalis*, 6(3–4), 459–75.
Mann, A., Sandu, G., and Sevenster, M. (2011) *Independence-Friendly Logic: A Game-Theoretical Approach* (Cambridge: Cambridge University Press).
Tarski, A. (1935) 'Der Wahrheitsbegriff in den formalisierten Sprachen', *Studia Philosophica*, 1, 61–405. In English in Tarski (1983).
—— (1983) *Logic, Semantics, Metamathematics*, 2nd edn (Oxford: Clarendon Press).

7
Against Relative Truth

Ilkka Niiniluoto

Jan Woleński and Peter Simons give in their article 'De Veritate' (1989) a vivid account of the 'Austro-Polish obsession with truth'. Another impressive proof of this obsession is the collection of Woleński's historical essays (Woleński, 1999a). After Kazimierz Twardowski's seminal paper in 1900, Polish logicians and philosophers have been staunch and passionate supporters of the absolute notion of truth. Woleński and Simons (1989, p. 422) even suggest that Alfred Tarski did not introduce relativization of truth to a model in the 1930s 'for fear it would lead to relativism'. On the other hand, what Tarski called the 'the classical Aristotelian conception of truth' is a relational notion, as shown by the medieval formula *veritas est adequatio rei et intellectus* (Woleński, 1993, p. 330). This relational character is compatible with the objectivity of truth. Indeed, in my view, Tarski's model-theoretical concept of truth amended with the notion of truthlikeness provides an adequate basis for a fallibilist critical scientific realism (see Niiniluoto, 1987; 1999a). While some ways of relativizing truth are 'innocent' from the viewpoint of a realist, who wishes to defend the epistemological status of science, it is important to show why relativists of various sorts are mistaken in their claims that all truths are relative to persons, communities, cultures, beliefs, paradigms, or perspectives.[1]

7.1 Knowledge and objective truth

In his campaign against the sceptics and the relativists, Plato developed an objective notion of knowledge. In his dialogue *Theaetetus*, Plato (152a) presented the sophist Protagoras as arguing that any given thing 'is to me such as it appears to me, and is to you such as it appears to you'.

Against such subjective relativism, Plato (201d) defined real knowledge (Gr. *episteme*) as true belief with a justification or explanation.

Still, it is natural to assume that objective knowledge has an epistemological subject. If we use the notation K_ap (a knows that p), B_ap (a believes that p), J_ap (a has justification for p), and Tp (it is true that p), where p is a statement or a proposition (cf. Hintikka, 1962), then the classical definition of knowledge can be expressed by

$$K_ap \equiv B_ap \& J_ap \& Tp \tag{1}$$

The grounds for the objectivity of knowledge could be sought in the impersonal nature of justification – Plato himself was impressed by the conception of mathematical proofs with rigorous standards. But Plato also defended the objectivity of truth (Gr. *aletheia*) with his metaphysical doctrine of the invariant domain of ideas or forms. In a more mundane fashion, Aristotle argued for the objectivity of truth: 'to say of what is that it is, and of what is not that it is not, is true' (cf. Tarski, 1944, p. 342). In this view, truth is correspondence or 'adequatio' between beliefs and mind-independent reality: the truth of a belief or statement p is constituted by the correlation of p with states of affairs or facts obtaining in the world W. Whether this relation obtains between p and W is independent of our beliefs and wishes. Thus, the truth predicate T is not relative to a person. Its objectivity is expressible by (a variant of) the Tarskian T-schema

$$Tp \equiv p, \tag{2}$$

that is, it is true that p if and only if p is the case. Factual truth in this objective sense has to be distinguished from the epistemological indicators or criteria (if any) that help us to recognize truths.

Ordinary beliefs may be true or false, so that they do not satisfy the success condition $B_ap \rightarrow Tp$. It follows that beliefs can be genuinely relative to different persons. It is indeed very common that two subjects a and b may have different, even conflicting beliefs:

It is possible that $B_ap \& B_b \sim p$ $\hspace{3cm}$ (3)

But by definition (1) knowledge presupposes truth, so that by (2) we have $K_ap \rightarrow p$. Hence, knowledge is impersonal in the sense that $K_ap \& K_b \sim p$ would entail the contradiction $p \& \sim p$.

7.2 Fallibilism

With many later followers, Plato and Aristotle advocated a strong notion of justification which gives complete or full certainty. However, with (2), the condition $J_a p \rightarrow p$ would render the truth condition Tp in (1) redundant. The ancient sceptical school recommended suspension of judgement (Gr. *epoche*) which can be expressed by the condition $\sim B_a p \, \& \sim B_a \sim p$. This cautious strategy helped them to avoid errors of falsity (that is, assertions of false statements) but at the same time they committed errors of ignorance (that is, failures to assert true statements). The alternative moderate view that belief and justification may have degrees was initiated already by the Academic sceptics. Cicero expressed this idea with the Latin terms *probabile* and *veri simile* which are key concepts in what Charles S. Peirce at the end of the nineteenth century called *fallibilist* epistemology (see Niiniluoto, 2000).

According to a weak form of fallibilism, beliefs may actually be true but this is always more or less uncertain for us: the probability of a hypothesis expresses its degrees of credibility on the available evidence. This is the basis of later probabilistic accounts of human knowledge. According to strong fallibilism, hypotheses are typically false but still may be close to the truth. This approach was developed in dynamic theories which observed that science may start from incorrect opinions, but on some conditions gradually approaches to the truth at least in the limit. Such views have been formalized in later theories of truthlikeness or verisimilitude which measure the distance of a theory from the complete truth (see Popper, 1972; Niiniluoto, 1987).

The school of American pragmatism developed fallibilist epistemology with an epistemic notion of truth (limit of inquiry, warranted assertability) which easily leads to a relativist position (see Section 3 below). An important feature of the weak and strong forms of fallibilism with realist notions of probability and truthlikeness is that they retain the ideal of objective truth and scientific progress without surrendering to the lure of relativism. Tarski's semantic definition of truth in 1935 was a revelation to fallibilists like Rudolf Carnap and Karl Popper (see Niiniluoto, 1999b): the former learned to distinguish objective truth from time-dependent and evidence-relative confirmation, and the latter rejoiced about the 'rehabilitation of the correspondence theory of truth'.[2]

Peirce's fallibilism is based upon the idea that the epistemological subject is not an individual person but rather a community C (cf. Niiniluoto, 2003a). The possibility of scientific progress presupposes critical discussion within a community of investigators. It turns out that

Popper's (1972) idea of 'objective knowledge without an epistemological subject' has a natural interpretation by the assumption that the subject is the scientific community.

Fallibilists may admit that persons or communities in different times and cultures have employed various standards of justification. This kind of historical relativity of J does not imply the relativity of truth T (Niiniluoto, 1999a, p. 241). Moreover, realist fallibilists reject the radical form of epistemological relativism which urges that 'there is no sense attached to the idea that some standards or beliefs are really rational as distinct from merely locally accepted as such' (Barnes & Bloor, 1982, p. 27). Indeed, it is important that the methods of science have improved with the progress of logical, mathematical, computational, and experimental techniques (cf. Boyd, 1984).

7.3   The lure of alethic relativism

Fallibilist accounts of human knowledge have also inspired alethic relativism about truth – but for bad reasons. The tentatively accepted results of scientific inquiry or the body of 'scientific knowledge' is historically changing and temporally variable: Newton's theory was a celebrated part of the accepted scientific doctrines until the late nineteenth century, but it was then replaced by relativity theory and quantum theory. As we are accustomed to thinking with Plato that knowledge implies truth, this change can be taken to indicate that scientific 'truth' has changed. Such considerations were influential in Hegel's dynamic account of 'truth as a process', and in the Marxist discussion about the 'dialectics of absolute and relative truth'.[3] But a more natural interpretation of scientific change is to acknowledge that the time-dependent system of "scientific knowledge" does not generally satisfy Plato's definition (1) with the success condition, but may include partial truths and truthlike falsities (cf. Niiniluoto, 1987; 1999a).

Similar remarks apply to cultural relativism. In the latter half of the nineteenth century, the emerging historical, ethnological, and social studies proved an amazing variation in human cultures. Even though these cultures include different cognitive systems, which may function to some extent in their local environments, we need not say that their tribal beliefs or world views are 'truths'. Sometimes the relativists formulate their position by stating that persons with different belief systems live in different 'worlds', where this notion is used in a non-objective sense. Echoes of this view can be found in Thomas Kuhn's thesis that scientists with different paradigms live in different worlds. However, what these claims reveal, instead of relativism about truth and reality, are only instances of the relativity of beliefs (cf. (3)). The same point can be made against some sociologists of science and contemporary social constructivists (Niiniluoto, 1999a, ch. 9).

Philosophically more interesting grounds for alethic relativism may arise from epistemic definitions of truth (cf. Niiniluoto, 1999a, ch. 4.6). Instead of defining knowledge by truth, such approaches reverse the analysis (1) by trying to give the best description of human knowledge seeking, for example, by coherence, empirical verification, proof, acceptance, provability, acceptability, warranted assertability, or community consensus. Then the results satisfying such conditions are deemed to be instances of 'knowledge' and 'truth'. This strategy has been characteristic of pragmatist and neo-pragmatist philosophers who think that our only access to truth is via justified beliefs: truths are not waiting out there to be discovered, as the realist assumes, but they have to be 'made' by our acts of verification.

While epistemic notions may serve as indicators of truth for a realist, it is clear that approaches which identify truth with epistemic concepts easily lead to alethic relativism. If truth is defined by means of our actual beliefs and actions, then truths are relative to their owners. Moreover, there are no unknown truths, but all truths up to now have to be true for someone. A more liberal version defines truth in terms of verifiability or acceptability, but argues that no 'recognition transcendent' truths should be allowed (see Dummett, 1978).

The traditional argument against the coherence theory of truth is that there are several mutually conflicting systems of coherent beliefs. Inspired by Otto Neurath's coherence theory, which rejected all attempts to relate statements with non-linguistic facts, Carl G. Hempel (1935) suggested that truth should be defined relative to 'the system which is actually adopted by mankind, and especially by the scientists of our cultural circle'.[4] Again there is the problem that scientific theories change over time. Therefore, in order to avoid alethic relativism, and to guarantee that truth is unique, not relative to time and place and actual persons, one should appeal to *ideal* epistemological conditions – like Peirce's ultimate limiting consensus of the scientific community. In the same spirit, Hilary Putnam (1981) claims that an 'ideally acceptable' theory cannot be false. Still, it is doubtful that such ideal conditions (if they could be specified) can be successfully recognized in a non-circular way: how could we know that the ideal theory has been reached without appealing to the notion of objective truth? If the ideal theory cannot be reached at all, then the pragmatist again faces the problem that truths are 'recognition transcendent'.

7.4 Definitions of relative truth

Twardowski argued that the subjectivist notion of 'truth for some person only' ultimately leads to a conflict with the fundamental laws of logic (Woleński, 1989, p. 48).[5] To explore this issue (cf. Niiniluoto, 2006), assume that personal truths are simply identified with personal beliefs. Thus, a statement p is *true for person a* (in symbols, $T_a p$) if and only if a believes that p:

$$T_a p \equiv B_a p. \tag{4}$$

Then p is *false for person a* if a believes that non-p, that is, $B_a \sim p$. A variant of this definition allows the alethic subject to be a community C. Thus, h is *true for community* C if and only if C as a group believes that h (Niiniluoto, 2003a). Both of these definitions are genuinely relativistic, since by (3) it is possible that for different persons a and b we may have $T_a p$ & $T_b \sim p$, and the same holds for different communities.

Depending on the logical properties of the belief-operator B_a, as defined by Jaakko Hintikka's (1962) doxastic logic, definition (4) would guarantee some typical conditions for the truth predicate T_a, given in G. H. von Wright's (1984; 1996) 'truth logic' (cf. Woleński, 1999b):

$$T_a(p \& q) \equiv (T_a p \& T_a q) \tag{5}$$

$$T_a p \to T_a(p \vee q) \tag{6}$$

$$T_a p \to \sim T_a \sim p \tag{7}$$

$$T_a p \to T_a T_a p \tag{8}$$

However, the following conditions do not hold by (4):

$$T_a(p \vee q) \to (T_a p \vee T_a q) \tag{9}$$

$$T_a p \vee T_a \sim p \tag{10}$$

$$T_a T_a p \to T_a p \tag{11}$$

Note that (7) assumes the belief system of the person to be consistent, and thus excludes paraconsistent truth-systems. Violation of (9) allows disjunction to behave non-classically. Violation of the principle of bivalence (10) would allow paracomplete truth-logics with truth-value gaps.

Personal truth need not be solipsistic: if I am the person a, I might believe that there are other persons b with their own truths. But, as far as I am aware of my beliefs (cf. (8)), ultimately definition (4) would make me omniscient in a peculiar way: I could not admit that there are some truths unknown to me or that some of my beliefs are false (Krausz & Meiland, 1982, p. 82; David, 2004, p. 375). Moreover, (4) does not admit or need any external constraint for attributions of truth and falsity. As a statement p does not have truth-conditions apart from my beliefs, formulas corresponding to the T-equivalence (2), which Tarski proposed as a condition for any adequate definition of truth (see Tarski, 1956, p. 188), that is,

$$T_a p \equiv p \tag{12}$$

do not make sense at all. Even if they did, they would not be valid, since formulas $B_a p \to p$ and $p \to B_a p$ are not accepted in Hintikka's doxastic system. Also von Wright's (1996) formula $T_a(T_a p \equiv p)$ fails to hold.

To these observations one can add a difficulty related to classical incoherence arguments concerning relativism (see Siegel, 1987). What does it mean to a relativist that she believes that p? If this statement about belief has absolute truth conditions, then at least the global relativist has lost the battle to the absolutist. Hence, as acknowledged by Martin Kusch (1991) in his defence of relativism, the claim that $B_a p$ should be understood relatively as the thesis that $B_a p$ is true for person a, that is, $T_a B_a p$ or $B_a B_a p$. But there is no end to this iteration of belief-operators or relative truth-operators. For example, Putnam (1981, p. 120) argues that according to Protagorean subjectivist relativism my utterance 'Snow is white' has to be understood as

I think that I think that I think that I [...] (with infinitely many   (13)

'I thinks') that snow is white.

This is not as such an inconsistency: iterations of belief-operators have been used in the definition of mutual group beliefs (cf. Niiniluoto, 2003a). But, to say the least, this makes it difficult for the relativists to communicate their position and for others to understand it. Similarly, David (2004, p. 377) observes that the T-condition for relative truth should have the form

The proposition that c is F is true for person a iff c is F for a,   (14)

but again it is difficult to understand iterated predications with the same person or different persons.

One alternative to (4) would be to define relative truth T_ap by the knowledge claim K_ap, but this is both too strong and circular, as the definition (1) of the knowledge-operator assumes the absolute notion of truth. Other more promising alternatives could explore various notions of justification, and define T_ap in terms of B_ap & J_ap. This leads to epistemic theories which replace truth-conditions with assertability-conditions.[6] For example, J_ap might state that proposition p is *provable* in an axiomatic system S available to person a.[7] It is known that this treatment leads to truth-systems satisfying the principles of intuitionistic logic. But Dummett's (1978) attempt to extend this account to the 'proof' of empirical propositions is problematic. It is easier to understand Dummett's semantical anti-realism as a theory of understanding and meaning rather than as a theory of truth (Niiniluoto, 1999a, p. 107; Niiniluoto, 2001a).

We have thus found reasons to conclude that attempts to define relative personal truth 'h true for a' in an interesting way, which is weaker than 'h is true' and different from merely reducing it to the condition 'a believes that h', are not successful (cf. Swoyer, 1982; Siegel, 1987; 2004).

7.5   Perspectivism

Instead of relativizing truth to persons and their beliefs, some formulations of epistemological relativism appeal to other factors, like world views, historical situations, traditions, paradigms, frameworks, points of view or perspectives (see Meiland, 1977; Swoyer, 1982). Many of them reduce truth to shared group beliefs, and thus are vulnerable to the criticisms given in the preceding sections.[8]

Perspectivism is a view that is often attributed to Friedrich Nietzsche: 'there are no facts, only interpretations' (cf. Danto, 1973, p. 37). Some of Nietzsche's statements in *Die fröhliche Wissenschaft* in 1882 have been interpreted as a defence of the utilitarian account of truth of some pragmatists.[9] But when rejecting the correspondence account of truth as a relation between beliefs and reality, Nietzsche also denied that 'there would be a world left over once we subtracted the perspectival'. Thus, from a relativist position he seemed to end up with complete scepticism about truth and reality.

A more interesting form of perspectivism is motivated by Albert Einstein's special theory of relativity, published in 1905. Einstein rejected the absolute notions of time and space of the Newtonian mechanics and admitted only relational movements which are relative to some other moving bodies. But his ultimate purpose was to show that the velocity of light and the basic laws of Maxwell's theory of electromagnetism are *invariant* with respect to all coordinate systems which move in constant velocity with respect to each other. The general theory of relativity generalizes this revolutionary step to all coordinate transformations, that is, to all perspectives of the observer.

As Eino Kaila argued in his works in the 1930s, the same idea of explaining the relativity of something to a perspective by deeper law-like invariances can be applied to human perceptions as well (see Kaila, 1979; cf. Niiniluoto, 2006). In fact, Kaila based his hierarchical concept of reality on the suggestion that degrees of reality correlate with degrees of invariance. For example, I can look at a coin D from many different angles A. By turning D in my fingers, it may appear round, oval, or flat. *Perceptual* statements of the form

The object D appears oval   (15)

are incomplete and without truth-values, but *perspectival* statements of the form

The object D appears as an ellipse E from a perspective   (16)

A at point m

are true or false in an objective sense. In Hintikka's logic of perception, such a claim can be formulated by a *seeing-as*-statement: I (situated at point m) see object D as an ellipse E (cf. Niiniluoto, 1982). Here the observed coin is a perspectivally identified object for Hintikka – and an object on the lower level of reality for Kaila. Statements of the form (16), with variable perspectives A and observed shapes E, can be derived by optical theories from a physical description of the form

The object D is located at point n, and its shape in the   (17)

three-dimensional physical space is a cylinder.

In Kaila's terms, (17) is a more invariant description of the physical situation than the various perspectival alternatives (16) it helps to explain, and the coin as a physical object is likewise more invariant and real than the observed coin. This account can be generalized from shapes to other observable physical properties – sometimes called 'secondary qualities' – like colours, sounds, and smells.

If an alethic relativist proposes that the notion of truth should be relativized to perspectives or viewpoints, then we may ask about the nature of statements of the form

Proposition *p* is true-from-perspective A.   (18)

Is (18) itself only true from a perspective, so that we end up with an infinite number of iterations of viewpoints? Following Kaila, a more plausible alternative is to treat (18) as a non-relative truth which can be explained by deeper invariances: the world *is* such that it *appears p* from the perspective A.

7.6   Truth-bearers and incomplete sentences

Doctrines of relative truth have sometimes been based on specific views about truth-bearers.[10] Max Kölbel has suggested that Gottlob Frege's 'Der Gedanke' in 1918 marked the birth of the modern theory of indexicals (Garcia-Carpintero & Kölbel, 2008, p. 2), but Twardowski in his 1900 essay had already defended absolute truth by emphasizing the distinction between sentences and the judgements they express on particular occasions (Woleński & Simons, 1989, p. 397; Woleński, 1989, p. 47). For example, the sentence 'It is raining' is incomplete or elliptical, but it can be completed with spatio-temporal indicators, for example, 'It is raining in Cracow on the 21st of October 2010.'[11] While it may seem that the incomplete sentence can be sometimes true and sometimes false, the complete judgement as a proper truth-bearer is absolutely true or false, depending on facts about the city of Cracow on that day. If true then, it is eternally true or true at all later times.

This treatment is today a standard way of assigning truth values to statements with indexicals (like 'we', 'here', 'now'). In this extension of Tarskian truth-conditional semantics, truth is determined by interpretation (meaning), world (model), and context (including agent, location and time). (Cf. also the step from (15) to (16).) Twardowski suggested further that the same approach can be applied to apparently relative expressions of taste and ethical rules.

These kinds of cases have recently been discussed by a new school of truth relativists (see the collection of essays edited by Manuel Garcia-Carpintero & Kölbel, 2008). Examples include statements about taste ('*a* is prettier than *b*'), epistemic possibility ('it might have been *a*'), epistemic justification ('*p* is justified by evidence *e*'), knowledge attributions ('*a* knows that *p*'), value statements ('*a* is good'), normative statements ('*a* ought to do *f*') and future contingents ('Spain is the World Champion of football in 2010', stated before 2010). The relativist strategy, largely inspired by David Kaplan, is to treat these statements as special kinds of propositions, comparable to temporally indefinite sentences, with a truth value changing relative to extra factors – such as standards of taste, state of knowledge, standards of justification, systems of morality, legal order, and time of utterance. But the moderate contextualist approach acknowledges that such statements involve hidden indexicals.[12] For example, Cappelen and Hawthorne (2009) treat predicates of personal taste (for example 'Skiing is fun') by indexical statements (for example 'Skiing is fun for me', 'Skiing is fun for all').

In my view, contextualism does not handle moral judgements in an adequate way. Modest (non-subjective) moral relativism can be defended by requiring that categorical value statements (for example, 'killing is wrong') have to be completed by reference to some axiological system (for example, the ethical principles valid in Christianity) or some community (for example, the legal code accepted by the Polish nation), see (Niiniluoto, 1999a, ch. 8.2; 2009). With such completion, which is not uniquely determined by the context of utterance, these statements have objective truth values. Hence, this kind of moral relativism need not and should not be construed as an instance of alethic relativism about truth.

7.7   Models and conceptual pluralism

Following the tradition of Franz Brentano and the Lvov-Warsaw school, Tarski rejected the idea of relative truth (Woleński, 1989). When he presented his semantical approach in Paris in 1935, Maria Kokoszyńska explained at the same congress that Tarski had defined an 'absolute' concept of truth (cf. Kokoszyńska, 1936).

With influence from his philosophical teacher Tadeusz Kotarbiński, whose 'reism' included only things but no facts or states of affairs in ontology (see Niiniluoto, 2002), Tarski was reluctant to use traditional phrases like 'correspondence to facts'.[13] In his groundbreaking paper in 1933/1936, he defined truth in terms of a satisfaction relation between sentences and sequences of objects – and nothing like facts were presupposed. (Tarski was more liberal than Kotarbiński in his allowance of sets among objects.) However, Tarski added, with reference to the 'Göttingen school grouped around Hilbert', that truth can also be defined relative to 'an individual domain' which is a subset of the class of all objects (Tarski, 1956, p. 199). Wilfrid Hodges (1986) has argued that Tarski did not have the concept 'truth in a structure' before mature model theory in the 1950s, and Woleński and Simons (1989, p. 422) agree that 'the relativization to a domain of individuals is not relativization to a model'. I think there are reasons to disagree with this conclusion (see Niiniluoto, 1999b). Tarski needed the concept of model in his 1935 paper on logical consequence, and his somewhat peculiar construction guarantees that each extra-logical constant (individual constant, predicate) has an interpretation in the domain of objects, (see Tarski, 1956, p. 471). An individual domain of objects, together with a language interpreted on this domain, indeed constitutes a structure or model in the sense of later model theory – and this notion was needed already before Tarski in the pioneering results by Löwenheim and Skolem. Tarski's later account of relational systems and models can be understood so that his set-theoretical constructions function as truth-makers of true sentences (see Niiniluoto, 2004).

A special feature of Tarski's early theory, rightly emphasized by Woleński and Simons (1989), is his assumption that truth applies only to an already interpreted language. When Kokoszyńska suggested in 1936 that the concept of truth is relativized to the concept of meaning, Tarski remarked that it would be simpler to relativize it to the concept of language (Woleński & Simons, 1989, p. 416; Woleński, 1993, p. 328). Tarski did not seem to realize that his conception of an interpreted language already presupposes something like meanings. In fact, he never made the interpretation explicit. In this respect, Carnap's version of logical semantics in the late 1930s and in his *Introduction to Semantics* (1942) was an advance in comparison to Tarski's papers in 1936 and 1944: like in later model theory, Carnap took a semantical system S to consist of a language as uninterpreted syntactical signs and a designation function Des such that names designate objects, predicates designate properties and relations of objects, and sentences designate propositions. Then

(C) Sentence s is true in S iff there is a proposition p such that s

designates p and p.

(Niiniluoto, 2003b). This is Carnap's counterpart to Tarski's T-schema which can be written in the explicit form as follows:

If p in metalanguage ML is the translation of sentence s in L,   (19)

then s in L is true iff p.

According to Kirkham (1992), when propositions are replaced by states of affairs, schema (C) expresses the 'essence' of the correspondence theory of truth.

In Tarski's mature model theory in the 1950s, the truth of a sentence s in language L in model or L-structure W has to be mediated by a third factor, viz. the interpretation function I from L to W. If we amend Tarski's account by making the interpretation function I from a language L to a domain W explicit, then truth and falsity in L are relative to the model W and the interpretation I.

Woleński (1999a, p. 107) states that Carnap and Arthur Pap gave an 'entirely wrong' interpretation of Tarski's semantic definition as 'a typical expression of relativism in the theory of truth', since it involves reference to language L and model M (cf. Woleński & Simons, 1989, p. 421). This claim is in need of clarification. Carnap (1942) did not complain about the schema (C), which is general in the sense that it can be applied to any language L. But he suggested further that there is in his terms an 'absolute' notion of truth besides the language-relative notion defined by (C): assuming that sentences in different semantical systems can designate the same proposition (for example, 'snow is white' and 'Schnee ist weiss'), then one can say that a proposition p is true. If, instead, propositions were simply identical with states of affairs in the world, then there would be no need to assign truth values to them. Carnap required that a definition of truth for propositions should have the following consequence:

Proposition p is true iff for every S and every s in S   (20)

if s designates p in S then s is true in S.

So far so good: if propositions are introduced as abstract entities and truth-bearers in the Fregean spirit, (20) expresses the sound semantical principle that we have access to propositions and their truth values only via sentences designating them. Thus, the truth of propositions is derivative of the truth of interpreted sentences.[14] But Carnap's next move is problematic, however. He gave another definition of absolute truth:

The proposition p is true $=_{df} p$ (Carnap, 1942, p. 90)   (21)

According to Carnap, this notion is not semantical, since it does not involve designation relations, so that it differs from Kokoszyńska's sense of 'absolute truth' (1942, p. 240). The problem with this proposal is that (21) taken as a *definition* of truth (that is, it is true that p iff p) makes truth redundant and leads to the deflationist notion of truth (cf. Horwich, 1990). When propositions are assumed as abstract entities, there are always several rival alternatives (for example, the propositions that snow is white, snow is red, snow is blue, etc.), and there must be something in the world as a truth-maker which decides which one of them is true. In this sense, even a proposition involves some sort of semantical relations to the world, and (20) is a way of making this connection explicit. This kind of account, compatible with (20), became available to Carnap in the 1950s when he developed the idea of propositions as functions from possible worlds to truth values (see Niiniluoto, 2003b).

Should one fear that the relativization of truth to a selected class of objects or a model leads to unacceptable relativism? In my view, the answer is negative, since such a relativization is a natural consequence of the relational character of truth as a semantical relation between language and world.[15] Model theory is an extremely powerful tool of semantics – and especially philosophically interesting when it is applied as 'possible worlds semantics' to any possible domains of objects.[16] As far as models are related to different worlds, it is not surprising that they have different and even conflicting truths. But if one's aim is to define the notion of *material* or *actual truth*, as Tarski was attempting to accomplish in 1933, there is a problem with the choice of 'individual domains' of objects: moving from the class of all actual objects to a subclass preserves all true universal sentences, but not necessarily all existential sentences. For example, if the domain is restricted to human beings, then the sentence 'There are penguins' is not true anymore.[17] One way of avoiding this trouble is to treat such a restriction as a kind of indexical: while 'There are penguins' is false in the domain of human beings, the sentence 'There are penguins but not among human beings' is actually true.

More generally, as the actual world W is one of the possible worlds, the notion of actual truth can be defined within model theory (see Niiniluoto, 1999a, pp. 220–6). Let the pair $K = (L, I)$ be an interpreted linguistic framework. The values of the function I for the extra-logical terms of L within the domain of actual objects in W constitute an L-structure W(K). This structure consists of the facts of the world W from the point of view of the framework K. Here W(K) is not an epistemic notion – it is not what we believe about the world W, but what the language L is able to tell about the world W if it were investigated via the framework K. Factual truth about W in language L interpreted by I (or with meanings specified by I) is defined by Tarski's model-theoretical definition of truth in structure W(K). We need not assume with metaphysical realists (cf. Putnam, 1981) that there is an ideal framework K which covers all of the variety of the actual world W so that $W = W(K)$. Instead, each conceptual framework K captures only a partial fragment of the inexhaustible reality W. But for each framework K truth is objective in the following sense: we may choose L and I, but world W decides the relevant truth values. Truth in W(K) is truth about the actual world. For other frameworks K' we have other truths in W(K'), but as descriptions of the same world W they cannot logically contradict the truths of K.[18]

The position outlined above is not relativistic. Rather, it expresses *conceptual pluralism*: the world can be described in alternative linguistic frameworks, and all of these frameworks may have interesting objective truths to offer us.

Notes

1. This paper summarizes and develops earlier work in (Niiniluoto, 1994; 1999a; 1999b; 1999c; 2000; 2002; 2003b; 2004), especially (Niiniluoto, 2006). I am grateful to Jan Woleński for his true friendship and absolutely enlightening discussions about truth over three decades.

2. Tarski's teacher Tadeusz Kotarbiński was able to combine physicalist realism and fallibilism already in 1929 (see Niiniluoto, 2002). In Poland, fallibilism was combined with a non-realist and relativist coherence account of truth by Edward Poznański and Aleksander Wundheiler in 1934 (see Kokoszyńska, 1936; Woleński, 1999a, p. 231).

3. British neo-Hegelian doctrines of 'degrees of truth' also confused errors of falsity with incomplete knowledge (see Niiniluoto, 1987). Similar ambiguities can be found in the Marxist discussion of absolute and relative truth.

4. After learning about Tarski's semantic definition, Hempel gave up his coherence theory (see Niiniluoto, 1999b).

5. It should be noted that the phrase 'true for person a' can be perfectly legitimate in the context of open formulas. For example, 'x is a logician' is satisfied by letting the variable x refer to Jan Woleński. In this sense, we may say that 'x is a logician' is *true for* Jan Woleński but *false for* Brigitte Bardot. But this notion is not a variant of truth (which is defined for closed sentences), and Tarski himself avoided this kind of unnecessary talk about relative truth by his notion of satisfaction. According to Tarski's schema, object b satisfies 'x is P' iff b is P*, where P* is the translation of P into metalanguage (Tarski, 1956, p. 192). Open formulas had an important role also in Jan Lukasiewicz's theory of probability (see Niiniluoto, 1998).

6. For a sophisticated but problematic attempt to build a bridge from inferential role semantics to truth-conditional semantics see (Brandom, 1994).

7. In mathematical systems with the soundness property, this notion of provability-of-p-in-S is sufficient to guarantee the Tarskian truth of statement p in the models of S. But Gödel's results show that, for all sufficiently rich theories S with a standard model W, there are truths in W not provable-in-S.

8. Bo Mou's (2009) 'substantive perspectivism' combines many roles and tasks of the theory of truth, including Tarski and Daoism, but without relativist implications. Antti Hautamäki's (1986) 'points of view' are in fact conceptual frameworks. In spite of his own inclination to relativism, Hautamäki's position can be interpreted in terms of conceptual pluralism (see Section 7.7). For

proposals that 'points of view' should be added to the truth relation between propositions and facts (see Meiland, 1977; Siegel, 1987, pp. 11–18).

9. Tarski followed Kotarbiński in his rejection of the utilitarian theory (see Tarski, 1956, p. 153). Hourya Benis Sinaceur (2009) concludes her illuminating account of Tarski's philosophical pluralism with the somewhat overstated conclusion that Tarski's 'philosophical final view' was 'effective pragmatism' accepting 'meaning as use'.

10. Hintikka (1973) has argued that in the Greek oral culture typical sentences were temporally indefinite, so that their truth was taken to presuppose an unchanging or permanent phenomenon (for example, 'snow is white'). In his defence of absolute truth, Tarski's teacher Stanisław Leśniewski rejected indefinite propositions (Woleński & Simons, 1989, p. 405).

11. Cappelen and Hawthorne (2009) defend, against the relativists, the contextualist view that the context of utterance of a declarative sentence determines a proposition which is absolutely true or false.

12. For my position on future contingents see (Niiniluoto, 2001b).

13. Woleński (1999a, p. 166, 174) suggests that Tarski was against the 'strong' correspondence theory which takes truth-bearers to be representations of reality by a similarity relation, and accepted only a weaker correlation between expressions and extra-linguistic entities (see also Woleński & Simons, 1989). However, Tarski's account of atomic sentences (for example, 'Jan loves Maria') is based on a strong form of correspondence, which resembles Wittgenstein's picture theory of language in the *Tractatus*, while this is not the case for his treatment of composite sentences like disjunctions and generalizations (see Niiniluoto, 1999b, p. 99; 2004).

14. Pap (1952) also favours the assignment of truth to propositions. On the basis of (19) he argues that Carnap treats '*s* is true' and '*p*' as logically equivalent, which he finds mistaken since the proposition '*p*' does not entail the existence of any sentences *s*. However, Pap ignores the fact that the equivalence given by (19) is conditional on the assumption that *s* designates *p*, so that it is also conditional on the existence of sentence *s*.

15. Cappelen and Hawthorne (2009) instead argue that truth and falsity *simpliciter* are monadic properties of propositions, more fundamental than the relational property of being *true at* a world.

16. Possible worlds semantics was formulated by Stig Kanger and Jaakko Hintikka in 1957 (see Woleński, 1999a, p. 241).

17. The same problem is faced by Jon Barwise's situation semantics, where situations are fragments of the world (see Barwise & Etchemendy, 1987).

18. This means that 'genuine relativism' in the sense of Kokoszyńska is avoided (see Woleński, 1999a, p. 166).

References

Barnes, B., and Bloor, D. (1982) 'Relativism, Rationalism and the Sociology of Knowledge', in M. Hollis, and S. Lukes (eds) *Rationality and Relativism* (Oxford: Blackwell), 21–47.

Barwise, J., and Etchemendy, J. (1987) *The Liar* (Oxford: Oxford University Press).

Boyd, R. (1984) 'The Current Status of Scientific Realism', in J. Leplin (ed.) *Scientific Realism* (Berkeley: University of California Press), 41–82.

Brandom, R. (1994) *Making it Explicit* (Cambridge, MA: Harvard University Press).

Cappelen, H., and Hawthorne, J. (2009) *Relativism and Monadic Truth* (Oxford: Oxford University Press).

Carnap, R. (1942) *Introduction to Semantics* (Cambridge, MA: Harvard University Press).

Danto, A. (1973) 'Nietzsche's Perspectivism', in R. Solomon (ed.) *Nietzsche: A Collection of Critical Essays* (New York: Anchor Books), 29–57.

David, M. (2004) 'Theories of Truth', in I. Niiniluoto, M. Sintonen, and J. Woleński (eds) *Handbook of Epistemology* (Dordrecht: Kluwer), 331–414.

Dummett, M. (1978) *Truth and Other Enigmas* (London: Duckworth).

Garcia-Carpintero, and M., Kölbel, M. (eds) (2008) *Relative Truth* (Oxford: Oxford University Press).

Hautamäki, A. (1986) 'Points of View and their Logical Analysis', *Acta Philosophica Fennica*, 41, (Helsinki: Societas Philosophica Fennica).

Hempel, C. G. (1935) 'On the Logical Positivists' Theory of Truth', *Analysis*, 2, 49–59.

Hintikka, J. (1962) *Knowledge and Belief* (Ithaca: Cornell University Press).

—— (1973) *Time and Necessity* (Oxford: Clarendon Press).

Horwich, P. (1990) *Truth* (Oxford: Blackwell).

Kaila, E. (1979) *Reality and Experience* (Dordrecht: D. Reidel).

Kirkham, R. L. (1992) *Theories of Truth: A Critical Introduction* (Cambridge, MA: MIT Press).

Kokoszyńska, M. (1936) 'Über den absoluten Wahrheitsbegriff und einige andere semantische Begriffe', *Erkenntnis*, 6, 143–65.

Krausz, M., and Meiland, J. W. (eds) (1982) *Relativism: Cognitive and Moral* (Notre Dame, Ind.: University of Notre Dame Press).

Kusch, M. (1991) *Foucault's Strata and Fields* (Dordrecht: Kluwer).

Meiland, J. (1977) 'Concepts of Relative Truth', *The Monist*, 60, 568–82.

Mou, B. (2009) *Substantive Perspectivism: An Essay on Philosophical Concern with Truth* (Dordrecht: Springer).

Niiniluoto, I. (1982) 'Remarks on the Logic of Perception', in I. Niiniluoto, and E. Saarinen (eds) *Intensional Logic: Theory and Applications*, *Acta Philosophica Fennica*, 35, (Helsinki: Societas Philosophica Fennica), 116–29.

—— (1987) *Truthlikeness* (Dordrecht: D. Reidel).

—— (1994) 'Defending Tarski against His Critics', in J. Woleński (ed.) *Sixty Years of Tarski's Definition of Truth* (Cracow: Philed), 48–68.

—— (1998) 'Induction and Probability in the Lvov-Warsaw School', in K. Kijania-Placek, and J. Woleński (eds) *The Lvov-Warsaw School and Contemporary Philosophy* (Dordrecht: Kluwer), 323–35.

—— (1999a) *Critical Scientific Realism* (Oxford: Oxford University Press).

—— (1999b) 'Tarskian Truth as Correspondence – Replies to Some Objections', in J. Peregrin (ed.) *The Nature of Truth – If Any* (Dordrecht: Kluwer), 91–104.

—— (1999c) 'Theories of Truth: Vienna, Berlin, and Warsaw', in J. Woleński, and E. Köhler (eds) *Alfred Tarski and the Vienna Circle* (Dordrecht: Kluwer), 17–26.

—— (2000) 'Scepticism, Fallibilism, and Verisimilitude', in J. Sihvola (ed.) *Ancient Scepticism and the Sceptical Tradition* (Helsinki: Societas Philosophica Fennica), 145–69.

—— (2001a) 'Information, Meaning, and Understanding', in L. Lundsten, A. Siitonen, and B. Österman (eds) *Communication and Intelligibility* (Helsinki: Societas Philosophica Fennica), 43–54.

—— (2001b) 'Future Studies: Science or Art?' *Futures*, 33, 371–7.

—— (2002) 'Kotarbiński as a Scientific Realist', *Erkenntnis*, 56, 63–82.

—— (2003a) 'Science as Collective Knowledge', in M. Sintonen, P. Ylikoski, and K. Miller (eds) *Realism in Action* (Dordrecht: Kluwer), 269–78.

—— (2003b) 'Carnap on Truth', in T. Bonk (ed.) *Language, Truth, and Knowledge: Contributions to the Philosophy of Rudolf Carnap* (Dordrecht: Kluwer), 1–25.

—— (2004) 'Tarski's Definition and Truth-Makers', *Annals of Pure and Applied Logic*, 126, 57–76.

—— (2006) 'The Poverty of Relative Truth', in T. Aho, and A. V. Pietarinen (eds) *Truth and Games: Essays in Honour of Gabriel Sandu* (Helsinki: Societas Philosophica Fennica), 165–74.

—— (2009) 'Facts and Values – A Useful Distinction', in S. Pihlström, and H. Rydenfelt (eds) *Pragmatist Perspectives* (Helsinki: Societas Philosophica Fennica), 109–33.

Niiniluoto, I., Sintonen, M., and Woleński, J. (eds) (2004) *Handbook of Epistemology* (Dordrecht: Kluwer).

Pap, A. (1952) 'Note of the "Semantic" and the "Absolute" Concept of Truth', *Philosophical Studies*, 3, 1–8.

Peregrin, J. (ed.) (1999) *The Nature of Truth – If Any* (Dordrecht: Kluwer).

Popper, K. (1972) *Objective Knowledge* (Oxford: Oxford University Press).

Putnam, H. (1981) *Reason, Truth and History* (Cambridge: Cambridge University Press).

Siegel, H. (1987) *Relativism Refuted: A Critique of Contemporary Epistemological Relativism* (Dordrecht: D. Reidel).

—— (2004) 'Relativism', in Niiniluoto, Sintonen and Woleński (2004), 747–80.

Sinaceur, H. B. (2009) 'Tarski's Practice and Philosophy: Between Formalism and Pragmatism', in S. Lindström *et al.* (eds) *Logicism, Intuitionism, and Formalism* (Dordrecht: Springer), 357–96.

Swoyer, C. (1982) 'True For', in M. Krausz, and J. W. Meiland (eds) *Relativism: Cognitive and Moral* (Notre Dame, Ind.: University of Notre Dame Press), 84–108.

Tarski, A. (1944) 'The Semantic Conception of Truth and the Foundations of Semantics', *Philosophy and Phenomenological Research*, 4, 341–75.

—— (1956) 'The Concept of Truth in Formalized Languages', in *Logic, Semantics, Metamathematics* (Oxford: Oxford University Press), 152–278.

Woleński, J. (1989) *Logic and Philosophy in the Lvov-Warsaw School* (Dordrecht: Kluwer).

—— (1993) 'Tarski as a Philosopher', in F. Coniglione, R. Poli, and J. Woleński (eds) *Polish Scientific Philosophy: The Lvov-Warsaw School* (Amsterdam: Rodopi), 319–38.

—— (1999a) *Essays in the History of Logic and Logical Philosophy* (Cracow: Jagiellonian University Press).

—— (1999b) 'Semantic Conception of Truth as a Philosophical Theory', in Peregrin (1999), 51–65.

Woleński, J., and Simons, P. (1989) 'De Veritate: Austro-Polish Contributions to Truth from Brentano to Tarski', in K. Szaniawski (ed.) *The Vienna Circle and the Lvov-Warsaw School* (Dordrecht: Kluwer), 391–442.

von Wright, G. H. (1984) 'Truth and Logic', in *Philosophical Papers III* (Oxford: Blackwell), 26–41.

—— (1996) 'Truth-Logics', in *Six Essays in Philosophical Logic* (Helsinki: Societas Philosophica Fennica), 71–91.

8
Truth without Truths? 'Propositional Attitudes' without Propositions? Meaning without Meanings?

Wolfgang Künne

I am proud of being able to say that Jan Woleński was my first friend in Poland and that he still is my best Polish friend. So I am glad that I can contribute a paper to his *Festschrift*. 'Künne likes to remain very close to natural language and its parlance', Jan wrote in his contribution to a symposium on a book on truth in *Dialectica* 62 (2008). Since I want to elucidate our workaday concept of truth, I do indeed like to stay close to natural language: after all, that's where this concept gets expressed in the first place. 'Sooner or later', Jan wrote in the same article, 'we encounter problems which require a clear decision concerning the metalogical properties of "is true." Is it a predicate or a modality? [...]' This is the main question that I want to clarify and to answer in this paper. Tackling this problem requires close attention to the syntactical and semantical status of *that*-clauses. The questions such clauses evoke when we brood on truth ascriptions ('It is true that *p*') reappear when we consider reports of propositional attitudes ('A *ϕ*s that *p*') and ascriptions of sentential meaning ('S means that *p*'), That's why I shall also discuss the second and the third issue which the verbose title of this paper alludes to. Arthur N. Prior saw the connection, and in each of these fields he opted for the same treatment of that-clauses. I shall argue that it is a mistreatment.

8.1 Truth predicate vs. truth connective

The expression '___ is true' is a sentence-forming operator on *singular terms*, – it is a truth *predicate*: insert a singular term like 'Alfred's most cherished belief', 'logicism' or 'this', and you obtain a sentence. By contrast, the expression 'It is true that ___' permits no such filling. It resembles the logicians' negation operator or their modal operators in being a unary sentence-forming operator on *sentences* – it is a truth *connective*: sentences like 'snow is white' and 'arithmetic is reducible to logic' as input deliver sentences as output.[1] Is one of these two truth locutions explanatorily more fundamental than the other, and if so, which one? On my view,[2]

(TC)   *It is true that* snow is white

is just a stylistic variant of

(TP)   That snow is white *is true*

and the latter is to

(TP+)   *The proposition* that snow is white *is true*

as 'Seven is prime' is to 'The number seven is prime'. In an attempt to elucidate our ordinary notion of truth we do well to focus on the truth *predicate* that is unmissable in construction (TP+) and discoverable in (TC). The proposition expressed by (TC) unobviously is what the proposition expressed by (TP+) obviously is: a proposition in which something is classified as true. Now sentences like 'Yesterday he met her in Cambridge' express different propositions in different contexts, some of which are true, some false. So let me formulate my priority thesis in such a way that sentences with semantically context-sensitive elements are covered as well:

(PROPERTY)
No matter which of the following three schemata a sentence instantiates,
   (tc) It is true that p
   (tp) That p is true
   (tp+) The proposition that p is true,
it express a truth in a context C, if, and only if, the proposition that is designated in C by the that-clause has the property that is signified by 'is true'.[3]

So whenever the sentence letter is replaced by the same sentence, the truth-conditions of the whole sentence (relative to C) are the same. Sameness of truth-conditions does not guarantee propositional identity: corresponding instances of (tc) and (tp) express the same proposition (relative to C), but the proposition expressed by their (tp+) counterpart is different. (PROPERTY) captures the sense in which I take the truth *predicate* to have priority over the truth connective in the order of explanation. (PROPERTY) implies that not only instances of (tp+) but also instances of (tp) and even those of (tc) contain a singular term that designates a proposition.[3] If I am right, an elucidation of the concept of truth as what is expressed by the truth predicate does not stand in need of supplementation by a separate account of the truth connective.

Arthur Norman Prior pleaded for the opposite view: 'the word [...] "true" in [its] primary use [is an] inseparable part of the adverbial phrase [...] "it is true that"'. (1967a, p. 229) If this is correct then the use of the general term 'true' as part of the truth predicate is *not* primary, the truth connective does *not* contain the truth predicate as a semantically relevant part, and (TC) contains *no* singular term that designates a proposition. If the truth connective is semantically indivisible, then nothing is classified as *true* in (TC), – only snow is classified as white. Quite generally, the truth-conditions of such sentences can be, and ought to be, specified without bringing the property of being true into the picture:

(OPERATION)
A sentence of the form 'It is true that *p*' expresses a truth in context C if, and only if: the operation signified by 'It is true that' yields a truth when applied to a truth and a falsehood when applied to a falsehood, and the embedded sentence expresses a truth in C.

If the use of 'true' as an inseparable part of the truth *connective* is primary, as Prior maintains, then an elucidation of the concept of truth that is not in the first place an account of the truth connective is *Hamlet* without the prince of Denmark.

I should at least mention that Prior goes two steps further.[4] From OPERATION it does not follow that the operation in question, when applied to a truth (falsehood), yields the *same* truth (falsehood). In his *first* additional step Prior endorses a weak redundancy thesis: in uttering (TC) we do not only talk *about* nothing but the stuff we would have talked about if we had uttered 'Snow is white', but we also *say* nothing but what we would have said if we had just uttered the shorter sentence. This claim is *not* implied by Prior's priority thesis. (A comparison may help to make this clear. If you contend that 'possible' in its primary use is an inseparable part of the one-place connective 'It is possible that', you'd better not go on and claim that in saying 'It is possible that *p*' one says nothing one would not have said if one had just uttered the plain '*p*'.) Prior's *second* additional step is to augment the redundancy thesis: not only (TC) but also (TP) and even (TP+) are used to say the same thing as 'Snow is white'.

Prior invokes Ramsey's authority for both additional steps, which is dubious,[5] but he could have invoked Frege's authority. In *Der Gedanke* (1918) Frege claims that (TC) and 'Snow is white' or rather, that 'It is true that I smell the scent of violets' and 'I smell the scent of violets' (in the same context) express the same proposition.[6] And in *Über Sinn und Bedeutung* (1892) he maintains that the same holds for the pair (TP+) and 'Snow is white', or rather, for 'The thought that 5 is a prime number is true' and '5 is a prime number'.[7] In this paper I shall assume that Frege's and Prior's redundancy claims are false. I will only be concerned with Prior's contention that basic truth talk makes use of a truth connective that is a semantically seamless whole. Recently Kevin Mulligan (2010) argued at length for this Priorean priority claim (without mentioning Prior in this connection),[8] and he holds the redundancy thesis in no higher esteem than I do.

8.2   Three truth-locutions

Often, if not always, sentences can be decomposed in more than one way: that is a point Frege repeatedly made. What Frege called *zerlegen*, Prior calls *parsing*. If the truth-connective parsing of (TC) is the only correct parsing of this construction, then it contains only a *spurious* occurrence of the clause 'that snow is white', for then the word 'that' belongs with the connective. By Prior's lights, (TC) no more contains a *genuine* occurrence of the that-clause than [Q], 'The woman who was married to Socrates was Greek', contains a genuine occurrence of the sentence 'Socrates was Greek'.[9] Now in [Q] putting the parsing line between 'Socrates' and 'was' is *obligatory*. But why should we take the truth-connective parsing of (TC) to be obligatory? On the contrary, the truth-predicate parsing is not only equally permissible, – it is more faithful to the way we understand (TC). Or so I shall argue.

It cannot seriously be denied that at least sometimes both connective- and predicate-parsing are *permissible*. After all, one can discover a truth connective even in (TP+), namely the sentence frame 'The proposition that ___ is true'.[10] The occurrence of the sentence 'snow is white' in (TP+) is certainly genuine. As regards parsing we have a lot of leeway. Consider 'Anna is witty, and Bella is pretty'. This compound sentence does not only contain a well-known two-place connective but also two unary connectives you presumably never dreamt of: '___, and Bella is pretty' and 'Anna is witty, and ___'. Clearly, these (extensional) operators deliver sentences as output for single sentences as input. But unlike the routine parsing these decompositions do not carve the object of understanding at the joints. Our understanding of the conjunction is based upon our comprehension of the conjuncts and of the two-place connective, – understanding the one-place connectives nowhere comes into play. In this sense the routine parsing is a *canonical* decomposition of our compound sentence, while the other two parsings are just *optional*. My contention, as against Prior, is this: our understanding of (TC) is based upon our understanding of the truth predicate and of the that-clause. (It is uncontroversial that our grasp of a genuine occurrence of a that-clause is in turn founded upon our understanding of the sentence to which the 'that' is prefixed and of the operation of this kind of sentence nominalization.) If PROPERTY is correct, then the parsing that finds the truth predicate and the that-clause in (TC) has got to be the canonical decomposition of (TC). I opt for affirming the antecedent.

Both in (TP+) and in 'The *number* seven is prime', there is a general term that expresses a concept which is not expressed in their unadorned counterparts, that is, in (TP) and 'Seven is prime'. Conceptual balance is a necessary condition of propositional identity. So in both pairs the verbose sentence and its sparing counterpart do not express the same proposition. Let us call the complex phrase that precedes 'is true' in (TP+) a *propositional description*.[11] According to PROPERTY, instances of (tc), (tp) and (tp+) contain singular terms that designate propositions, so this principle implies that propositional descriptions are singular terms. What kind of a singular term is a propositional description? It is not a *standard* definite description, for if it were, the definite article would be followed by a complex general term, but 'the unique x such that x is *a proposition that snow is white*' makes no sense. Does my *analogia proportionalitis* help here, too? 'The number 7' certainly does not mean: the unique x such that x is *a number 7*, but we can treat it as meaning: the unique x such that x is a number and x is 7. If one were to apply this strategy to the phrase 'the proposition that p',[12] instances of 'x is that p' would have to be meaningful. But a string like *'Logicism is that arithmetic reduces to logic' does not seem to be grammatically acceptable.[13] Apparently, we must declare instances of 'the proposition that p' to be definite descriptions *sui generis*.[14]

An instance of 'the proposition that p' is not a propositional description *unless* the phrase 'the proposition' is followed by a that-clause. This restriction might seem to be entirely superfluous, but a string of words does not wear the property of being a that-clause on the sleeves. If a part of a sentence S is a that-clause, then what follows 'that' expresses in S, in context C, a (complete) proposition. Hence in 'The proposition *that Ann wrote on the blackboard yesterday* is the Pythagorean Theorem' the italicized string is *not* a that-clause: it is a relative clause, and the subject term of the sentence is a standard definite description. Sometimes a sequence of words within a sentence S expresses a proposition under one reading of S, and it does not do so under another reading. Thus 'The proposition that Ann wrote on the blackboard yesterday is true' can be understood as containing a that-clause or as containing a relative clause.[15] Sometimes one might call an embedded sentence a that-clause in spite of the *absence* of 'that' on its syntactical surface. In 'Ben believes Ann is ill', for instance, the embedded sentence can be prefaced with 'that' without affecting the meaning of the whole. No wonder that the complementizer pops up as soon as you insert a parenthesis after the main verb: 'Ben believes, as he told me, that Ann is ill'.

According to PROPERTY, naked that-clauses are singular terms, too, and this is a highly controversial claim.[16] I start my attempt at defending it by reflecting on two pairs of sentences. Consider

(1a) That nobody is infallible is incompatible with the dogma of papal infallibility.
(1b) Fallibilism is incompatible with the dogma of papal infallibility.

Surely, the expression 'is incompatible with' has the same meaning in both sentences, and it is a first-order two-place predicate, a sentence-forming operator on (pairs of) singular terms. If a dyadic predicate is not saturated by two singular terms, then at least one of the saturating expressions must be a quantificational expression (like 'everything', 'something', 'every F' or 'some F'). Since the that-clause in (1a) is not a quantificational phrase,[17] it has to be a singular term. Or take

(2a) That there are abstract entities entails that not everything is a particular.
(2b) Platonism entails the negation of particularism.

The verb 'entails' has the same meaning in both sentences, and it is a first-order two-place predicate. Since none of the expressions that flank this predicate in (2a) or (2b) is a quantificational phrase, it is flanked in both sentences by singular terms.

What the that-clauses in (1a) and (2a) designate are things of the same sort as what is designated by the names 'fallibilism', 'platonism' and 'particularism' and by the description 'the dogma of papal infallibility', – they all designate propositions.[18] And propositions are what we quantify over when we derive the conclusions

(C_1) Something is incompatible with something

$\exists x\ \exists y$ (x is incompatible with y)

(C_2) Something entails something

$\exists x\ \exists y$ (x entails y)

from one or the other of our four sample sentences. Quantification into the position of the that-*clauses* in (1a) and (2a) is objectual quantification into the position of singular terms.

Now this argument shows at best that in environments like (1a) and (2a) that-clauses are singular terms. If PROPERTY is correct, then that's what they also are in sentences like (TC). Let me approach this issue indirectly. Is there any semantical difference between (TC) and (TP)? This is one of the few points in this area on which I agree with Prior (1971, p. 11): he also regards them as stylistic variants of each other. Mulligan disagrees. He (2010, pp. 569–70) suggests the following view: – By uttering (TP) with assertoric force one incurs an ontological commitment that one would not incur if one had assertively uttered (TC). In stating that it is true that snow is white, one is *not* obliged to agree that there is at least one truth, to wit, that snow is white, whereas in stating that (TP) one does incur this obligation. – I find this hard to swallow. In classical Latin both (TC) and (TP) would have to be rendered by a sentence composed of '*verum est*' and '*nivem albam esse*', and due to the accusative-*cum*-infinitive construction the Latin translation does not allow for a truth-connective parsing.[19] Were the Romans, as far as truth talk is concerned, bound to incur an ontological commitment that speakers of English can easily avoid? I don't think that this is a reasonable question. But let us leave the Romans to their fate. Suppose Ann and Ben are looking at photographs of former schoolmates. Ann claims that Tom and Dick were both very dull, but Ben protests, 'It is true that *he* [+ Tom's picture] was dull, but that *he* [+ Dick's picture] was dull is not true at all.' I cannot see that there are 'differences in ontological commitment' between the two parts of Ben's protest. Can you incur an ontological commitment, or shake it off, just by adding or removing a pleonastic pronoun and changing the word order in the vehicle of your assertion? Isn't such a variation too frail a reed to bear such a weight? (In Section 3 I shall explicitly pose the question I have just answered in passing when I called the pronoun 'it' in (TC) pleonastic.)

If PROPERTY is correct, then not only (TP+) and (TP) but also (TC) implies that something is true. Isn't the consequent of this conditional intuitively plausible? Suppose an excited would-be philosopher (or a post-modernist *philosophe*) exclaims, 'But is *anything* true?', and you try to cool him down by saying, 'Well, it is true, for example, that snow is white, so *something is true*, namely that snow is white.' Syntactically, both the insertibility of 'for example' in the premiss of your response and the namely-rider appended to its conclusion show (*contra* Prior) that the word 'that' in the premiss is not chained to 'It is true'. The existential generalization here appears to be of the same type as in the case of the inferences from (1a) or (2a) to (C_1) resp. (C_2): quantification into the position of the that-clause in (TC) very much looks as if it were objectual quantification into the position of a singular term. If appearances are not deceptive, the conclusion in your response can be rendered as '$\exists x\ (x$ is true)'.

Admittedly, the ordinary-language quantifier 'something' is transcategorial: 'The Moon is round' implies not only 'Something is round' but also 'The Moon is something (namely round)', and only the former is a quantification into the position of a singular term. But the logical role of the quantifier in '*Something is true, for example,* ___' is the same, no matter whether 'logicism', 'Frege's most famous claim in the philosophy of mathematics', 'the proposition that arithmetic reduces to logic' or the naked that-clause fills the gap. After all, these are four ways of giving the same (rather infelicitous) example.

What is the semantical relation between corresponding instances of (**tc**), 'It is true that p', (**tp**), 'That p is true' and (**tp+**), 'The proposition that p is true' on the one hand and the embedded sentence on the other? If lack of truth (in the case of a truth-candidate) coincides with falsity, then each de-nominalizing biconditional of the forms 'It is true that p, if, and only if, p', 'That p is true iff p' and 'The proposition that p is true iff p' expresses a truth. But if bivalence does *not* hold, then an instance of (tc), (tp) or (tp+) may express a falsehood though the embedded sentence does not. The definite description in [**R**], 'Socrates's first book was a great success', does not designate anything. Suppose we follow Frege and Strawson and take that to deprive the proposition expressed by [R] of any truth-value. Then the [R]-instances of (tc), (tp) and (tp+) express falsehoods while [R] does not.[20] But even without endorsing bivalence one can maintain quite generally: provided that 'p' expresses a truth, (1) it is true that p, <u>because</u> p, (2) that p is true <u>because</u> p, and (3) the proposition that p is true <u>because</u> p. Furthermore, if 'p' expresses a truth then the proposition that p is true *because* (that p is true) *and* (that p is a proposition), just as the number seven is prime *because* seven is prime *and* seven is a number: in both cases the truth of what the longer sentence expresses is *partly due to* the truth of what the shorter sentence expresses.

8.3   Some reflections on 'it'

What is the role of the pronoun in (TC), 'It is true that snow is white'? It does not play here the role of a dummy subject, as in

(S1) It is dawn.
(S2) It is snowing.

For there is a striking syntactical difference: If we delete the first word in (TC), we can restore sentencehood by an inversion that gives us (TP), 'That snow is white is true', but this is not possible if we delete the first word in (S1) and (S2). Now one may be tempted to say that there also obtains the following semantical difference: – While the pronoun in (S1) and (S2) certainly does not serve to single out *something that is dawn or *something that is snowing, the pronoun in (TC) does pick out something that is true, and it does so by anticipating the subsequent that-clause which explicitly gives us what is true according to (TC). – Here are two authors who yielded to this temptation. Paul Horwich wrote:

> In light of the loc[u]tion, 'It is true that p', it might be thought that a theory of the truth *predicate* would have to be supplemented with a separate theory of the truth *operator* [that is connective]; but this is not so. We can construe 'It is true *that p*' on a par with 'It is true, *what Oscar said*' as an application of the truth *predicate* to the thing to which the initial 'It' refers, which is supplied by the subsequent noun phrase, '*that p*'. (1990, p. 17) [my underlining]

And I followed suit:

> Isn't the modest account inapplicable to the truth [connective]? Bolzano and Horwich have pointed towards a solution to this problem. Sometimes pronouns are used *cataphorically*. Consider the role of 'he' . . . in 'He was wise, the man who drank the hemlock'. . . Similarly, we can treat the pronoun in 'It is true that p' as cataphoric. . . [21] (*CT*, p. 351)

Mulligan (2010, pp. 571–2) is right as against Horwich and Künne vintage 2003: we should not assimilate the role of 'it' in (TC) to that of the anticipatory pronouns in 'It is true, what Oscar said' and 'He was wise, the man who drank the hemlock', which designate what the phrase after the comma designates.[22] A comparison may help to drive home this point. At the very beginning of Schiller's *Wilhelm Tell* a fisherboy sings in his boat:

> *Es lächelt der See, er ladet zum Bade.*[23]

What are the pronouns doing here? The *'er'* in the second sentence is anaphoric: it designates what its 'antecedent' *('der See')* designates in the boy's song, to wit, Lake Lucerne. By contrast, the *'es'* in the first sentence is an *expletive* pronoun: it is not in the business of designating something, it is semantically vacuous.[24] My mistake in *CT* resembles that of mistaking Schiller's (Ex) for its cousin (Cat):

(Ex)　*Es lächelt der See*
(Cat)　*Er lächelt, der See.*

In (Cat) the pronoun (note the gender!) is indeed cataphoric: it gets its semantic content from the subsequent noun phrase.[25] Similarly, in

(Cat$_1$)　It is true, what Oscar said
(Cat$_2$)　It is true, the proposition that p

the first word is cataphoric: it receives its semantic content from what follows the comma. The German translation of (TC) begins with *'es'*, whereas that of (Cat$_2$) begins with *'sie'* – because of the gender of the noun, *'die Proposition'*.

Both Bolzano and Horwich are convinced that (TC) expresses the same proposition as (TP), but, contrary to what I insinuated in my book, the answer that my first and favourite witness gives to the question *why* this is so differs from the answer Horwich gives. Here is what Bolzano wrote:

> Sometimes the word *it* seems to be quite superfluous [*überflüssig*], as in the expression: It is true that etc. For after all, this is completely equivalent [*durchaus gleichgeltend*] with: 'The proposition that etc. has truth'. Similarly, the idiom: It's fine weather today [*Es ist heute schönes Wetter*] means the same as [*eben so viel heißt als*]: The weather is fine today etc.[26] (1837, vol. I, p. 216)

Far from ascribing reference to the first word in 'It is true that p', Bolzano declares it to be 'superfluous'. Interestingly, some grammarians call expletives 'pleonastic pronouns'.[27] If a certain part, X, of a sentence S is propositionally superfluous, then S *minus* X expresses what is less economically expressed by S. In this sense the third word in 'Socrates was indeed wise' is superfluous.[28] Sometimes, after removal of the superfluous expression an inversion is required, as in the case of Schiller's (Ex). By Bolzano's lights, the transition from (TC) to (TP) is of the same kind as the move from (Ex) to *'Der See lächelt'*. That seems exactly right to me.[29] Unfortunately, in the passage quoted above Bolzano actually talks about (tp+) rather than (tp). But one may very well wonder whether it is his considered view that the sense of the prefixed noun phrase 'the proposition' is part of the sense of an instance of (tc).

The first word in (TC), like the *'es'* in its German translation, is a pleonastic or expletive pronoun: it contributes nothing to the content of an utterance of (TC). If you look at the translation of (TC) into Italian, you see that the Italians renounce the luxury of an expletive: *'È vero che la neve è bianca.'*

8.4   Interlude: negation and 'It is not true that'

Consider a question–answer sequence whose first member is a yes/no interrogative ('Was Socrates stupid?', 'Are all Greeks philosophers?', 'Are some Greeks fond of Angela Merkel?') and whose second member is a 'No'. By squeezing the message of such a question–answer sequence into the confines of a single sentence we obtain the negation of the declarative sentence that corresponds to the first part of the question–answer sequence. Unlike the sequence the result of the condensation can be embedded. In English as spoken outside logic classes there is no *systematic* way of obtaining the required result: we slide a 'not' between copula and general term, or we prefix it to 'all Fs', or we transform 'some Fs' into 'no F', etc. And we run into a bit of trouble when the yes/no interrogative is a compound sentence (even in as simple a case as 'Is Anna witty, and is Bella pretty?'). Logicians have invented an unbreakable one-place connective that can be prefixed to any declarative sentence to obtain the required result, and in their informal prose they have forced a long-winded natural-language expression into the same service. They are to be congratulated for the invention, and they should be mildly criticized for the act of force.

In his late essay *'Die Verneinung'* Frege (1919, p. 148, original pagination) formulates the double negation of the truth that the Schneekoppe is higher than the Brocken as follows:

(DN) *It is not true that* the Schneekoppe is *not* higher than the Brocken.

The 'inserted (*eingefügt*)' negation sign is atomic, but the double negation of that geographical truth is not expressed by the string. *'The

Schneekoppe is not not higher than the Brocken', for that is an ungrammatical stutter. If one puts 'not' in front of the internally negated sentence, the result is not grammatically acceptable either. As an emergency solution for this annoying grammatical problem Frege prefixes a wordy connective to that sentence, clearly intending this connective to perform the same job as the inserted 'not'. Here is what Tarski says about his own use of the same connective for the same purpose:

> For stylistic reasons we sometimes use, instead of the word 'not', the expression 'it is not true that'. In doing so we treat the whole expression as a single word, without ascribing to its parts, especially to the word 'true' it contains, any independent meaning. (1935, §2, note 17) [my translation]

Frege would surely have endorsed this. In his paper '*Gedankengefüge*' he takes as his target language a *regimented* version of German. In a similarly regimented English, (DN) would be replaced by

(DN$_R$) Not (not (the Schneekoppe is higher than the Brocken)).

The negation connective of *regimented* English is just a typographical variant of the small vertical stroke in Frege's *Begriffsschrift*, of Peano's tilde, of Łukasiewicz's 'N' and of Gentzen's hook, etc.[30]

Tarski *stipulates* that the long-winded connective is to be understood as if it were as unstructured as the word 'not' and the logicians' negation operator. One could visually mark this intention by using hyphens: 'It-is-not-true-that'. If the hyphenated connective is an indissoluble unit, then comprehending it is not founded upon understanding the words it ostensibly consists of. (We do not grasp the meaning of the word 'attic' in virtue of understanding 'at' and 'tic'.) As we all know, the hyphenated negation connective, '¬' for short, can be explained as follows: a sentence of the form '¬p' expresses a truth in context C iff (the operation signified by '¬' yields a truth when applied to a falsehood and a falsehood when applied to a truth, and the embedded sentence expresses a falsehood in C). The typographical presence of 'true' in the hyphenated connective may serve as a reminder of this explanation, but that is only a mnemotechnical role, not a semantical one. So there is a vast difference between 'It is not true that' and its hyphenated opposite number. I do not dream of maintaining that the latter (or the logicians' negation operator or our good old 'not') is secondary to any predicative construction. But I do maintain this for the pristine connective 'It is not true that'.

What I said in Section 2 about the interrelations between 'snow is white', (TC), (TP) and (TP+) also applies, *mutatis mutandis*, to the interrelations between 'snow is not blue' and

(NC)   It is not true that snow is blue.
(NP)   That snow is blue is not true.
(NP+) The proposition that snow is blue is not true.

Each of these three sentences expresses a truth if, and only if, the proposition designated by 'that snow is blue' does not have the property that is signified by 'is true'. So they all have the same truth-conditions. The 'it' in (NC) is an expletive pronoun, hence (NC) expresses the same proposition as (NP). The latter is to (NP+) as 'Seven is prime' is to 'The number seven is prime'. Each N-sentence is about what (NP+) unmistakably is about, namely a proposition (and only indirectly about snow). Hence none of the N-sentences expresses the same proposition as 'Snow is not blue' or its Loglish counterpart, for the latter sentences are certainly not about a proposition.[31] So the word 'not' in 'Snow is not blue' and the logicians' negation operator are not at all in the same boat as the wordy one-place connective in (NC).

What is the semantical relation between corresponding instances of (**nc**), 'It is not true that p', (**np**), 'That p is not true' and (**np+**), 'The proposition that p is not true' on the one hand and the negation of the embedded sentence on the other? If lack of truth (in the case of a truth-candidate) coincides with falsity, then each de-nominalizing biconditional of the forms 'It is not true that p, if, and only if, $\neg p$', 'That p is not true iff $\neg p$' and 'The proposition that p is not true iff $\neg p$' expresses a truth. But if bivalence does *not* hold, then an instance of (nc), (np) and (np+) may express a truth though the negation of the embedded sentence does not. Suppose, once again, that the proposition expressed by [R], 'Socrates's first book was a great success', is neither true nor false. Then what 'Socrates's first book was not a great success' and '$\neg$ (Socrates's first book was a success)' express also falls into the truth-value gap, whereas the [R]-instances of (nc), (np) and (np+) all express truths.[32] But even without endorsing bivalence one can maintain quite generally: if '$\neg p$' expresses a truth, then (1) it is not true that p, *because* $\neg p$, (2) that p is not true *because* $\neg p$, and (3) the proposition that p is not true *because* $\neg p$.

Although the victims of our elementary logic courses are hardly ever cautioned at this point, pronouncing the negation operator as 'It is not true that' (or as 'It is not the case that') is potentially misleading, for in our language the latter expression isn't an indivisible connective (though it *contains* an atomic negation sign). So again I disagree with Prior who says about a close relative of (NP):[33]

> When, instead of saying simply 'Snow is not blue', we say 'That snow is blue is not the case', we construct from the sentence 'Snow is blue' what looks like a name ('That snow is blue') and then complete the sentence with what looks like a verb ('is not the case'). We are not, however, *very* strongly tempted to treat this name, in this sentence at any rate, as genuinely denoting an object, and its 'verb' as genuinely describing an activity of this object — it is sufficiently obvious that the whole complex 'That ___ is not the case' simply has the force of the adverb 'not', appropriately placed. (Prior, 1963a, p. 193)

Elsewhere Prior says about 'It is not the case that *p*' what he here says about 'That *p* is not the case' (1967b, p. 458 and 461; 1971, p. 11 and 19). As regards the weak temptation that he wants us to resist, his description of those who yield to it is a bit of a caricature. I have never come across anyone who thought of sentences like (NP) as saying of an object (be it a state of affairs or a proposition) that it performs an activity called 'not being the case' or 'not being true'. Let that pass as a bit of propaganda. What I mainly object to is Prior's claim that the one-place connectives 'That ___ is not the case (not true)' and 'It is not the case (not true) that ___' are semantically unbreakable.

8.5   Some substitution failures

It has to be admitted that in truth talk instances of 'the proposition that *p*' (*propositional descriptions*) and corresponding instances of the unadorned schema 'that *p*' (*that-clauses*) are not always exchangeable *salva congruitate*, that is, without loss of grammaticality. (The same point has often been registered with respect to 'propositional attitude' reports.[34]) There are substitution failures in both directions.

Replacing the that-clause in (TC) by the corresponding propositional description results in loss of grammaticality:

(TC) It is true that snow is white.
(X)   *It is true the proposition that snow is white.

Of course, if we insert in (X) a comma after 'true', grammatical law and order will be restored, but then we do (a tiny bit) more than just replace the clause by its adorned counterpart, and we transform the pleonastic or expletive pronoun in (TC) into a cataphoric or anticipatory pronoun.

(Recall that the same operation in German has an ill-formed result, *'*es ist wahr, die Proposition, dass Schnee weiß ist*', since '*die Proposition*' calls for an anticipatory '*sie*'.)

Is the ungrammaticality of the ill-starred formulation evidence for the claim that the clause in (TC) does not designate the proposition that snow is white? I don't think so. Since the first word in (TC) is an expletive, (TC) expresses the same proposition as

(TP) That snow is white is true,

and in (TP) the exchange *is* possible *salva congruitate*. So the fact that in (TC) the clause is not replaceable without further ado by the propositional description has got nothing to do with the *content* of (TC): it is due to a semantically insignificant grammatical constraint. Consider a similar case. In 'He kissed Ann', as uttered in context C, the name cannot be replaced without violence to grammar by a 'she' that designates Ann in C,[35] but this substitution does preserve well-formedness in 'Ann was kissed by him', – in a sentence that expresses in the same context the same proposition as its active counterpart.[36] So the non-exchangeability in the first sentence is just a caprice of grammar.

Here is an example of a (less obvious) substitution failure in the other direction. Sometimes a propositional description cannot be replaced *salva congruitate* by an unadorned that-clause. Thus

(S3) She wonders whether the proposition that snow is white is true
(S4) If the proposition that snow is white is true, then snow is not blue

are fine, while the strings

(S3*) *She wonders whether that snow is white is true
(S4*) *If that snow is white is true, then snow is not blue

are grammatically unacceptable.[37] Again, I don't think that this shows that the plain that-clause does not serve the same purpose as its adorned counterpart, namely to single out a proposition. It only shows that in certain positions that-clauses need a crutch. Consider the sentence frame 'The Polish philosopher ... admired Bolzano'. We can insert the name 'Kazimierz Twardowski' but not the co-designative definite description 'the founder of the Lwów-Warsaw School'. The latter needs some add-on before it can enter the slot: '*who was* the founder of the LWS'.[38] This failure of *exchangeability without further ado* will hardly be regarded by anyone as evidence for the claim that the name and the definite description do not serve the same purpose, namely to pick out a certain man. In both cases the non-exchangeability is just a whimsy of grammar.[39] It is part of the grammarian's job to explain such restrictions on substitution.

In a regimented version of English (S3*) and (S4*) could be improved for example by using brackets: '...whether (that snow is white is true)', 'If (that snow is white is true), then ...'. Not all grammatical features of the vernacular matter for the elucidation of the concept of truth. In English all substitution-instances of the schema '*p* iff it is true that *p*' are grammatically well-formed. But unfortunately the German translation of this schema, that is '*p genau dann, wenn es wahr ist, dass p*', has no grammatically acceptable substitution instance in which the predicate isn't a full verb. If you have a close look at

(W) *Schnee ist weiß genau dann, wenn es wahr ist, dass Schnee weiß ist*

you will notice that this is not really an instance of the schema. In the case of classical Latin you need not look closely. The translation of 'If it is true that snow is white, then snow is white' is '*Si nivem albam esse verum est, nix alba est*', where you have an accusative-*cum*-infinitive construction rather than an embedded sentence in the antecedent. Although I am rather fond of my mother tongue, I sympathize with those of my readers who will exclaim at this point, 'So much the worse for German (Latin)!' In a suitably regimented version of German, call it 'Logman', (W) gets transformed into

(W$_R$) *(Schnee ist weiß) genau dann, wenn (es ist wahr, dass (Schnee ist weiß))*.

The sentence schema '*Wenn p, dann p*' has not a single well-formed substitution instance. Does it follow that in German the schema is not universally valid? Certainly not. In Logman '*Wenn (Schnee ist weiß), dann (Schnee ist weiß)*' is grammatically impeccable, and we know how to translate from German into Logman, and vice versa. Frege explicitly registered the same kind of problem, if it deserves this name, in his fragment '*Logische Allgemeinheit*'. He was clearly aware of, and justifiably unmoved by, the fact that syntactically precise rules of derivation are only applicable to a *regimented* version of the vernacular.[40]

8.6 'Propositional attitudes' without propositions?

That-clauses were Prior's *bête noir*: he fought them wherever he met them. I now want to discuss his moves in two other areas where that-clauses are employed, before I return to the topic of truth talk in Section 8.9. The first site is the embattled field of 'propositional attitude' reports. The second area is occupied by certain meaning ascriptions, and it is largely unexplored.

Prior pleaded for a *non-relational* account of ascriptions of propositional mental acts-or-states and of indirect speech (1955, pp. 33–4; 1963b, pp. 147–50; 1971, pp. 16–21).[41] It runs as follows. When we make a statement of the form 'A ϕs that p', we don't claim that a relation between individual A and the proposition that p obtains. Sentences like 'Alfred says (believes) that snow is white' should not be parsed as 'Alfred/says (believes)/that snow is white' but rather as 'Alfred/says (believes) that/snow is white'. By thus moving the second parsing line a bit to the right we see the sentence as formed from the indissoluble operator '___ says (believes) that ___' by inserting a singular term at the front and a sentence at the rear. Since such operators are like predicates on the left and like connectives on the right, I dubbed them 'prenectives'.[42] Unlike two-place predicates, prenectives do not signify relations, so the *prenective* in a report of the form 'A ϕs that p' does not signify a relation between A and the proposition that p. If the prenective parsing is obligatory, ascriptions of propositional acts-or-states do not require propositions.

In *Word and Object* Quine (1960, p. 216) toyed with the idea of decomposing ascriptions of beliefs and other 'propositional attitudes' in such a way that 'the verb "believes" ceases to be a term and becomes part of an operator "believes that" '. In his *Philosophy of Logic* he (1970, p. 32) nicknamed such operators 'attitudinatives'. (This label is not apposite to the analogous components of reports of mental or illocutionary *acts* like 'concludes that' and 'asserts that'.) Because of the rather casual remark in *Word and Object,* Quine received a somewhat devious compliment by Prior (1971, p. 20): "This is ... one of the two points in the philosophy of logic on which Quine seems to me to be dead right." (Prior forgot to tell us what the second point is.) I don't know whether Quine ever heard of this compliment. At any rate, he did not accept it for long:

> There had been little point in my citing the attitudinative in the first place. I preferred the first alternative all along: 'believes' as transitive verb and 'that p' as noun phrase. (Quine, 1995, p. 244)

I share Quine's preference. But calling the verb in 'A believes that p' transitive, though grammatically correct, can be philosophically misleading (as we shall see in Section 8.7), and one may reasonably wonder whether that-clauses should be classified as noun phrases: I, for one, would like to say that in 'She doubts the existence of intelligent extraterrestrials' we use a noun phrase *instead of* a that-clause.

The prenective parsing of ascriptions of propositional acts-or-states is by no means obligatory. As I said before, we have a lot of leeway as regards decomposition. Even from a conjunction like 'Anna is witty, and Bella is pretty' we can cut out an (extensional) prenective by deleting the first name and the second sentence: '___ is witty, and ___'. (For that matter, we can also isolate a 'connecticate', an operator that is like a connective at the front and like a predicate at the rear: '___, and ___ is pretty'.[43]) Now in a case like this, Prior (1971, p. 18) admits that such a decomposition is 'a little odd ... and not in itself very illuminating'. We see that this is a bit of an understatement as soon as we ask *why* such decompositions are not illuminating. Our understanding of the conjunction is based upon our understanding of the conjuncts and of the two-place connective: understanding the odd prenective nowhere comes into play (nor does understanding the connecticate). Now I think the same holds *mutatis mutandis* for reports of the form 'A ϕs that *p*': our understanding of such reports depends on our comprehension (of the instances) of 'A', of 'ϕs' and of the that-clause. Prior's parsing does not carve the objects of understanding at their joints: it is only an optional decomposition.

Here are three reasons for rejecting his contention that prenectives of the form '___ ϕs-that ___' are semantically indivisible. Consider the argument 'Anna believes that Mt Etna is still active, and so does Bella. So there is something they both believe, namely that Mt Etna is still active.' Here we have quantification into the position of a that-clause, and the namely-rider appended to the conclusion shows that the word 'that' is not chained to the verb.[44]

A prenective cannot possibly have the same sense as a predicate, and if the prenective is semantically atomic (as Prior insists), then it contains no genuine component that has the same sense as a predicate. Consequently, for those who take the Prior line on prenectives it will be extremely hard, to put it mildly, to explain the irresistible impression that the following argument is formally valid:

(p1) Anna *asserts that* a vixen is a female fox.
(p2) That a vixen is a female fox is Casimir Lewy's favourite proposition. So,
(c) Anna *asserts* Casimir Lewy's favourite proposition.

On Prior's reading this argument has to be classified as enthymematic: no parsing of conclusion (c) will bring a prenective to light, – as regards such sentences the predicative parsing is the only game in town. By contrast, it is very easy to explain the impression of formal validity if one allows for the predicative parsing of (p1) and takes (p2) to be an identity statement, for then the impression is veridical: the argument exemplifies a pattern that is universally valid in the predicate calculus, namely $aRb, b = c \therefore aRc$.

Finally, in some languages, Prior's prenective parsing, far from being obligatory, is not even possible. In classical Latin 'Alfred says (believes) that snow is white' would be rendered by '*Alfredus dicit (existimat) nivem albam esse*': the accusative-*cum*-infinitive construction forecloses Prior's decomposition.[45]

Declaring a mode of decomposition to be merely optional is not to condemn it as illegitimate. One can consistently deny that Prior's connective parsing is obligatory and concede that it is permissible. After all, every instance of 'A ϕs that p' contains a genuine occurrence of a sentence.[46] It is no objection against epistemic or doxastic logic that it employs Prior's way of decomposing ascriptions of knowledge and belief, but neither does the elucidatory power of such logics show that a non-relational account of knowledge- and belief-ascriptions is correct.[47] In any case, the limits of the elucidatory power of such calculi (quite apart from the idealizations they involve) become manifest when we consider an argument like '(p1), (p2), therefore (c)' in which a prenective and a predicate interact. If we regard the prenective in (p1) as a semantically indissoluble unit, the impression of formal validity at once evaporates.

As against Prior, I plead for (a version of) the traditional *propositionalist relational* account of ascriptions of propositional acts-or-states.[48] Schematically, one can formulate the view I want to uphold as follows:

(PR) A *de dicto* report of the form 'A ϕs that p' expresses a truth in context C
iff the content of A's Φ is the proposition designated by 'that p' in C.

Note that a report that is *de re* rather than *de dicto* is not always recognizable by its form: one would not want to apply (PR) to 'Father Paolo believes that the kitschy picture of the Madonna in the oratory is a great piece of art'. Because of clauses with context-sensitive elements, for example 'that it is raining', the relativization to context is needed. (I shall henceforth take it as understood.) The Greek majuscule in (PR) is a placeholder for verbal nouns (like 'belief', 'statement', 'hope') that correspond to the verbs ('believe', 'state', 'hope', etc.) for which the Greek minuscule is a dummy. My use of 'content' is meant to line up with the classification of the clauses in instances of 'A ϕs that p' as (declarative) *content clauses*, which is due to the Danish linguist Otto Jespersen. The relations that make this view a *relational* account are signified by dyadic predicates that are instances of the predicate schema 'the content of x's Φ is y'. It is a *propositionalist* account, since 'y' is taken to be a dummy for expressions that designate propositions – and not, say, sentences, utterances or mental acts-or-states.

8.7 More substitution failures

The (PR) account does not imply that in reports of the form 'A ϕs that p' the that-clause and the corresponding propositional description are always interchangeable. This is all to the good, since more often than not, prefixing 'the proposition' to the clause has a detrimental effect. (Prior and Geach registered one aspect of this phenomenon in the sixties, and already half a century earlier Bertrand Russell had observed it – and shrugged it off.[49]) We have to reckon with two kinds of damage. One is the failure to preserve truth-value:

(1) Sometimes, as in instances of 'A fears/explains that p', the clause can be replaced *salva congruitate* but not *salva veritate* by the corresponding propositional description.[50]

The other kind of mischief is the failure to preserve grammaticality:

(2) Sometimes, as in instances of 'A says/hopes that p', the clause is not even replaceable *salva congruitate* by the corresponding propositional description.[51]

Do these observations show that (at least in such cases) that-clauses do not designate the proposition which is designated by the corresponding propositional description? I don't think so. As regards point (1), consider this substitution failure: if you replace the subject term in 'The author of *Die Leiden des jungen Werther* owed his early fame mainly to this novel' by 'The author of *Die Wahlverwandtschaften*', you turn a truth into a falsehood.[52] Nevertheless, these expressions *are* co-designative. Or take

(RTL) Richard the Lionheart was so-called because of his bravery.

The subject term of (RTL) cannot be replaced *salva veritate* by the term 'Richard I of England', although they both designate one and the same man.[53] So, why should the non-replaceability registered in (1) be a conclusive reason for maintaining that that-clauses do not designate here what the corresponding propositional descriptions designate?

As regards point (2), consider the following substitution failure: in (RTL) the grammatical subject cannot be replaced *salva congruitate* by 'the king who was held captive in an Austrian castle', since the anaphoric 'so-called' requires a *name* as antecedent. (A name N that designates an object x introduces x into discourse as being called by the name N, whereas a definite description D that designates x introduces x into discourse as having the property that is signified by the predicate from which D is built.) The grammatical obstacle to substituting for 'Richard the Lionheart' that definite description does not prevent these singular terms from being co-designative. Or take the subject term in 'Young Goethe fell in love with Charlotte Buff'. In this context the name 'Goethe' can only be substituted for, *salva congruitate*, by another *name*. Nevertheless, 'Goethe' and, say, 'the author of *Götz von Berlichingen*' designate the same man.[54] So, why should the grammatical fact that the that-clause in a type-(2) report can only be replaced *salva congruitate* by another *that-clause* be a conclusive reason for denying that it designates what the corresponding propositional description designates?

Let us call the transitive verb that occupies the 'ϕ'-position in a sentence S of the form 'A ϕs ...' *objectual* (as used in S) if, and only if, a grammatically permissible answer to the question 'Whom or what does A ϕ?' can be given by filling the gap with a name or a standard definite description, and let us call such a verb *clausal* (as used in S) iff it is not objectual. (In case the verb is not objectual, the gap can only be filled by a that-clause.)

If a transitive verb has both a clausal and an objectual use, we get type-(1) cases. Using 'p_1' as abbreviation of '*Mt Etna will soon erupt again*', we can say: 'Anna fears that p_1' expresses a truth if, and only if, the relation that is signified by 'x is in a state of fear whose *content* is y' obtains between Anna and the proposition that p_1. This relation is not the relation that obtains between Joan and Jack the Ripper if 'Joan fears Jack' expresses a truth, for what follows the verb here specifies the *intentional object* of her state. If Anna were to fear the proposition that p_1, a proposition thus being what her fear is a fear *of*, she would suffer from a special kind of eidophobia (*vulgo* nominalism).[55] The clausal verb ($\approx$ 'reckons anxiously with the possibility') does not have the same meaning as the homophonous objectual verb ($\approx$ 'is scared of'). This is not a case of accidental homonymy like that of 'bank' if 'fear' in '*x* is in a state of fear whose content is *y*' and '*x* is in a state of fear whose intentional object is *y*' is univocal. Similarly, if we say, (U) 'She explains that the lecture had been cancelled', we specify the *content* of the act she performs, but if we say, (V) 'She explained the proposition that tachyons are spinless', we indicate what her explanation is an explanation *of*, in other words: we specify the *intentional object* of her act. Once again, the clausal verb in (U), ≈ 'lets it be known', does not have the same meaning as the homophonous objectual verb in (V), ≈ 'gives an explanation of'. Note that in the predicate schema '*x* performs an act of ϕing / is in a state of Φ whose ___is *y*' the *y*-position is hospitable to propositional descriptions. No matter whether we fill the blank with 'content' or with 'intentional object', predicates of this form signify a relation that may obtain between a person and a proposition.

When we utter (RTL) with assertoric force, we are right just in case the king referred to was known as Richard the Lionheart because of his bravery. If we reformulate our message accordingly, we obtain

(RTL+) Richard the Lionheart was called 'Richard the Lionheart' because of his bravery,

and here the subject term *is* replaceable *salva veritate* by 'Richard I of England' (and by 'the king who was held captive in an Austrian castle'). (RTL+) specifies the truth-conditions of (RTL) and tells us what having the property ascribed to the king in (RTL) consists in. Similarly, when we say 'Anna fears that Mt Etna will soon erupt again', we are right if, and only if, her fear has a certain content. If we reformulate (and abbreviate) our message along these lines, we obtain 'That p_1 is the content of Anna's fear', and here the clause *is* replaceable *salva veritate* by 'the proposition that p_1'. So the fact that sometimes a clause, though replaceable *salva congruitate* by the corresponding propositional description, cannot be replaced by it *salva veritate* does not show that a person's being in a state of, or performing an act of, type (1) does not consist in her or his being related in a certain way to the proposition designated by the that-clause.

If the verb in an instance of 'A ϕs that *p*' has *only* a clausal use, we get type-(2) cases. The report that Alfred says that snow is white is true just in case the relation that is signified by '*x* performs an act of saying whose *content* is *y*' obtains between Alfred and the proposition that snow is white. And the report that Zeno hopes that p_1 is true iff the relation that is signified by '*x* is in a state of hope whose *content* is *y*' obtains between Zeno and the proposition that p_1. A proposition cannot be the intentional object of acts or states of this kind. But the fact that sometimes the clause cannot even *salva congruitate* be prefaced with 'the proposition' does not show that a person's being in a state of, or performing an act of, type (2) does not consist in his or her being related in a certain way to a certain proposition.

Every instance of the following schema expresses a necessary a priori truth:

(CLAUSAL)   A ϕs that p iff that p is the content of A's Φ.

The right-hand side of a biconditional that instantiates CLAUSAL tells us what, in a given case, ϕing that p consists in. Admittedly, in cases of kind (2) we cannot move from 'A ϕs that p' to 'A ϕs some proposition' or 'There is a proposition which A ϕs',[56] but that is simply due to the fact that such an inference requires an objectual verb. However, from a sentence that specifies the truth-conditions of our premiss, that is from 'The content of A's Φ is that p' we can infer 'There is a proposition which is the content of A's Φ', for on the RHS of CLAUSAL 'that p' can be replaced *salva veritate*, and hence *salva congruitate*, by the corresponding propositional description. So the failure of the first inference does not provide us with a good reason for denying that a that-clause is a singular term. Look again at (RTL).[57] We cannot infer *'There is a man who was so-called because of his bravery' from (RTL), since the anaphoric 'so-called', which requires an antecedent, is left dangling in the second sentence. Nobody would take the failure of the first inference to be a good reason for doubting that 'Richard the Lionheart' is a singular term.[58] From a sentence that specifies the truth-conditions of our premiss and tells us what having the property ascribed to the king in the premiss consists in, that is from (RTL+), we can infer 'There is a man who was called "Richard the Lionheart" because of his bravery'.

Here is the opposite number of CLAUSAL. Every instance of the following schema expresses a necessary a priori truth:

(OBJECTUAL)   A ϕs the proposition that p iff
          the proposition that p is the intentional object of A's Φ.[59]

In many cases the message of the RHS of such biconditionals can be conveyed by a sentence of the form: *A's Φ is a Φ of the proposition that p.* At this point I can explain, at last, why I kept on putting the widely used Russellian label 'propositional attitude' between scare quotes.[60] One reason for having reservations is fairly obvious: no propositional act, and hardly any propositional state, is an *attitude*. (In central cases, the attitude one has towards X involves one's being well-disposed, or ill-disposed, towards X.) The second reason for disliking that label is less obvious but more important.[61] An attitude is always *towards* something, and this very feature of the notion of an attitude tends to suggest to those who use the Russellian tag that every propositional act-or-state is towards (or 'directed at') a proposition. But that suggestion should be rejected: a 'propositional attitude' is not 'towards a proposition' unless it can correctly be reported by means of a sentence that has truth-conditions of the form represented by OBJECTUAL. So even if we put aside all animadversions against the misuse of the word 'attitude', a *belief that p* – ever since Russell the most prominent example in philosophical debates about 'propositional attitudes' and their reports – is *not* 'directed at' a proposition.[62]

From (U), 'She explained that the lecture has been cancelled', as well as from (V), 'She explained the proposition that tachyons are spinless', we can correctly infer 'She explained something'.[63] But since the clausal verb 'explains' in (U) does not have the same meaning as the objectual verb in (V), we really have *two* conclusions: 'She explained$_{cls}$ something' and 'She explained$_{obj}$ something'. What we infer from (U) is tantamount to 'For some y, y is the content of her explanation'. By contrast, what we infer from (V) comes to the same thing as 'For some y, y is the intentional object of her explanation' or 'For some y, y is what her explanation is an explanation of'. A sentence like

(Z1) She explained that the lecture has been cancelled and the proposition that tachyons are spinless

is a *zeugma* (lit. a yoke): one word is forced into the conflicting services of a clausal and of a non-clausal verb.[64] Compare a more familiar zeugma[65] in

(Z2) She took counsel and tea

the nouns 'counsel' and 'tea' are yoked together by the verb 'to take' to form a conjunctive predicate in which the verb resists coherent interpretation. Combined with 'counsel', it does not have the meaning that it has when combined with 'tea'. [If you translate 'to take (counsel)' and 'to take (tea)' into German, you get something like *'(Rat) einholen'* and *'(Tee) zu sich nehmen'*.] The interpretative impasse produced by the zeugma is overcome as soon as (Z2), with its conjunctive predicate, is dissolved into a *conjunction*: (C2), 'She took counsel, and she took tea'. Admittedly, this is not a snippet of brilliant prose, but in (C2) the meaning of the verb is first partly determined by its combination with the first noun and then by its combination with the second.[66] Similarly, if we transform (Z1) into the conjunction of (U) and (V):

(C1) She explained that the lecture has been cancelled, and she explained the proposition that tachyons are spinless,

we are out of the hermeneutical deadlock: in the first conjunct the meaning of the verb is partly determined by its combination with a naked that-clause, in the second conjunct it is partly determined by its combination with a propositional description. A naked that-clause which designates a proposition P introduces P into discourse as a possible *content* of an act or a state, while a name, a standard definite description or a propositional description that designates P introduces P into discourse as a possible *intentional object* of an act or a state. The former kind of introduction requires the 'lets it be known' interpretation of 'explains', the latter demands the 'gives an explanation of' reading.

Let me complete my survey of the pertinent substitution failures. In some ascriptions of propositional acts-or-states we have non-exchangeability *salva congruitate* in the other direction:

(3) Sometimes, as in instances of 'A endorses / rejects the proposition that p', the propositional description cannot be replaced *salva congruitate*, let alone *salva veritate*, by the corresponding that-clause.[67]

No matter whether the general term in the prefixed noun-phrase is 'proposition' or a less catholic term like 'claim' or 'theorem',[68] if you delete the prefix, the result is ungrammatical. In type-(3) cases the expression 'the (NOUN) that p' specifies the *intentional object* of the ϕing. It is noteworthy that in Fregean ascriptions of thinking in the style of 'A grasps (*fasst*) the thought that p' the expression 'the thought' cannot be deleted *salva congruitate*. So this kind of report singles out the intentional object of the act rather than its content.[69]

Let us briefly take stock. What follows a clausal verb in (an instance of) 'A ϕs that p' always designates the content of A's act or state. What follows the objectual verb in (an instance of) 'A ϕs the proposition that p' always designates the intentional object of A's act or state, and the same holds if the verb in (an instance of) 'A ϕs that p' only seems to be clausal.

8.8   Meaning without meanings?

According to Prior, we also use an unbreakable prenective when we say of a sentence S that it means that *p*.[70] Let us consider an example. In

(M_S) *'nix alba est'* means that snow is white

a (Latin) sentence is mentioned, and its translation into the metalanguage (into English) is used to say what the quoted sentence means. Unsurprisingly, Prior regards the prenective parsing of (M_S) as obligatory. In his own example, which is taken from medieval discussions about the *significatum propositionis*, the metalanguage contains the object-language:

> The expression 'means that' ... constructs a sentence not out of two names but out of a name and a sentence — 'a man is a donkey' means-that a man is a donkey. So we need not ask what is named by the clause 'that a man is a donkey'; the word 'that' does not belong here but with the 'means' that precedes it, and what is left, 'a man is a donkey', names nothing because it is not a name but a (subordinate) sentence. (Prior, 1962, pp. 137–8)

If this is correct, such meaning ascriptions do not require meanings. Against this view one can turn variants of arguments I used against Prior's claim that psychological and illocutionary prenectives are atomic. Our practice of existential generalization shows that here, too, the that-clause is a semantical unity: *'nix alba est'* means that snow is white, and so does *'śnieg jest biały'*, so there is something both sentences mean, namely that snow is white. And in some languages Prior's parsing is not even possible: if Cicero wanted to say what the Greek sentence *'chiōn leukē estin'* means, he had to use the accusative-*cum*-infinitive construction: ... *significat nivem albam esse*. Once again Prior's carving knife fails to cut at the joints.

In his famous paper 'Truth and Meaning' Davidson urges us to "sweep away the obscure "means that"" in statements like (M_S) by moving to crystal-clear Tarskian biconditionals like ' "*nix alba est*" is true iff snow is white'. I agree with Terence Parsons' comment on this passage: 'Davidson [...] is not sweeping away a genuine semantic unit; he is sweeping away artificially combined portions of different units.' (Parsons, 1993, pp. 445–6)

Certain ways of saying what a *word* means provide us with further evidence against Prior's insistence on the prenective parsing of (M_S).

'What does "*sapiens*" mean?' you are asked by a child who is trying hard to make sense of a Latin sentence, and you dutifully answer,

(M_W) '*sapiens*' means wise.

That's a good answer, but (M_W) is grammatically garbled unless the final word plays a non-standard role: it designates its own meaning (which is identical, let's assume, with that of '*sapiens*').[71] Meaning ascriptions like (M_W) should not be confused with statements of synonymy. If one knows that an expression e_1 means μ, one understands e_1, but one can know that e_1 means the same as e_2 without understanding either expression. By looking up a Latin-German dictionary a monoglot Russian might come to know that '*sapiens*' and '*weise*' are synonyms: he does not thereby come to understand the Latin word or its German translation. (The same hold *mutatis mutandis* for (M_S) and its ilk.)

No parsing can isolate a prenective in (M_W), so here the verb signifies a relation. Now one can hardly deny that this verb makes the same contribution to what is said by (M_W) and to what is said by (M_S). But that would be impossible if, in (M_S), it were just a fragment of the allegedly atomic prenective '___ means-that ___'. So 'means', not only in (M_W) but also in (M_S), is a two-place predicate that signifies the relation in which an expression x stands to y iff y is the, or a, meaning of x.

Actually, it would be better to delete 'that' in (M_S) and to formulate ascriptions of sentence-meaning, too, in the style of (M_W). When somebody says on Monday

(T) That today it is raining in Berlin is true, I am afraid,

the that-clause designates a proposition, and when the same words are uttered on the following day, it designates a *different proposition*. But no matter whether I answer the question 'What does '*Heute regnet es in Berlin*' mean?' on Monday or on Tuesday, – in saying

(M1) '*Heute regnet es in Berlin*' means that today it is raining in Berlin.

I ascribe *one and the same* linguistic (lexico-grammatical) *meaning* to that sentence. So the that-clause plays very different roles in (T) and (M1). If I had answered the linguistic question by saying

(M2) '*Heute regnet es in Berlin*' means Today it is raining in Berlin,

my reply would have been less misleading.[72] The word sequence (M2) is not a well-formed sentence unless the sentence I underlined plays a non-standard role: in this context it designates its own meaning, – it functions as a singular term. The verb 'means' in (M2) is not a fragment of a prenective but a two-place predicate. So, my proposal to expel the word 'that' from meaning ascriptions is certainly not grist to Prior's mill.[73]

8.9 Once again: is the truth connective atomic?

Prima facie Prior's affirmative answer to this question looks hopeless. After all, the presence of 'true' in 'It is true that' does not seem to be an orthographic accident as its presence in 'obstruent'.[74] There can be no doubt that the truth connective is *syntactically* divisible. Just look at the way we shed ballast in moving from 'Some people deny that the mayor is corrupt, but it is true that he is corrupt' to 'Some people deny that the mayor is corrupt, but it is true.' The possibility of stripping off 'that *p*' shows that it is a genuine syntactical constituent of 'It is true that *p*'. If the word 'that' were chained to 'true', it should be possible to end with 'it is true that'. Furthermore, while insertion of a parenthesis before 'that' is possible: 'It is true – as you know only too well – that snow is white', the result of such an insertion immediately after 'that' is ungrammatical. So 'that' goes with the sentence which follows it to form a genuine syntactical unit.[75]

These arguments do not yet show that the indivisibility claim as intended by Prior is false. They can be easily adapted to show that the truth predicate, like any other predicate of the form '__ is F', is not syntactically atomic either.[76] Parenthesis insertion is possible: 'Casimir Lewy's favourite proposition is – as I needn't tell you – true', and so is stripping off: 'Hardly any statement in that paper is true, but the last one is'. So Quine is wrong when he says (1950, §34; 1974, §39) that a predicate like '__ is true' and '__ is witty' is "an indissoluble unit in which [the general term] stands merely as a constituent syllable" comparable to the 'tic' in 'attic'. Not even a predicate like '__ smokes' is unbreakable. In 'Ann smokes, but Ben does not' we strip off the verb-stem. (The latter plays in 'ϕs' the role that the general term plays in 'is F', so we might call the final 's' and 'smoke', as they occur in the predicate, copula and general term.) But ever since Frege most philosophers tend to regard predicates like 'is witty' and 'smokes' as semantically atomic. So they will not take the arguments from parenthesis insertion and stripping-off as demonstrations of *semantical* divisibility.[77]

Furthermore, the case of *idioms* shows that syntactical structure does not guarantee semantic structure. In the slangy death-notice 'Ben kicked the bucket'[78] the verb 'kicked' does not occur as a two-place predicate, nor does the noun phrase 'the bucket' occur here as a singular term, for we cannot reasonably conclude from that rude statement that there is something Ben kicked, and we cannot sensibly ask whether the bucket he kicked is the same as the bucket that was kicked the day before by poor old Tom. So, unlike the predicate in 'Ann kicked the *ball*', the predicate in the death-notice, though *syntactically* structured, does not differ semantically from the predicate in 'Ben died'. Similarly, Prior could argue, what precedes '*p*' in 'It is true that *p*' is only syntactically structured.

Mulligan tries to weaken resistance against the Priorean atomicity claim by comparing the truth connective with connectives like 'It is possible (probable, certain, . . .) that' which have a *one-word* adverbial, or rather adsentential, counterpart, (2010, p. 576): 'Possibly (probably, certainly, . . .).'[79] Can we regard 'truly' as the indivisible counterpart of the more verbose truth connective? Two uses of 'truly' should be distinguished here. In one of these uses it modifies a *verbum dicendi* or *cogitandi*. Following Tadeusz Kotarbiński's footsteps Prior makes a very interesting point about this use:[80]

> 'X says (believes) truly that *p*' means 'X says (believes) that *p*; and *p*.'[. . .] What is [. . .] defined [. . .] is not a property of anything, but rather what it is to say with truth that something is so; it is an account of the adverbial phrase 'with truth' rather than of the adjective 'true'. (Prior, 1967a, pp. 229–30; cp. 1971, p. 98)

'Truly' in the sense of 'with truth' is really an ad*verbial*. (Of course, unlike 'solemnly' for example, it doesn't characterize the *manner* in which the act of saying is performed: '*x* said truly that *p*' implies 'what *x* said is true', whereas '*x* said solemnly that *p*' does not imply 'what *x* said is solemn'.[81]) But there is also a use of 'truly' in which it modifies a whole sentence, as in the centurion's exclamation 'Truly, this was the Son of God'.[82] Suppose the adsentential 'truly' as used in that sentence is related to the wordy truth connective as 'probably' is related to 'it is probable that'.[83] Can the semantical indivisibility of this, or any other, one-word adsentential really lend support to the claim that the corresponding long-winded connective is semantically atomic, too? The general term in 'Ermelyn is a vixen' is semantically and syntactically unbreakable, but although 'Ermelyn is a female fox' serves to say the same thing, nobody would conclude from this that the general term in the second sentence has no semantically relevant parts.

The adsentential 'truly' is a genuine semantical constituent of instances of 'Truly, p'. It signifies an operation that yields a truth when applied to a truth and a falsehood when applied to a falsehood. As the *canonical* decomposition of instances of that schema shows, they really have the structure that was wrongly ascribed to instances of 'It is true that p' by the principle OPERATION in section 1. Wrongly, if I am right in taking the expression 'It is true that' to be a hybrid sentence fragment consisting of an expletive, a predicate and a complementizer. If PROP-ERTY is correct, that expression is *not* a genuine semantical constituent of instances of 'It is true that p' but the result of artificially combining portions of different semantical constituents, the product of a merely *optional* parsing.

In *CT* I argued that a certain anaphoric use of the adverb 'also' pro-vides evidence for the claim that 'It is true that p' contains a genuine occurrence of the truth predicate (and hence against the view that the truth connective is semantically indivisible). Consider the follow-ing ironic comment on a dispute between Jane and Jim about their supervisor:

(An) It is true that he is brilliant, but Jim's critical remark about him is also true.

With such an example in view, I argued, (*CT*, p 351):

> We can make literal sense of the 'also' if it is preceded by another application of the predicate 'is true' in the first half of the sentence. But on the [truth-connective] reading we can find no predication of 'is true' there.

I am no longer satisfied with this argument.[84] Consider a variant of (An),

(An$_0$) He *is* brilliant, no doubt, but Jim's critical remark about him is also true.

Surely, the coherence of (An$_0$) must be explainable without performing the impossible feat of finding another application of the predicate 'is true' in the first conjunct. (Or take 'When he left Hamburg it was raining cats and dogs, but when he arrived at Palermo the weather was *also* very bad': there is no predication of 'bad' in the first part, and yet the anaphora is comprehensible.) So the coherence of (An) cannot *depend* on the availability of my explanation either, and this explanation lends no support to the claim that the predicate '___ is true' is present in 'It is true that p'.

We get a better argument for this claim if we broaden the view. Truth talk does not always contain a that-clause. So far I have followed Prior and Mulligan in focussing on *'expressive'* truth talk, that is, on sentences in which a truth locution is applied to something that expresses – or that contains a part that expresses – a truth candidate.[85] As regards 'expressive' truth talk, redundancy theorists have a point: what we say is strongly equivalent with something we could say without using *any* truth locution. But an elucidation of our concept of truth cannot afford to put aside *'non-expressive'* truth talk like

–: Whatever follows from a truth is true.
–: Everything the Pope says *ex cathedra* is true.
–: The dogma of papal infallibility is true.
–: Her favourite hypothesis is true.
–: Fallibilism is true.

No parsing of such sentences will bring a truth connective to light, – as regards *this* kind of truth talk the predicative parsing is obligatory. Now an elucidation of our concept of truth cannot afford tunnel vision. It must survey 'expressive' *and* 'non-expressive' truth talk, and it must throw light on their interaction within the confines of one and the same argument. A sentence-forming operator on sentences cannot possibly have the same sense as a predicate, and if the truth connective is a semantically seamless whole, it contains no genuine component that has the same sense as the general term which the copula transforms into the truth predicate. Consequently, if one takes the Prior line on the truth connective, it will be extremely hard, to put it mildly, to explain the irresistible impression that the following argument is formally valid:

(P1) *It is true that* every even number greater than 2 is the sum of two primes.
(P2) That every even number etc. is Goldbach's Conjecture. So,
(C) Goldbach's Conjecture *is true*.

Under Prior's reading of (P1) this argument will have to be classified as enthymematic. By contrast, it is very easy to explain the impression of formal validity if one allows for the predicative parsing of (P1) and takes (P2) to be an identity statement, for then the impression is veridical: the argument exemplifies a pattern that is universally valid in the predicate calculus, namely $Fa, a = b \therefore Fb$. We cannot adequately account for our multi-faceted use of 'true' without bringing truths (with an 's') into the picture, that is, things to which the truth predicate applies.

As I had occasion to emphasize before, declaring a mode of decomposition to be merely optional is not to condemn it as illegitimate. One can consistently deny that Prior's connective parsing is obligatory and concede that it is permissible. After all, every instance of 'It is true that *p*' contains a genuine occurrence of a sentence.[86] In Georg Henrik von Wright's 'truth-logics' the vocabulary of classical propositional logic is embellished with one new symbol, the operator 'T', that we are asked to read as 'it is true that'. Thanks to the introduction of 'T' one can distinguish *within* those calculi mere lack of truth (in a truth candidate) and falsity, the former being expressed by '¬T*p*', the latter by 'T¬*p*'.[87] Of course, it is no objection against the calculi von Wright constructs that they employ an atomic truth connective, but neither does the elucidatory power of such calculi show that our comprehension of sentences of the form 'It is true that *p*' is not based upon understanding the truth predicate. As a matter of fact, von Wright (1984; 1986 and especially 1996) makes no claim about explanatory priorities.[88] In any case, the limits of the elucidatory power of such T-calculi becomes manifest when we consider an argument like '(P1), (P2), therefore (C)' in which truth connective and truth predicate interact. If we use von Wright's 'T' to formalize (P1), the impression of formal validity at once evaporates.

This completes my defence of the claim that the truth predicate has explanatory priority over the truth connective.

8.10   Postlude: Frege on 'Abstract Noun Clauses'[89]

In his paper 'On Sense and Reference' Frege calls that-clauses in ascriptions of propositional acts-or-states *'abstracte Nennsätze* (abstract noun clauses)'. A *'Nennsatz* (noun clause)', he says, is a clause that constitutes a complex name (1892, 41c), and an *'abstract* noun clause, introduced by "that"' (37a) can be regarded as 'a proper name (*Eigenname*) of a thought' (39c), that is, as a singular term that designates a proposition.[90] It will not have escaped the reader that in this respect I am with Frege.

Why does Frege call that-clauses *abstract*? His translator Max Black tries to explain this in a footnote, and this footnote has been preserved in all reprints of his translation:[91]

Frege probably means clauses grammatically replaceable by an abstract noun-phrase; for example 'Smith denies that dragons exist' = 'Smith denies the existence of dragons'; or again, in this context after 'denies', 'that Brown is wise' is replaceable by 'the wisdom of Brown'.

This is very farfetched, I think, and it founders on 'Smith says / thinks that dragons exist (that Brown is wise)'. We come closer to a plausible interpretation if we look for the intended contrast. What is a *concrete* noun clause? Unfortunately Frege never uses the title '*concreter Nennsatz*', but he gives an example that can lay claim to it. The *German* sentence (W), I contend, begins with such a clause:

(W) *Der die elliptischen Planetenbahnen entdeckte, starb im Elend.*
The one person who discovered the elliptic form of the planetary orbits died in misery.[92]

The grammatical subject of the German sentence is a headless relative clause.[93] It could be replaced *salvo sensu* by the definite description 'the discoverer of the elliptic form of the planetary orbits' (and *salva veritate* by the name 'Kepler'). Frege classifies it as a noun clause (41b). Why would it be reasonable to characterize it also as concrete? The answer is not far to seek: Because it designates a concrete entity. If that is correct, we also have a natural answer to our question why Frege classifies the that-clause in 'Copernicus believed that the planetary orbits are circles' (37b) as an abstract noun clause: Because it designates an abstract entity. Thus understood, the distinction between concrete and abstract noun clauses is a distinction between the kinds of objects they designate (if they are not empty). In the same manner Quine (1950, §34; 1974, §39) classifies singular terms like 'piety' ['the author of *Waverley*'] as abstract [concrete] because they 'purport to refer to abstract [concrete] objects'.[94] To be sure, the adjectives 'abstract' and 'concrete' do not belong to Frege's official theoretical vocabulary. But he does not hesitate (1903, §74) to call his 'logical objects', adopting Cantor's terminology, '*abstracte Gegenstände*'. Why should he have any qualms to apply this term also to propositions?[95]

As we saw, Frege makes the *general* claim that abstract noun clauses can be regarded as designators of propositions (*Eigennamen eines Gedankens*). But as a matter of fact, he invokes it only when he is concerned with ascriptions of propositional acts-or-states in which the sentence preceded by 'that' cannot always be replaced *salva veritate* by another sentence with the same truth-value. Obviously sentences of the form 'It is true that *p*' or 'That *p* is true' are not exposed to that risk. But this also holds for substitution-instances of 'If anyone were to believe / assert that *p*, he or she would be right in so believing / in asserting that'. Are they exceptions to the rule that a that-clause preceded by 'believes' or 'asserts' designates a proposition? That would be a rather uncomfortable position. Furthermore, expressive truth talk and reports of propositional acts-or-states interact, as in the argument

(Π1) *Plato believes that* Socrates is wise.
(Π2) *It is true that* Socrates is wise. So,
(Γ) Plato believes something that *is true.*

This very much looks like a formally valid argument, and that's what it is if we regard both premises as predications about propositions; for then the argument instantiates a schema that is universally valid in classical predicate logic, namely *Fa, Ga ∴ ∃x (Fx & Gx).* So there are good reasons for applying Frege's thesis about that-clauses also to expressive truth talk, thereby treating the latter as having a predicational structure. As we saw in Section 1, Frege would not agree, for he thinks that corresponding instances of *'p'* and *'It is true that p'* express one and the same proposition (in the same context). But here we should part company with him.[96]

Notes

1. In labelling the second expression 'truth connective' I grudgingly yield to the logicians' odd practice of classifying a sentence-forming operator on sentences as a connective even if it is does not connect anything with anything. The expression 'one-place connective' is, as Prior (1971, p. 19) put it, 'a reasonable logical barbarism'.
2. Künne (2003), henceforth: *CT*, 350–1.
3. In my usage, only singular terms *designate* anything, and demonstratives, proper names and definite descriptions, for example, do that in different ways. Predicates, on the other hand, *signify* (or connote) properties. So, the property of being wise is signified by 'is wise', whereas it is designated by 'wisdom'. Connectives, too, do not designate anything, – they signify operations.
4. Prior (1967a, pp. 229–30; 1971, pp. 5–6, 11–13, 20).
5. Cp. *CT* 339–41.
6. Frege (1918, original pagination 61). He upholds this kind of identity thesis in *NS* 153 (*PW* 141), 211 (194), 271 (251).
7. Frege (1918, original pagination 34). He repeats this kind of identity thesis in 'Logic in Mathematics' (1914), in *NS* 251 (233).
8. My paper incorporates, often in a thoroughly revised form, some parts of my reply to Mulligan (Künne, 2010c), but this time the canvas is much larger.
9. Cp. Geach (1955, p. 229; 1965, p. 110).

10. A truth connective can be uncovered even in semantical truth talk where truth (in L), or the property of expressing a truth (in L), is ascribed to a sentence: a Tarski-style truth ascription like 'The sentence "snow is white" is true (expresses a truth)' contains the one-place connective 'The sentence "___" is true'.

11. I borrow this term from King (2002; 2007, ch. 5).

12. As I did in *CT* 10 n.12, 255–7.

13. Mulligan emphasized this in (2010, p. 573). Glock (2011, p. 151) is less severe with (*): he declares it to be 'less common and perspicuous than' what results from (*) if 'that' is replaced by a colon. More on this issue in Section 7 below. – Throughout this paper the *asterisk* is to signalize lack of grammaticality.

14. A definite description need not be appositive in order to be non-standard. Terms of the form 'the set of all and only those things that are F' cannot be treated along Russellian lines either, for 'the unique x such that x is *a* set of all etc.' makes no sense.

15. This point is due to Cartwright: see *CT* 256. In many languages, for example in German, there is no such ambiguity. In *'die Proposition, die ...'* and *'das Theorem, das ...'* we have relative clauses, whereas in *'die Proposition, dass ...'* and *'das Theorem, dass ...'* we have (German versions of) that-clauses.

16. Some of its post-Priorean opponents are listed in Rosefeldt (2008, n. 3). Note especially Moltmann (2003). Rosefeld and Pryor (2007) also belong on this list. Glock (2011, p. 150) sees 'no need' to join either party. Most opponents focus on that-clauses in 'propositional attitude' reports. In Section 7 I shall enter this minefield.

17. Recanati disagrees: 'Can that-clauses be considered as quantified phrases? Why not? [...] We can [...] analyse (1) "John believes that grass is green" as (2) "John believes *something that is true iff grass is green*".' (2004, p. 231) [numerals added] I don't find this plausible. Firstly, the putative analysans contains an uncovered cheque: (1) entails (2), but (2) entails (1) only under a *very* strong reading of 'iff' that (admittedly) still awaits explanation. As it stands, (2) provokes a question that (1) does not provoke: 'And what is it that John believes?' Secondly, one can attach the rider 'namely that grass is green' to (2), but appending it to (1) makes no sense. Doesn't that disqualify (2) as analysans? (To be sure, there are quantifications that do not tolerate such a namely-rider, for example, 'There is a snowflake that is never individually referred to', but Recanati's putative analysans is not of this type.) Thirdly, the application of Recanati's analysis to 'That snow is white is true' has the consequence that the alleged analysans contains more occurrences of the truth predicate than the analysandum: 'something that is true iff snow is white is true'. Does one invoke the concept of truth twice when one says, 'That snow is white is true'?

18. Both in (1a) and in (2a) the that-clause can be replaced *salva veritate* by the corresponding propositional description.

19. Only in medieval Latin is the construction *'verum est'* + *'quod nix alba est'* also available: Kretzmann (1970, p. 776). In ancient Greek both (TC) and (TP) can be rendered with an accusative-*cum*-infinitive (*'alēthés estin chióna leukēn eînai'*) as well as with a clause (*'alēthés estin, hóti chiōn leukē estin'*).

Note for future reference that there is no counterpart to the English (German) expletive 'it' (*'es'*) either in Latin or in Greek. In *Cat.* 13b18–19 Aristotle, or rather his translator John Ackrill, writes: 'Neither "Socrates is sick" nor "Socrates is well" will be true if Socrates himself does not exist at all'. The Greek construction is: definite article (*'tò'*) + a.c.i. The article suggests Ackrill's translation, and again and again Aristotle takes declarative sentences to be truth-bearers. But the a.c.i. points into the direction of 'It will neither be true *that* Socrates is sick nor *that* he is well', and the definite article might be short for *'tò'* + *verbum dicendi*: 'It will neither be true *to say* that Socrates is sick nor true to say that he is well'. I am grateful to Klaus Corcilius for an illuminating exchange on this issue. He pointed out to me that on the whole Aristotle prefers an *oratio obliqua* construction, lit. 'To say that *p* is true' where the Greek counterpart of 'that *p*' is either an a.c.i. or a clause. Compare, for example, *An. Pr.* 52a8–12, *De caelo* 282a29f and (elliptically) *De sensu et sensibilibus* 438a12–14. Aristotle employs the same construction in his famous definition of truth in *Met.* IV, 7; for an attempt to make sense of it, see *CT* 95 ff.

20. As Dummett correctly maintained, this makes for a tension in Frege's theory. I rejected Dummett's point in *CT* 37–42, and I defend and develop it in Künne (2010d).

21. Repeated in my (2010a, pp. 558–9). Compare the retrospective use of 'she' in (1) 'If Ann has time, she will join us' and its prospective or anticipatory use in (2) 'If she has time, Ann will join us': in (1) we have anaphora, in (2) we have cataphora. Both pronouns designate what 'Ann' designates.

22. One can find the point already in Rundle (1979, p. 319).

23. Unfortunately, in an English translation the first pronoun is bound to disappear: 'The lake smiles, it invites to bathe'.

24. German Lieder and poems abound with this construction: *'Es ist ein Ros entsprungen'* (Christmas hymn), *'Es schlug mein Herz, geschwind zu Pferde!'* (Goethe), *'Es bellen die Hunde, es rasseln die Ketten'* (Müller/Schubert), *'Es klappert die Mühle am rauschenden Bach'* (Volkslied).

25. The difference between expletive and cataphoric pronouns is less obvious, though equally real, if we move from *'Es lächelt das Mädchen'* to *'Es lächelt, das Mädchen'*, where the gender of the noun allows us to use *'es'* in both cases.

26. 'Equivalent (*gleichgeltend*)' in Bolzano means *extensionally equivalent*, but in the case under discussion he maintains propositional identity, as the addition of 'completely' suggests and the claim about the next pair shows.

27. From Greek *'pleonasmós* (abundance)'.

28. Cp. Bolzano (1837, vol. I, p. 123) on *'in Wahrheit'* and *'wirklich'*.

29. In Kretzmann (1970, p. 777) Bolzano's claim is repeated (and turned against Prior): (TC) 'is an expletive construction, the nonexpletive version of which is' (TP). By definition, expletive pronouns are semantically vacuous, but there are expletives and expletives. The *'es'* in (Ex) tends to disappear when (Ex) is negated, and it disappears when (Ex) is embedded in a conditional and when it is transformed into a yes/no interrogative. Furthermore, it has no counterpart in the English translation of (Ex). The 'it' in (TC) survives those operations, and it does have a counterpart in the German translation of (TC).

30. Frege (1923) orig. 41, *et passim*.
31. Here I find myself in complete agreement with Napoli (2006) 245. Cp. also Bolzano's careful distinction between 'Socrates is not stupid' and 'The proposition that Socrates is stupid is false': (1837) II, 16, 44 ff, 63, 269, 419.
32. Cp. my (2010a, 558–9).
33. I took the liberty of replacing his example by mine.
34. See Section 8.7. There I will argue that a naked that-clause that designates a proposition *P* introduces *P* into discourse as a possible *content* of an act or a state, whereas a propositional description that designates *P* introduces *P* into discourse as a possible *intentional object* of an act or a state.
35. The ungrammaticality of this kind of substitution falsifies the principle '(P2*)' in Rosefeldt (2008, p. 309). The vagaries of pronoun inflection (declination) provide Oliver with one of various types of examples by means of which he refutes the tenet that co-designative singular terms are always exchangeable *salva congruitate*: Oliver (2005); cp. my (2010a, 303–4). Crispin Wright dubbed this tenet 'The Reference Principle'. In spite of its title, Dolby (2009) is not a defence of this principle but an attempt to replace it by a more defensible one. So I am surprised that Glock (2011, p. 151) still maintains that 'we can substitute co-referential expressions *salva significatione* even in intensional contexts'. [Presumably he wanted to say '... *salva congruitate* ... ', for surely such a substitution will normally affect the meaning (*significatio*) of the sentence.]
36. In a regimented version of English 'He kissed her' and 'She was kissed by him' would be reformulated in such a way that they become instances of the schemata for pairs of predicates that signify converse relations, '*R* (*a*, *b*)' and '*Ř* (*b*, *a*)', and (I presume that) the reformulations could be treated as representations of the deep-structure of the originals.
37. I owe this datum to Mulligan (2010, p. 574) who contrasts (S3*) and (S4*) with the (TC)-variants of (S3) and (S4).
38. This kind of argument from apposition is due to (Schiffer, 2003, 93–5). Cp. my (2010a, pp. 303–4).
39. The same holds for the pair 'Logicism is the proposition that arithmetic reduces to logic' and (*) 'Logicism is that arithmetic etc.'
40. In his 1923/24, p. 281 Frege considers this example: If you instantiate the argument schema '*p. Wenn p, dann q. Also q.*', the second premiss will often not be grammatically acceptable: **'Wenn Napoleon ist ein Mensch, dann Napoleon ist sterblich.'* Nobody will conclude from this that in German Modus Ponens is not universally valid.
41. More recently, this account was also endorsed in Recanati (2000).
42. CT 68.
43. Recanati used this nickname for what I call prenectives.
44. In his 2000 Recanati rejects Davidson's account of *oratio obliqua* because "it blatantly violates" the constraint that discrepancies between grammatical form and logical form should be kept to a minimum: On Davidson's account, 'John believes that S' consists of two sentences, but "it is clear, on grammatical grounds, that that is not the case" (2000, 28–9). I think it is equally clear, on grammatical grounds, that 'John believes that S' contains a clause as semantical unit – and not the atomic prenective 'believes-that'. So Recanati's

criticism of Davidson backfires on his adoption of 'Prior's Adverbial Analysis'. In (Recanati, 2004) this tension is implicitly acknowledged.

45. In medieval Latin the clausal construction '*dicit (existimat) quod nix alba est*' is also available.

46. As opposed to the merely typographical occurrence of 'Socrates was Greek' in [Q]: see sect. 8.2, 1st par.

47. Similar problems arise with respect to alethic modal logic (and deontic logic). Suppose we have only indissoluble one-place modal connectives at our disposal when it comes to formulating necessity claims. Then our language is too poor to capture the point of statements like 'All logical laws are necessary', 'Some a posteriori truths are necessary' and 'Not all a priori truths are necessary'. Halbach and Welch (2009) argue for a way out that was hinted at in (Schiffer, 2004, p. 97): adding a truth *predicate* (!) to the language of modal logic and replacing the predicate '___is necessary' by the predicate '___is necessarily true'.

48. The handy classificatory terminology is borrowed from Schiffer (2006).

49. For type (1) below see Russell (1918, p. 218) ('It seems natural to say one believes a proposition and unnatural to say one desires a proposition, but as a matter of fact that is only a prejudice [...] You may desire to get some sugar tomorrow and of course you may [...] believe that you will'), (Prior, 1971, p. 16) and (Geach, 1967, p. 168). For types (1)–(3), or most of them, see Rundle (1968; 1979, ch. 7; 2001), King (2002), repr. in his 2007, pp. 137–63 with additional note on pp. 142–3 (Pryor, 2007; Rosefeldt, 2008; Forbes, 2011).

50. Kind (1) includes *appreciate (consider / demand / desire / discover / expect / feel / forget / hear / indicate / mention / notice / predict / recall / recognize / recommend / remember / see / suggest / suspect / understand) that p*.

51. Kind (2) includes *agree (be amazed / answer / bet / boast / complain / conclude / confess / gather / be disappointed / be glad / guess / have the impression / insist / lament / object / protest / rejoice / be relieved / remark / reply / request / suppose / be sure / surmise / be surprised) that p*. Rundle tends to blur the difference between type-(1) and type-(2) cases, since his absurdity verdicts do not distinguish glaring falsity from ungrammaticality. To be sure, the assumption that Ben desires the proposition that Labour will win is perfectly absurd, but that does not make the sentence 'Ben desires the proposition that etc.' ungrammatical. By contrast, *grammar* frowns upon *'Jim is glad the proposition that etc.' – How is *believe that p* to be classified? According to Rundle (2001, p. 145), it belongs to type (1). The reason he gives looks plausible to me: Take any '*p*' you like, believing *the proposition that p* is having a belief *about* this proposition, namely the belief that it is true, whereas one can believe *that p* without having a belief about the proposition that *p*. (As for the other direction, one can hardly have the meta-belief without having its first-order counterpart.) Let me at least mention a different position. Forbes (2011) tries to show that every instance of 'A believes the proposition that *p*' is necessarily false but some are figuratively true. He does not explicitly answer the question whether an instance of that schema can be figuratively false while its counterpart with the naked that-clause is true (or vice versa).

52. In my 2010a, pp. 302–3 I also used a frivolous variant of an example I had found in Kiteley (1991): Replacing the definite description in 'The director's secretary is paid very well for her services' by the co-designative term 'the direcor's mistress' might affect the truth-value.
53. Forbes (2011) employs Quine's better-known 'Giorgione/Barbarelli' example for the same dialectical purpose, but not in the same way as I shall use it.
54. Singular terms with attributive adjectives belong to the battery of examples by means of which Oliver (2005) refutes the principle that co-designative singular terms can always replace one another *salva congruitate*.
55. Cp. *CT* 254–61 and my (2010a, pp. 303–4).
56. Rosefeldt's main objection to the view that that-clauses are singular terms is based on this observation: see his 2008, pp. 309–10; cp. Pryor (2007, p. 227).
57. More advisable, I dare say, than watching *RTL* on German TV.
58. The restriction Rosefeldt subsequently imposes on his principle '(P3)' does not save it from this counterexample: see his 2008, p. 309 and 330, fn. 12.
59. Using a neo-Davidsonian semantics for ascriptions of acts *and* states as framework, Forbes (2011) spells out the RHS of biconditionals instantiating CLAUSAL and OBJECTUAL. He gives good reasons for the linguistic thesis that the surface structure of the ('Davidsonized') RHS is the deep-structure of the LHS of such a biconditional. He shows that treating a clausal verb as if it were objectual ('transitive'), or mistaking a clausal verb for its objectual ('transitive') counterpart amounts semantically to assigning to a proposition the role of intentional object ('theme') when it really has the role of content. I stick by the Husserlian terminology I used in *CT* 258 ff. to highlight the same difference, and I prefer 'objectual' to 'transitive'. Judged by the standard criteria of transitivity (in the grammarians' sense), the verb in a sentence S of the form 'A ϕs that p' *is* transitive: S allows for passivization, and the that-clause in S can be used to answer the question '(Whom or) what does A ϕ according to S?'
60. The main source, I gather, is Russell's Harvard Lectures, published as *An Inquiry into Meaning and Truth*, London 1940.
61. Cf. Rundle (2001, p. 143).
62. Perhaps a belief that p does have an intentional object, but whatever that object might be, it is *not* the proposition that p. In this respect, Husserl's Fifth *Logical Investigation* is more insightful than most analytical writings about intentionality. I explain and discuss his view in my (2011).
63. King (2002, pp. 362–4) takes this fact to be embarrassing for the propositionalist relational view he argues for. I agree with Forbes (2011) that he shouldn't, and as opposed to Rosefeldt (2008, pp. 315–16), I think that King's worry can be dispelled *without* assuming that that-clauses do not designate anything and that 'something' in the conclusion is a non-nominal quantifier.
64. *Contra* Pryor (2007, p. 230).
65. The comments on zeugmata in Kiteley (1991, pp. 380–2) are very illuminating. (Z2) is extracted from Alexander Pope's lines about Hampton Court and Queen Anne: "Here Thou, great Anna! whom three Realms obey, / Dost sometimes Counsel take – and sometimes Tea" (*The Rape of the Lock*, 1714, Canto III, ll. 7–8). (For the sake of the rhyme we must assume that Pope did not pronounce 'obey' or 'tea' in the way we do.)

66. As noted in Kiteley (1991), this creates trouble for compositionality.

67. Kind (3) includes *advance (attack / contradict / criticize / debate / defend / dismiss / embrace / entertain / evaluate / misunderstand / refute / repudiate / ridicule) the* (NOUN) *that p.* Cp. *CT* 260.

68. Some more examples: *allegation, axiom, belief, conjecture, contention, dogma, hypothesis, judgement, law, principle, report, speculation, statement, suggestion, tenet, thesis, thought.* The verbal nouns in this list can be used to refer to acts or states, but when inserted in 'A ϕs the ___that p' they don't: cp. *CT* 249–52. In my (2010b) I used this battery of examples in order to explain the term 'proposition' as the most general one that can be inserted in that schema.

69. In 2010a, pp. 514–19 I explain why I find this unfortunate.

70. Because of what precedes 'means' this verb is here the counterpart of the German '*bedeutet*' (in its colloquial, non-Fregean use), not of '*meint*'. (At this point German may be less prone to mislead than English.)

71. A sentence of this sort can be ambiguous: ' "*nihil*" means nothing' is false, if understood as a verdict of meaninglessness, and it is true as an ascription of a certain meaning. It's the other way round with 'The phrase "asa nisi masa" in Fellini's *Otto e mezzo* means nothing'. The intonation pattern tends to disambiguate.

72. There is an additional reason for deleting 'that': the format of (M_S) is at best suited for saying what a *declarative* sentence means. But if I had been asked, 'What does "*Regnet es?*" mean?', I could have obliged, too: 'It means Is it raining?' If prefaced by 'it means', *any* (meaningful) expression manages to designate its own meaning.

73. As regards meaning ascriptions I almost totally disagree with Schiffer (2003, pp. 100–5).

74. *CT* 351.

75. Both arguments are due to Kent Wilson (1990, pp. 23–4).

76. See my (2010c, n. 61).

77. But then, in an inference like 'Ann is witty, but Ben is not; so, she is something he is not (namely witty)' we quantify into the position – not of a predicate but – of a general term, as the survival of the copula and namely-rider show. And we can also quantify into the position of a verb-stem: 'He does something she doesn't, namely smoke'. Doesn't this indicate *semantical* divisibility? Cp. the homage to the copula in my (2006, §§I and V) and the references to Wiggins and Strawson contained therein.

78. Cp. the analogues of 'to kick the bucket' in German: *den Löffel abgeben* (lit. 'to give the spoon away'), *ins Gras beißen* (lit. 'to bite into the grass') or, as is most appropriate in this volume, in Polish: *kopnąć w kalendarz* (lit. 'to kick the calendar').

79. If one applies his argument from restrictions on embedding (574) to such pairs, the odd result is that one of them is, as he puts it, 'at best less than a fully-fledged sentence', since instances of 'I wonder whether possibly (probably,...) p' or 'If possibly (probably,...) p, then q' are not well-formed sentences whereas their 'it is [adjective] that p' counterparts are fine. (Timothy Williamson pointed this out to me.)

80. On Kotarbiński's anticipation see *CT* 343–6. The German counterpart of the definiendum is '*x sagt wahrheitsgemäß (wahrheitsgetreu, der Wahrheit*

entsprechend), dass p'. Nowadays the word *'wahr'* is always used as an adjective. In the late eighteenth century, though, it was also used adverbially: *'Redest Du wahr?'*, Wilhelm asks Barbara (Goethe, *Wilhelm Meisters Lehrjahre*, VII/8).

81. White (1968).

82. Matthew 27: 54 (King James Bible), where 'Truly' corresponds to *'Wahrlich, ...'* in Luther's translation, to *'Vere ...'* in the Vulgata and to *"Ἀληθῶς ...'* in the original.

83. *Pace* Mulligan (2010, p. 576): '"truly" (like German "*wahrlich*" ...), as far as I can see, connote sincerity or truthfulness or express attitudes such as surprise.' As for the alleged connotations, they seem to be absent in the centurion's sentence, and as for the attitudes, a speaker can also use 'it is true that *p*' to give voice to her surprise at the fact that *p*, and neither expression seems to be standardly used for this purpose.

84. Neither is Mulligan, but I don't think his machinery is needed for rejecting the argument: cp. his (2010, pp. 578–9).

85. Cp. *CT* 52–3 *et passim* on 'propositionally revealing' *vs.* 'propositionally unrevealing' truth talk.

86. Again in marked contrast to the merely typographical occurrence of 'Socrates was Greek' in [Q], see Section 8.2.

87. von Wright (1984), (1986) and esp. (1996).

88. As a matter of fact, he uses a truth *predicate* and applies it to propositions whenever he explains his axioms. We are told, for example, that the axiom '$Tp \leftrightarrow T\neg\neg p$' which belongs to the core of each of his 'truth-logics' says that 'a proposition is true if, and only if, its negation is false', or that '[the axiom '$Tp \rightarrow \neg T\neg p$' of one of his calculi] says that if a proposition is true then it is not (also) false': cp. (Wright, 1996, §§6–7). The latter axiom is the earmark of his calculus TL that accommodates the possibility of a truth-value gap.

89. This postscript is based on Künne (2010a, pp. 291–2, pp. 417–18).

90. In a letter to Russell (13.11.1904, in *WB* 246) Frege puts forward the same view. Herbert Feigl translates *'Nennsatz'* as 'subjective clause', which is bad, and *'abstracter Nennsatz'* as 'abstract clause', which simply ignores *'Nenn-'*. In the above I followed Max Black's rendering. [I should add that at other places Feigl's translation is far better than Black's: cp. my (2010a, pp. 821–2, *s.v.* 'Black', 'Feigl'.]

91. The only change that recent editors deemed necessary consisted in replacing Black's reasonable translation of *'Bedeutung'* as 'reference' either by 'meaning' (which makes Frege devote several pages to the utterly silly question whether sentences have meaning) or by the non-translation *'Bedeutung'*.

92. The headless relative clause in (W) is surprisingly difficult to translate. Obviously my translation does not preserve headlessness. Black's rendering does preserve it: "Whoever discovered [...]." But this translation seems bad, for as far as I can see (which may not be very far) it is either wrong or ambiguous. It is *wrong* if 'whoever' means any person who (as in 'Whoever wrote that is a fool'). Thus understood, the English sentence would express a truth if a team every member of which died in misery had made the discovery. But Frege does certainly not want his sentence to be understood as a universal generalization. The translation is *ambiguous* if 'whoever' also has a reading

under which it means the one person who. Surely this is what Frege intends: he needs a clause whose *Bedeutung* is Kepler. Feigl's translation doesn't fare much better: 'He who discovered [...]', for this construction can also serve as a restricted universal quantifier (as can be seen from proverbs like 'He who hesitates is lost' and 'He who fails to prepare, prepares to fail'). [Note how Goethe moves in 'Mignon's Song' from a general *'wer'*-clause to the kind of *'der'*-clause we find in Frege's example: *'Nur wer die Sehnsucht kennt, / Weiß, was ich leide. /.../ Ach! der mich liebt und kennt, / Ist in der Weite. |...* (Only somebody who knows longing / Understands what I suffer! / ... / Alas! the one person who loves and knows me, / Is far away. /...)'.]

93. We have just stumbled over another translation problem. In his long (and sadly neglected) discussion of the semantical roles of 'subordinate sentences (*Nebensätze*)' (op. cit. 36–50), Frege employs the terminology of German grammar books (36c). In these books subordinate sentences are regarded as 'representatives of parts of [sc. simple] sentences' (ibid.), hence they inherit their names from them: *Nennwörter* (nouns) / *Nennsätze* (noun clauses), *Beiwörter* (adjectives) / *Beisätze*. Feigl translates *'Beisatz'* as 'relative clause'. This is bad for two reasons: it makes the correlation with a certain class of words invisible, and it neglects the fact that Frege calls the headless relative clause in (W) a *Nennsatz*. For both reasons Black's translation is to be preferred: 'adjective clause'.

94. Similarly John Stuart Mill, *System of Logic* (1843, I, 2, §4).

95. A long time ago Alonzo Church suggested this interpretation (1953, p. 93). This reading of Frege's text is supported by the things grammarians of the nineteenth century say under the heading *Abstracte und concrete Nennsätze*, – for instance, Max Wilhelm Götzinger in *Die deutsche Sprache und ihre Literatur*, vol. 1, pt. 2: 'Satzlehre', §§123–30, Stuttgart (1839), repr. Hildesheim (1977).

96. I have benefited from discussions of various parts of this material in Frankfurt/M, Hamburg, Kraków, Padova and Saarbrücken. I am especially grateful to Kevin Mulligan who gently forced me to reconsider some claims and arguments in *CT*. It is largely his fault that half a page of that book has now been replaced by a fairly long paper.

References

Ackrill, J. L. (1963) *Aristotle's "Categories" and "De Interpretatione"*, (Oxford: Clarendon Press).

Bolzano, B. (1837) *Wissenschaftslehre*, vol. 4, (Sulzbach). Partially translated by R. George *Theory of Science*, (Berkeley: University of California Press), 1972 and B. Terrell *Theory of Science* (Dordrecht: Riedel), 1973. Complete translation by P. Rusnock, and R. George (forthcoming).

Church, A. (1953) '[Review of] P. Geach and M. Black, *Translations from the Philosophical Writings of Gottlob Frege'*, *Journal of Symbolic Logic*, 18, 92–3.

Dolby, D. (2009) 'The Reference Principle: a Defence', *Analysis*, 69, 286–96.

Forbes, G. (2011) 'Content and Theme in Attitude Ascriptions', in *Workshop in Philosophy and Linguistics* (University of Michigan), Spring, [presented paper]. Available at: http://web.eecs.umich.edu/~rthomaso/lpw11/forbes.pdf, date accessed 18 October 2012.

Frege, G. (1892) 'Über Sinn und Bedeutung', reprinted with original pagination in Patzig, G. (ed.) *Funktion-Begriff-Bedeutung* (Göttingen: Vandenhoeck and Ruprecht,) 2008, translated by
 (a) M. Black (1948), reprinted in *CP* and in M. Beaney (ed.) *The Frege Reader*, (Oxford: Blackwell), 1997;
 (b) H. Feigl (1949) in H. Feigl, and W. Sellars (eds) *Readings in Philosophical Analysis*, (New York: Appleton-Century-Crofts) and in J. Garfield, and M. Kiteley (eds) *Meaning and Truth: Essential Readings in Modern Semantics* (New York: Paragon House), 1991.

—— (1918) 'Der Gedanke', translated in *CP*, reprinted with original pagination and commentary in Künne (2010a).

—— (1919) 'Die Verneinung', translated in *CP*, reprinted with original pagination and commentary in Künne (2010a).

—— (1923) 'Gedankengefüge' translated in *CP*, reprinted with original pagination and commentary in Künne (2010a).

—— (1923/24) 'Logische Allgemeinheit', translated in *PW*, reprinted with original pagination and commentary in Künne (2010a).

—— (NS) *Nachgelassene Schriften*, (Hamburg: Meiner), 1969; translated as Frege, G. (PW) *Posthumous Writings* (Oxford: Blackwell), 1979.

—— (CP) *Collected Papers on Mathematics, Logic, and Philosophy*, in B. McGuinness (ed.) (Oxford: Blackwell), 1984.

—— (WB) *Wissenschaftlicher Briefwechsel* (Hamburg: Felix Meiner Verlag), 1976.

Geach, P. T. (1955) 'Class and Concept', reprinted in Geach (1972), 226–35.

—— (1965) 'Logical Procedures and Expressions', reprinted in Geach (1972), 108–15.

—— (1967) 'The Identity of Propositions', reprinted in Geach (1972), 166–74.

—— (1972) *Logic Matters* (Oxford: Blackwell).

Glock, H. J. (2011) 'A Cognitivist Approach to Concepts', *Grazer Philosophische Studien*, 82, 131–63.

Halbach, V., and Welch, P. (2009) 'Necessities and Necessary Truths', *Mind*, 118, 71–100.

Horwich, P. (1990) *Truth* (Oxford: Blackwell).

King, J. C. (2002) 'Designating Propositions', *Philosophical Review*, 111, 341–71.

—— (2007) *The Nature and Structure of Contents* (New York: Oxford University Press).

Kiteley, M. (1991) 'Subjectivity's Bailiwick', in Garfield and Kiteley (1991), 372–95.

Kretzmann, N. (1970) 'Medieval Logicians on the Meaning of the *propositio*', *Journal of Philosophy*, 67, 767–87.

Künne, W. (CT) *Conceptions of Truth* (Oxford: Oxford University Press), 2003.

—— (2006) 'Properties in Abundance', in P. F. Strawson, and A. Chakrabarti (eds) *Universals, Concepts and Qualities: New Essays on the Meaning of Predicates* (Burlington, VT: Ashgate), 249–300.

—— (2010a) *Die Philosophische Logik Gottlob Freges* (Frankfurt am Main: Klostermann).

—— (2010b) 'Circularity Worries: Reply to Paul Boghossian', *Dialectica*, 64, 585–97.

—— (2010c) 'Truth Without Truths? Reply to Kevin Mulligan', *Dialectica*, 64, 597–615. French translation in: *Philosophiques*, 38 (2011), 195–217.

—— (2010d) 'Un Conflitto Interno alla Teoria di Frege: Risposta ad Andrea Sereni', in M. Carrara, and V. Morato (eds) *Verità* (Milano: Mimesis), 98–109.

—— (2011) ' "Denken ist immer Etwas Denken." Bolzano und (der frühe) Husserl über Intentionalität', in *Abhandlungen der Akademie der Wissenschaften zu Göttingen*, NF Bd. 14, (Berlin), 78–99.

Moltmann, F. (2003) 'Propositional Attitudes Without Propositions', *Synthese*, 135, 77–118.

Mulligan, K. (2010) 'The Truth Predicate *vs* the Truth Connective. On Taking Connectives Seriously', *Dialectica*, 64, 565–84.

Napoli, E. (2006), 'Negation', *Grazer Philosophische Studien*, 72, 233–52.

Oliver, A. (2005) 'The Reference Principle', *Analysis*, 65, 177–87.

Parsons, T. (1993) 'On Denoting Propositions and Facts', *Philosophical Perspectives*, 7, 441–60.

Prior, A. N. (1955) 'Berkeley in Logical Form', reprinted in Prior (1976), 33–8.

—— (1962) 'Some Problems of Self-reference in John Buridan', reprinted in Prior (1976), 130–46.

—— (1963a) 'Is the Concept of Referential Opacity Really Necessary?', *Acta Philosophica Fennica*, 16, 189–99.

—— (1963b) 'Oratio obliqua', reprinted in Prior (1976), 147–58.

—— (1967a) 'Correspondence Theory of Truth', in P. Edwards (ed.) *The Encyclopedia of Philosophy*, vol. 2, (New York: Macmillan), 223–32.

—— (1967b) 'Negation' in P. Edwards (ed.) *The Encyclopedia of Philosophy*, vol. 5, (New York: Macmillan), 458–63.

—— (1971) *Objects of Thought* (Oxford: Clarendon Press).

—— (1976) *Papers in Logic and Ethics* (London: Duckworth).

Pryor, J. (2007) 'Reasons and That-Clauses, *Philosophical Issues*, 17, 217–44.

Quine, W. V. O. (1950) *Methods of Logic*, 3rd edn (La Salle, IL: Open Court), 1974.

—— (1960) *Word and Object* (Cambridge: The MIT Press).

—— (1970) *Philosophy of Logic* (Cambridge: The MIT Press).

—— (1995) 'Reactions', reprinted in D. Føllesdal, and D. B. Quine (eds) *Quine in Dialogue* (Cambridge: The MIT Press), 2008, 235–50.

Recanati, F. (2000) *Oratio Obliqua, Oratio Recta* (Cambridge: The MIT Press).

—— (2004) 'That-Clauses as Existential Quantifiers', *Analysis*, 64, 229–35.

Rosefeldt, T. (2008) 'That-Clauses and Non-Nominal Quantification', *Philosophical Studies*, 137, 301–33.

Rundle, B. (1967/68) 'Transitivity and Indirect Speech', *Proceedings of Aristotelian Society*, 58, 187–206.

—— (1979) *Grammar in Philosophy* (Oxford: Oxford University Press).

—— (2001) 'Object and Attitude', *Language and Communication*, 21, 143–56.

—— (1918) 'The Philosophy of Logical Atomism', reprinted in B. Russell (1956) *Logic and Knowledge* (London: Allen and Unwin), 177–281.

Schiffer, S. (2003) *The Things We Mean* (Oxford: Clarendon Press).

—— (2006) 'Propositional Content', in E. Lepore, and B. C. Smith (eds). *The Oxford Handbook of Philosophy of Language* (Oxford: Clarendon Press), 267–94.

White, A. R. (1968) ' "True" and "Truly" ', *Noûs*, 2, 247–51.

Wilson, W. K. (1990) 'Some Reflections on the Prosentential Theory of Truth', in M. J. Dunn, and A. Gupta (eds) *Truth and Consequences* (Dordrecht: Kluwer), 19–32.

von Wright, G. H. (1984) 'Truth and Logic', in *Philosophical Papers*, vol. III (Oxford: Blackwell), 26–51.

—— (1986) 'Truth, Negation, and Contradiction', *Synthese*, 66, 3–14.

—— (1996) 'Truth-Logics', *Six Essays in Philosophical Logic*, 60 (Helsinki: Societas Philosophica Fennica), 71–91.

9
Formal Concepts

Kevin Mulligan

9.1 Introduction

It may seem that there is a big difference between concepts such as the concepts of something, of logical consequence and of disjunction, on the one hand, and concepts such as the concepts of yellow, of emotion or of inflation, on the other hand. Attempts to specify or understand the difference between (some) formal or logical concepts (properties, objects, expressions) and non-formal or non-logical concepts (properties, objects, expressions) loom large in Austrian philosophy. Bolzano, Husserl and Wittgenstein all took the project seriously.

Perhaps the two most important Polish contributions to this Austrian project are Tarski's (1986) invariance-theoretic account of the logical constants and Stanisław Jaśkowski's (1934) discovery of the method of natural deductions from arbitrary suppositions. The independent discovery of this method in the same year by the German logician Gerhard Gentzen led the latter to make the following doubly hesitant claim:

> The introductions are so to speak the "definitions" of the relevant signs and the eliminations are in the end only consequences thereof. Gentzen (1969, p. 14).

Gentzen's claim stands at the beginning of many influential research programs. But scepticism has often been expressed about the possibility of distinguishing between what is and is not logical. In the course of distinguishing between propositions which are logically analytic and propositions which are analytic but not logically analytic Bolzano writes of the distinction between concepts which do and do not belong to logic:

This distinction has something unstable (*Schwankendes*) about it, for the domain of concepts which belong to logic is not so sharply delimited as to rule out controversy. (Bolzano, 1929, §148 (3)).

Tarski writes in very similar terms in 1936:

Underlying our whole construction [of the concept of logical consequence] is the division of all terms of the language discussed into logical and extra-logical. This division is certainly not quite arbitrary [...] On the other hand, no objective grounds are known to me which permit us to draw a sharp boundary between the two groups of terms.[1] (Tarski, 1956, pp. 418–19).

In what follows I provide a brief sketch of the different ways in which Husserl and Wittgenstein sought to understand the differences between formal and non-formal concepts, in particular of their use of the notion of an operation to understand form, logical and arithmetical.

9.2 Formal vs. non-formal

'My fundamental thought', Wittgenstein says, 'is that the "logical constants" do not represent' (TLP 4.0312). The thought had been formulated, but not endorsed, by Husserl:

The *a* and the *the*, the *and* and the or, the *if* and the *then*, the *all* and the *none*, the *something* and the *nothing*, the *forms of quantity* and the *determinations of number* etc. – all these are meaningful propositional elements (*Satzelemente*), but we should look in vain for their objective correlates (if such may be ascribed to them at all) in the sphere of real objects [...].[2] (Husserl, LU VI §43) [underlining by KM].

Constants cannot represent logical objects, Wittgenstein thinks, because 'there are no "logical objects"'.[3] (TLP 4.441)

9.2.1 Lists

Wittgenstein may have arrived in Cambridge with views about logical form in his rucksack. In a letter dated 26 January 1912, Russell writes that Wittgenstein had proposed 'a definition of logical *form* as opposed to logical *matter*'.[4] In the *Tractatus*[5] he distinguishes between formal concepts and concepts proper (*eigentliche Begriffe*) and says that he introduces the former expression in order to make clear the confusion of formal concepts with concepts proper, a confusion "which pervades all of the old logic" (TLP 4.126). The confusion was one Husserl took himself to have avoided 30 years earlier.

One can quite correctly call the concepts *something, one, multiplicity* and *cardinal number* form-concepts (*Formbegriffe*). They are general and the poorest in content of all concepts. They are not concepts of contents of a definite kind but in a certain way comprehend every and each content [...] There are other similar relation concepts such as *identity* and *difference* [...] (Husserl, 1970, pp. 84–5).

Ten years later, in his *Logical Investigations*,[6] Husserl distinguishes sharply between formal (pure) and material (*sachhaltige*) concepts or categories, and between two types of formal concepts, logical and ontological (objectual, *gegenständlich*) concepts or categories. His examples of material concepts are the concepts *house, tree, colour, tone, space, sensation, emotion* (LU III §11). As examples of 'categorical forms in statements' which are 'meaningful propositional elements (*Satzelemente*)', he mentions *being* as well as

[t]he *a* and the *the*, the *and* and the *or*, the *if* and the *then*, the *all* and the *none*, the *something* and the *nothing*, the *forms of quantity* (*Quantitätsformen*) and the *determinations of number* etc. (LU VI §43).

Amongst the meaning categories are

the concepts of the elementary forms of connection (*Verknüpfungsformen*), [...] e.g. the conjunctive, disjunctive, hypothetical connection of propositions to [make] new propositions. Further, the forms of combination (*Verbindung*) of lower (-order) meaning elements to [make] simple propositions [...]; the different subject-forms, predicate-forms, the forms of conjunctive and disjunctive combination, the plural form [...]. (LU P §67).

One of his lists of formal, non-logical concepts is introduced as follows:

In [...] connection with [...] the *categories of meaning* [...] stand other correlative concepts such as Object, State of Affairs, Unity, Plurality, Number, Relation, Connection (*Verknüpfung*) etc. [which are] the pure, the *formal object categories* (LU P §67).

Another list of "formal-ontological categories" is

Something or one (*Eins*), object, attribute (*Beschaffenheit*), connection, plurality (*Mehrheit*), cardinal number, order, ordinal number (*Ordnungszahl*), whole, part, magnitude (*Grösse*) (LU III §11).

According to Wittgenstein, the concepts of object (thing (*Ding*), entity (*Sache*)), complex, fact, function and number are formal concepts (TLP

4.1272). And since a fact is the obtaining or non-obtaining of a state of affairs we may perhaps ascribe to him the view that the concept of a state of affairs is a formal concept.[7] The concept of member of a formal series is a formal concept (TLP 4.1273). The propositional sign is a fact (TLP 3.14). So perhaps the concept of a propositional sign is also formal.

It is striking that the formal concepts explicitly mentioned as such by Wittgenstein are what Husserl calls formal ontological concepts. They seem to agree that the concepts of object, complex, fact, number, series, and state of affairs are formal concepts.

Wittgenstein refers not only to formal concepts and concepts proper but also to pseudo-concepts. The concept of object is a pseudo-concept (TLP 4.1272). Husserl, on the other hand, thinks that both formal and material concepts are concepts. Neither Husserl nor Wittgenstein ask whether there are concepts which are neither formal nor material.

In one of his lists of formal categories Husserl mentions the category of facts:

> [T]he pure truths of logic are all the ideal laws which have their whole foundation in the 'sense', the 'essence' or the 'content', of the concepts of Truth, Proposition, Object, Property, Connection, Law, Fact (*Tatsache*) etc. (LU P §40).

'Fact' here probably refers not to the category of obtaining states of affairs but to the much narrower category of contingently obtaining states of affairs, to matters of fact, a category Husserl often refers to by using Hume's English expression. Husserl refers to the category of proposition as a 'categorial idea' (LU IV §13) and says:

> I understand [...] by pure logical laws all the ideal laws which are grounded in the sense (in the 'essence', 'content') of concepts [...] propositions [...]. (LU P §37).

Within the family of formal-ontological categories, one category, Husserl says, is fundamental: they are 'all grouped around the empty idea of object or something as such'. The category of proposition plays a similar role in the family of formal-logical categories. Logic, Husserl and Wittgenstein agree, stands on its own:

> Pure logic [...] does not presuppose [anything]. (Husserl, 1979, p. 232).
>
> Logic must take care of itself. (TLP 5.473).

And the starting point of logic, Husserl says, is the proposition:

[...] a notion so fundamental as that of an (ideal) propositional meaning, the ultimate point of unity to which all things logical must be traced back [...] (LU V §40).

Husserl could, then, have said, with Wittgenstein (1979, 22.1.15): 'My *whole* task consists in explaining the nature of the proposition.' But although Husserl says, as we have seen, that the concept of object is the most basic formal ontological concept he also says that, just as

the general theory of the *forms of propositional meanings* includes the theory of the *forms of all meanings* [...], so too the general theory of the *pure forms of states of affairs* includes the theory of *all objective categorial forms*. (LU VI §63).

For

[...] there is no categorical form which could not become a compo-nent (*Bestandstück*) of the form of a state of affairs. (Husserl, 1901, LU VI §63).

Wittgenstein thinks that

It is essential to things that they can be constituents (*Bestandteil*) of a state of affairs. (TLP 2.011).

Does Wittgenstein, too, allow for a *Sachverhaltsform*? Apparently:

The way objects hang together in the state of affairs is the structure of the state of affairs (TLP 2.032). Form is the possibility of structure. (TLP 2.033).

Similarly, the 'general form of the proposition is: Things stand thus and so (*Es verhält sich so und so*)'. (TLP 4.5) As Husserl puts it: 'If I assert something, I think of things, and that things stand thus and so (*dass sich die Sachen so und so verhalten*): this is what I express, and perhaps also know'. (LU VI §67) But if Husserl can agree with the second claim made by Wittgenstein when he says:

The proposition [Satz] *shows* how things stand (*wie es sich verhält*), *if* it is true. And it *says*, *that* they do so stand (TLP 4.022),

he rightly rejects claims of the following sort:

The simplest proposition, the elementary proposition, asserts the obtaining of a state of affairs. (TLP 4.21).

He argues as follows:

> We note [...] that just as a nominal meaning by itself says nothing to the effect that its object exists, does not predicate this, so too a proposition does not predicate that its state of affairs is a really obtaining state of affairs. This would lead to an infinite regress. If we say "S is P", then we say about S that it is P, but we do not say that the state of affairs that S is P really obtains; otherwise the last statement would in turn state that the state of affairs that the state of affairs that S is P really obtains really obtains, and so on *in infinitum*. (Husserl, 2001, 96, cf. 91).

Husserl's distinction between propositions, which are true or false, and states of affairs, which obtain or do not obtain, develops gradually and becomes more consistent over the course of his *Investigations*. He says at one point: '[...] a certain state of affairs obtains or does not obtain'. (LU P §6) We have noted his use of 'fact' to refer to contingently obtaining states of affairs. But he also seems to use the term 'fact' with a wider meaning, to refer to 'the objective state of affairs, the fact' (LU V §33, cf. V §36), to contingent and non-contingent facts. He does not, as far as I can see, define facts in terms of obtaining or non-obtaining states of affairs or in terms of the obtaining or non-obtaining of states of affairs. A year after the publication of Husserl's *Investigations*, another Austrian philosopher, Meinong, argued, very plausibly, that Husserl's term 'Sachverhalt' should be rejected because of its connotation of factuality. Meinong's alternative, technical term is: 'objective'. 'An objective, which obtains', he says, 'is also called [a] "fact"'. (Meinong, 1977, p. 69)

One of the most striking applications of the formal concept of fact by Wittgenstein is the opening of the *Tractatus*:

> The world is all that is the case. The world is the totality (*Gesamtheit*) of facts, not of things. The world is determined by the facts, and by their being *all* the facts [...]. What is the case, the fact, is the obtaining of states of affairs. (TLP 1–2).

Some 20 years earlier Husserl had asserted:

> The world is nothing more than the unified objective totality (*gesamte*) corresponding to, and inseparable from, the ideal system of all factual truth. (LU P §36(6)).

A 'factual truth' is a contingent truth. To each such truth there corresponds a contingent state of affairs. But by 1913 Husserl had either changed his mind or forgotten his own good idea: 'The world is the totality of objects that can be known through experience [...]'.[8] (Husserl 1977, §1, p. 46)

Form, variables and operations, Husserl and Wittgenstein believe, belong together. 'The general propositional form is a variable', says Wittgenstein (TLP 4.53); 'every variable is the sign of a formal concept' (TLP 4.1271) and 'the variable name "x" is the proper sign of the pseudo-concept *object*' (TLP 4.1272). This is because

> every variable represents a constant form, which all its values possess and which can be taken to be a formal property of these values. (TLP 4.1272).

According to Husserl some 20 years earlier:

> The task of an accomplished science of meanings would be to [...] track down the primitive meaning-patterns (*Bedeutungsgestalten*) and their inner structures and, in connection with this, to establish the pure categories of meaning which circumscribe the sense and extension of the indeterminates (*Unbestimmten*) – the variables, in a sense quite close to that of mathematics – that occur in such laws.[9] (Husserl, LU IV §13).

In their discussions of the relations between variables and form they employ overlapping terminologies which have not found favour with later logicians and philosophers. Thus Husserl's reflections on Bolzano's method of varying the components of propositions led him to oppose formalization to materialization (deformalization, particularization) and generalization to specialization:

> Formalisation consists in [...] replacing material (*sachhaltige*) determinations (*Bestimmungen*) by indeterminate determinations which are taken to be unlimited variables. (LU III §12)

> we sharply distinguish material concepts or propositions, from merely formal concepts of propositions, which are free of all [...] matter. (LU III §11)

> the so to speak materialisation (*Materialisierung*) of this form [*this S is p*], its particularisation (*Besonderung*) in definite propositions, is possible in infinitely many ways.[10] (LU IV §10)

In 1913 Husserl sets out a terminology designed to capture the difference between the 'extensions' of what is formal and of what is material:

> One must sharply distinguish the relations of generalisation (*Generalisierung*) and specialisation (*Spezialisierung*) from the essentially

different relations of [formalisation], the universalisation (*Verallge-meinering*) of what is material into the formal-logical, or conversely, the materialisation (*Versachlichung*) of the logically formal. Gener-alisation is something totally different from formalisation [...] and specialisation is something totally different from deformalisation (*Entformalisierung*). (Husserl, 1977, §13).

Similarly, in 1914, Wittgenstein notes:

If the completely general proposition is not completely demateri-alised (*entmaterialisiert*), then a proposition does not get demate-rialised at all through generalisation, as I used to think.[11] (1979, 23.10.14).

In 1932 Wittgenstein is said to gloss what looks very like Bolzano's method of variation with the help of a distinction very like Husserl's,

The expression of form is the result of transforming the constant parts of the proposition into variable parts. This transformation into variables is quite different from generalisation (*Verallgemeinern*).[12] (Wittgenstein, 1984, p. 224).

'Form', he says, 'is no generalisation and no property common to a class of propositions'. (Wittgenstein, 1984, p. 220) 'The property fx is a gen-eralisation of the property fx. gx. Generalising leads from one property to another'. (1984, p. 224)

9.2.2   Nodality

If formal concepts are not concepts proper, then they cannot occupy nodes in trees, whether these are genus-species trees (nouns), determinable-determinate trees (adjectives) or even verbal trees. In the *Tractatus* we are told:

In logic there is no side by side (*Nebeneinander*), there can be no classification.

In logic, there cannot be the more general and the more special (*Allgemeineres und Spezielleres*). (TLP 5. 454).

Around 1932 Wittgenstein is said to have repeated these claims: 'Forms have nothing to do with generality. A form is neither general nor spe-cial.' 'A form cannot fall under another form (there is superordination and subordination only for concepts).' (1984, p. 225) These claims were rejected, Wittgenstein says, by Frege and Russell: 'The whole Frege-Russell logic is based on the confusion of concept and form.' (1984,

p. 224) The general point of view rejected by Wittgenstein is formulated very clearly by Husserl in 1913 and earlier. Husserl agrees with Wittgenstein that formal concepts (essences, species) do not occupy nodes in trees in the way peculiar to material concepts. But they do, he thinks, occupy nodes in trees of a distinct type. Forms *are* more or less general and there *is* subordination and superordination in the case of forms:

> Each essence, whether material or empty (thus, purely logical) ['formal', says Husserl in an annotation], has its place in a hierarchy of essences, in a hierarchy of generality (*Generalität*) and specificity (*Spezialität*). Descending, we arrive at the *infimae species* or, as we also say, the *eidetic singularities*; ascending through the specific and generic essences we arrive at a *highest genus* [...] In this sense, in the purely logical domain of significations (*Bedeutungen*) the highest genus is "signification as such", each determinate propositional form and each determinate form of a propositional member (*Satzgliedform*), is an eidetic singularity; the proposition as such is an intermediate genus. In the same fashion, cardinal number as such is a highest genus. Two, three, etc. are its *infimae species* or eidetic singularities. In the material sphere, thing as such, sensory quality, spatial form and experience as such, for example, are highest genera; the essential make-up (*Wesensbestände*) belonging to a determinate thing, sensory quality, spatial form or experience is a material, eidetic singularity. (Husserl, 1977, §12).

Husserl's first example in this passage of a formal tree is combined with his Platonic view of propositions and thus will not be allowed by a philosopher who, like Wittgenstein, does not allow such creatures. But one may well think that, on any view of propositions and meanings, the category of meaning divides into the categories of proposition and of non-propositional parts of propositions. Husserl's second example of a formal tree, the view that cardinal number is a genus of which 2 and 3 are lowest differences is to be found in his *Philosophy of Arithmetic* and in his *Investigations* (LU P §46). It is a view Wittgenstein specifically rejects: 'Numbers are not concepts. One does not obtain them by generalisation.' (1984, p. 224.) An early example of logical classification given by Husserl is this:

> The form of inference represents a class concept under which falls an infinite manifold of propositional connections all with the constitution with which this form sharply stamps them. (LU P §7, cf. §70).

Husserl thinks that even Bolzano failed to see the difference between material and formal trees. Does Husserl's claim that material trees and formal trees differ in kind involve the claim that the relations of height in the two trees differ or only that the terms of these relations differ in kind? In 1929 he asserts that there are two kinds of relation of height:

> Mathematicians [...] generalise (*verallgemeinern*) mathematically mathematical objects, particularise (*besondern*) generalities (*Allgemeinheiten*); not by following the differentiations of species and genus in the sense of the Aristotelian tradition, which are here meaningless, but in conformity with the superordinations and subordinations that the domain of the formal presents.[13] (Husserl, 1974, §30).

But he does not give any justification for this claim.

9.2.3   Formality vs. Internality

Throughout his *Investigations* Husserl regularly distinguishes but always *en passant* inner and outer relations and properties.[14] Wittgenstein refers to many internal relations and to inner relations, such as inner similarity (4.0141) or connections (5.1311) and to external relations. Husserl does not identify the formal and internal nor does he identify the material and the external. Parthood is a formal relation but, in the most basic cases, is not any internal relation. Wittgenstein seems to think that internality is or entails formality.

One internal relation of great interest to our philosophers is described by Husserl as follows:

> It is not an accidental fact that a propositional thought (*Satzgedanke*), here and now, fits a given state of affairs. The relation (*Verhältnis*) concerns the identical propositional meaning and the identical state of affairs (*Sachverhalt*). (LU P §51).

Wittgenstein talks of 'the [...] internal relation which obtains between language and world' (TLP 4.014) and asserts that

> The proposition [...] must [...] hang essentially together (*zusammenhängen*) with the situation (*Sachlage*). (TLP 4.03).

As Husserl had put the point in 1894:

> *Satz* and *Sachverhalt* [...] stand in an objective, law-governed connection (*Zusammenhang*) [...] they are ideally related to one another and inseparable from one another. (1979, p. 341).

What Wittgenstein and Husserl call the form of an object (TLP 2.0141) and the object's matter (Husserl) or content (Wittgenstein) are approached in radically different ways. Wittgenstein seems to think that the form of an object is or comprehends its internal properties (TLP 2.0231, 2.024, 2.025) and asserts that "space, time and colour (colouring, colouredness) are forms of objects" (TLP 2.0251). According to Husserl, as we have seen, spatiality and the different qualities are paradigm material properties.

Their distinct accounts of formality and internality derive in part from the differences between their essentialisms. Husserl's is heavy,[15] Wittgenstein's lite.[16] Wittgenstein's essentialism, unlike Husserl's, does not explicitly countenance essences (species) nor a fortiori instances thereof. (One interpretation of Wittgenstein's simple objects does, it is true, provide a possible counter-example to this claim). As we have already noted, Husserl says

> I understand [...] by pure logical laws all the ideal laws which are grounded in the sense (in the 'essence', 'content') of the concepts truth, proposition, object, attribute, relation, connection, law, fact etc. (LU P §37).

Similarly, he claims

> From the formal Urcategory 'object' further formal categories such as state of affairs, attribute (*Beschaffenheit*), relation, connection, multiplicity or set, series, cardinal number, series, ordinal number, magnitude etc. follow. (Husserl, 1988, p. 9).

> [T]he general logical laws divide into several groups: into laws which have their roots in the concept of proposition, into laws which have their roots in the concept of concept, into laws which have their roots in the concept of object. (Husserl, 2001a, p. 241).

Amongst the many things which, according to Husserl, are explained by or grounded in the essences or natures of objects or concepts are operations and the laws governing operations:

> basic propositional forms [...] the primitive types of complex propositional unity, conjunctive, disjunctive and hypothetical, the distinctions between universality and particularity, on the one hand, and singularity, on the other hand, the syntaxes of plurality, negation and modality, the results of compounding [...] these according to operational laws (*Operationsgesetze*) – [they] are all rooted in the essence of meanings. (LU IV §14)

> the operational laws of compounding meanings are grounded in the primitive meaning structures, the primitive types of members and connections. (LU IV §14).

[There are] operations which are *grounded* in the idea of cardinal numbers. (Husserl 1970a, p. 435)

9.2.4   Form, Operations and Rules

In 1900 Husserl argued not only that the deontic propositions which express logical rules or norms – One ought not to judge that *p* and that not-*p* – must be sharply distinguished from the propositions of logic, since the latter are purely theoretical in nature, but also that the former are partially grounded in the latter (LU P §§13–16). From his *Philosophy of Arithmetic* on he regularly returns to the relation between the rule-followings he calls operations, on the one hand, and formal concepts, expressions, objects and properties, on the other hand.

If Husserl and Wittgenstein are to be believed, form and operations are inseparable. As Husserl says in 1929, summing up over 30 years work on the matter, 'the concept of an operation' is 'the guiding concept in the investigation of forms'. (1974, §13(c)) The fundamental feature of operations to which Husserl (from 1891 on) and Wittgenstein attach so much importance is formulated by Husserl as follows:

> Every construction by operations (*operative Gestaltung*) of a form from another form has its law, and in the case of genuine operations this is such that the result can be subjected to the same operation again. (1974 §13(c)).

The distinction between logical rules or norms concerning what one may (and may not) think or assert, and logical theses, differs from the distinction between rules for using signs and theses. The two distinctions are special cases of the distinction between theses and rules. In 1900 Husserl introduces the second distinction, as follows:

> [There is a] remarkable duplication of all concepts of pure mathematics, and in particular those of arithmetic, such that the general arithmetical signs which at first stand for the correlated number-concepts, with which definition has connected them, subsequently function as pure operational signs, that is, as signs whose meaning (*Bedeutung*) is exclusively determined by the external operation forms (*Operationsformen*), each sign counting as a mere something-or-other (*Irgendetwas*) with which one may operate in this or that way, in these definite forms, on paper. (LU P §54).

Husserl also refers in 1900 to a 'much favoured comparison of mathematical operations to rule-governed games e.g. chess', a comparison which was to become even more favoured throughout the twentieth century, particularly by those under the influence of the claim by Gentzen quoted above:

> Bits of ivory and wood become chessmen through the game's rules which give them their fixed game-meaning (*Spielbedeutung*). And so arithmetical signs have, beside their original meaning, their so to say game-meaning [...] Signs taken in a certain operational meaning or game-meaning do duty for the same signs together with their arithmetical meaning.[17] (LU I §20).

Two fuller formulations of this point are:

> Calculatory signs (*Rechenzeichen*) or arithmetical signs as algorithmic signs have their meaning (*Bedeutung*) (connotation in Mill's sense) exclusively in the rules of connection, separation, replacement, in short in the operational rules which together make the algorithm an algorithm. It is in the capacity to be subject to different rules in such connections, the permission to replace connections of one definite sort by connections of another definite sort, and the prohibition against connections and replacements which contradict the rules that the general meaning lies which gives the signs their general meaning [...]. Meaning is then here constituted exclusively by completely external prescriptions [...]. (Husserl, 1979, p. 393, cf. p. 8)

> [T]o every principle corresponds a certain rule for operating with the signs, and every derived proposition is obtained by mere step-wise subsumption under these rules. Thus there is a one-to-one corresponding parallelism between the game system and its rules, on the one hand, and the number-system and its laws, on the other hand. Hence there is no mechanically-symbolically derivable proposition that does not have its counterpart in the domain of arithmetic. (Husserl, 2001a, p. 311).

Frege's view of the relation between theses and rules resembles Husserl's:

> Let us seek to make the essence of formal arithmetic more clear. How does it differ from a mere game? [...] If one were to go back to the meanings, then the rules would find in these meanings their justification (*Begründung*). (Frege, 1998 II, §90; cf. §91).

Both Husserl and Frege argue, against formalist theories of arithmetic, that sense or meaning understood as constituted by rules or prescriptions is *grounded*, at least in part, in meaning which is not so constituted. But 'meaning' in the passages from Husserl just quoted does not mean what 'meaning' means in the passage from Frege. Frege's 'meaning' is the semantic value of the sense of expressions. Husserl's 'meaning' corresponds to what Frege and Husserl call 'sense'.

The relation between arithmetical operations and the rules these instantiate, on the one hand, and arithmetical theses, on the other hand, resembles, argues Husserl, the relation between another type of operation and logical theses. Under the heading 'The laws of the compounding of meanings and the pure logico-grammatical theory of forms' Husserl writes, in a passage from which we have already quoted:

> The task of an accomplished science of meanings would be to investigate the law-governed, essence-bound structure of meanings and the laws of connection [...] which are grounded in this structure and to reduce such laws to the least number of independent elementary laws. In order to do this it would of course be necessary to track down the primitive meaning-patterns (*Bedeutungsgestalten*) and their inner structures and [...] to establish the pure categories of meaning which circumscribe the sense and extension of the indeterminates (*Unbestimmten*) – the variables, in a sense quite close to that of mathematics – that occur in such laws. What formal laws of connection (*Verknüpfungsgesetze*) may achieve can be made fairly plain by arithmetic. There are definite forms of synthesis, through which, quite generally or in certain definite conditions, two numbers give rise to new numbers. The 'direct operations' $a + b$, ab, a^b etc. yield unrestrictedly numbers as results, the 'inverse operations', $a - b$, a/b ... yield numbers as results only under certain conditions. That this is the case must be laid down by an *existential proposition* or rather an existential *law*, and perhaps proved from certain primitive axioms [...] [I]t is clear that there are similar laws governing the existence or non-existence of meanings which obtain in the meaning sphere and that in these laws the meanings are not free variables but are restricted to the extension of different categories which are grounded in the nature of the sphere of meaning. (LU IV §13).

Husserl goes on to identify a family of operations which might be called *sense operations* since their task is to fix the boundary between sense and nonsense (*Unsinn*). One type of such operations concerns the combining (*Verbindung*) of sub-propositional meanings to make either other sub-propositional meanings or propositional meanings, the other the connections between propositions. One of his examples of the latter is:

> Any two propositions, when connected in the form *M and N*, yield another proposition [...] To any two propositions, *M*, *N*, there

belong, likewise, the primitive connective forms, *If M, then N, M or N,*
so that the result is again a proposition. (LU IV §13).

Husserl's sense operations differ from Wittgenstein's truth operations,
although each type of operation is supposed to illuminate the notion
of logical form. According to Wittgenstein, the structures of propositions stand in internal relations to one another (TLP 5.2). As we have
seen, structure is to form as actuality is to possibility. 'We can bring
out these internal relations in our mode of expression', says Wittgenstein, 'by presenting a proposition as the result of an operation which
produces it from other propositions (the bases of the operation' (TLP
5.21). An 'operation is what must happen to a proposition in order to
make another proposition out of it' (TLP 5.23). 'And that will of course
depend on their formal properties, on the internal similarity of their
forms.' (TLP 5.231) 'The internal relation which orders a series is equivalent to the operation by which a member arises out of another member'
(TLP 5.232). It is not clear to me just what Wittgenstein took the relation between this account of the link between operations and logical
form and his account of truth-operations to be. The main claims of the
latter account are:

> The truth-functions of elementary propositions are the results of
> operations which have the elementary propositions as bases (I call
> these operations truth-operations). (TLP 5.234)

> The truth-operation is the way in which a truth-function arises from
> elementary propositions. (TLP 5.3).

Wittgenstein, like Husserl, notes that there are internal relations both
between numbers and between propositions:

> Series which are ordered by (durch) *internal* relations I call series of
> forms (formal series, *Formenreihen*).

> The number series is not ordered according to an external but
> according to an internal relation.

> Likewise the series of propositions "aRb", "(Ex):aRx.xRb", "(Ex,y):aRx.
> xRy.yRb" etc [...]. (TLP 4.1252, cf. 5.131).

And, as we have seen, thinks that each such internal relation is
"equivalent" to an operation.

In 1891 Husserl described the internal relations between numbers in
the language of his heavy essentialism:

> Whether we consider ordinal numbers or cardinal numbers (both
> terms taken in their genuine sense) we always recognize that the

ordering of the series (*Reihenordnung*) is grounded in the nature of these concepts. In the case of the cardinal numbers, for example, the ordering principle consists in the fact that every successive number in the series is one more (*um Eins mehr*) than the preceding number. (Husserl, 1970a, p. 176).

We now understand some of the background to the claims by our two Austrians:

Number is the exponent (*Exponent*) of an operation. (TLP 6.021).

Numbers are measures of operations (*Operationsetalons*) in a definite domain of operations [...]. (Husserl, 1970a, p. 475).

But perhaps the most surprising twist in our tale is Wittgenstein's exploration in the early 1930s of the idea of inferring from and according to suppositions.[18]

Notes

1. Tarski also notes that 'the division of terms into logical and extra-logical [...] plays an essential part in clarifying the concept analytical' (Tarski, 1956, p. 419). He also writes: 'sometimes it seems to me to be convenient to include mathematical terms, like the ε-relation, in the class of logical ones, and sometimes I prefer to restrict myself to terms of "elementary logic"'. (Tarski, 1987, p. 29)
2. Cf. Meinong (1977, p. 404). On concepts which have no object, cf. (Bolzano, 1929 §78 n. 2). Cf. Mulligan (2004).
3. On logical objects, cf. Bolzano (1929, §223).
4. Russell's letter is quoted by Monk (1990, p. 41).
5. References to the *Tractatus* (Wittgenstein, 1998) are given in the text in brackets which contain 'TLP' followed by a section number.
6. References to the *Logical Investigations* (Husserl 1975; 1984; 1984a) are given in the text in brackets containing 'LU' followed by the number of the *Investigation* followed by the relevant section. 'P' refers to the Prolegomena to the *Investigations* (Husserl, 1975).
7. Cf. Favrholdt (1964, p. 122).
8. Cf. Mulligan (2009).
9. Husserl seems to use 'term', 'variable' and 'indeterminate' interchangeably even when the variables are propositional. (Cf. LU, IV §13)
10. Cf. LU, III §24; LU P §8, §§67–72.
11. Cf. 17.12.14. On logical vs. material relations, cf. Ramsey (1931, p. 146). On formal vs. material cf. Johnson (1922, pp. 138–45).
12. What Husserl had called 'Generalisieren' is here called 'Verallgemeinern' by Wittgenstein.
13. Husserl uses here 'Verallgemeinern' to refer to what he had earlier called 'Generalisierung'.
14. For example, LU VI §4, §24 §48, §50, §59.

15. Cf. Mulligan (2004).
16. Cf. Wittgenstein (1979, p. 98), TLP 2.011, 2.0123, 3.1, 3.1431, 4.016, 5.47, 5.471, 6.1232.
17. Cf. Husserl (2001, p. 30f).
18. (Wittgenstein, 2003, p. 208). On Husserl on operations, cf. Centrone (2010). On Wittgenstein on operations, cf. Marion (1998), ch. 2, Marion (2000) – My sketch of the relations between what Husserl and Wittgenstein say about formal concepts omits several points of comparison: their views about the relation between formality and exactness, materiality and vagueness (cf. Husserl, LU P§21), between formality and simplicity or indefinability, between formal concepts and second-order concepts and marks; the relation between Husserl's account of the way in which predicative logical form determines a class of propositions and of the contrast between this kind of generality and that of the quantifiers (LU II §16(c), II §42 (c)) and Wittgenstein's account of the way propositional functions determine a class of propositions (TLP 5.522–3). – The category of an 'arbitrary supposition' (Jaśkowski, 1967) or 'Annahme' (Gentzen) received its first philosophical elucidation in the accounts of suppositions or assumptions (*Annahme*) given by Husserl in his *Logical Investigations* and, a year later, by Meinong in *Über Annahmen*. In the latter work Meinong indeed introduces the idea of an inference from an assumption (*Annahmeschluss*) and wonders about the relation between this and judgements on the basis of an assumption. Similarities between Bolzano's account of 'because', his theory of the relation he calls *Abfolge*, and Gentzen's account of natural deduction have often been noted. It is therefore of some interest that the logician and philosopher Paul Hertz, who is known to have influenced Gentzen, was familiar with the philosophy of Bolzano, his theory of *Abfolge*, with Meinong's work and the theory of psychological suppositions (cf. Hertz, 1923; 1931; 1935, in particular 234–5).

References

Bolzano, B. (1929) (1837) *Wissenschaftslehre*, 4 vols., in J. E. von Seidel, Sulzbach, A. Höfler, and W. Schultz (eds) (Leipzig: Meiner).

Centrone, St. (2010) *Logic and Philosophy of Mathematics in the Early Husserl* (Dordrecht: Springer).

Favrholdt, D. (1964) *An Interpretation and Critique of Wittgenstein's Tractatus*, (Copenhagen: Munksgaard).

Frege, G. (1998) *Grundgesetze der Arithmetik I/II* (Hildesheim: Georg Olms Verlag).

Gentzen, G. (1969) *Untersuchungen über das logische Schliessen* (Darmstadt: Wissenschaftliche Buchgesellschaft). (1934/1935 *Mathematische Zeitschrift*, 39, 176–210, 405–31).

Hertz, P. (1923) *Über das Denken und seine Beziehung zur Anschauung* (Berlin).

—— (1931) 'Vom Wesen des Logischen, insbesondere der Bedeutung des modus barbara', *Erkenntnis*, 2, 369–92.

—— (1935) 'Über das Wesen der Logik und der logischen Urteilsformen', *Abhandlungen der Friesschen Schule*, 6, 225–72.

Husserl, E. (1970) *Logical Investigations*, 2 vols., English translation of Husserl (1975; 1984; 1984a), (London: Routledge & Kegan Paul).

—— (1970a) *Philosophie der Arithmetik. Mit ergänzenden Texten (1890–1901)*, Husserliana XII, L. Eley (ed.) (The Hague: Nijhoff).

—— (1974) *Formale und transzendentale Logik. Versuch einer Kritik der logischen Vernunft. Mit ergänzenden Texten*, Husserliana XVII, P. Janssen (ed.) (The Hague: Nijhoff).

—— (1975) *Logische Untersuchungen*, Vol. I, Husserliana XVIII (The Hague: Nijhoff).

—— (1977) *Ideen zu einer reinen Phänomenologie und phänomenologischen Philosophie. Erstes Buch: Allgemeine Einführung in die reine Phänomenologie*, Husserliana III-1, K. Schuhmann (ed.) (The Hague: Nijhoff).

—— (1979) *Aufsätze und Rezensionen (1890–1910)*, Husserliana XXII, B. Rang (ed.) (The Hague: Nijhoff).

—— (1984) *Logische Untersuchungen*, Vol. II, Part 1, Husserliana XIX/1 (The Hague: Nijhoff).

—— (1984a) *Logische Untersuchungen*, Vol. II, Part 2, Husserliana XIX/2 (The Hague: Nijhoff).

—— (1988) *Vorlesungen über Ethik und Wertlehre 1908–1914*, Husserliana XXVIII, U. Melle (ed.) (Dordrecht: Kluwer).

—— (2001) *Logik. Vorlesung 1902/03*, Husserliana, Materialienband II, E. Schuhmann (ed.) (Dordrecht: Kluwer).

—— (2001a) *Logik. Vorlesung 1896*, Husserliana, Materialienband I, E. Schuhmann (ed.) (Dordrecht: Kluwer).

Jaśkowski, S. (1967) 'On the Rules of Supposition in Formal Logic', in S. McCall (ed.) *Polish Logic 1920–1939* (Oxford), 232–58, (1934 *Studia Logica*, I, 5–32).

Johnson, W. E. (1922) *Logic, Part II* (Cambridge: Cambridge University Press).

Marion, M. (1998) *Wittgenstein, Finitism and the Foundations of Mathematics* (Oxford: Clarendon Press).

—— (2000) 'Operations and Numbers in the *Tractatus*', *Wittgenstein Studies*, 2, 105–23.

Meinong, A. (1977) (1902) *Über Annahmen*, Gesamtausgabe, Vol. IV (Graz: Akademische Druck- u Verlagsanstalt).

Monk, R. (1990) *Ludwig Wittgenstein. The Duty of Genius* (New York: The Free Press).

Mulligan, K. (2004) 'Essence and Modality. The Quintessence of Husserl's Theory', in M. Siebel, and M. Textor (eds) *Semantik und Ontologie. Beiträge zur philosophischen Forschung* (Frankfurt: Ontos Verlag), 387–418.

—— (2009) 'Tractarian Beginnings and Endings. Worlds, Values, Facts and Subjects', in G. Primiero, and S. Rahman (eds) *Acts of Knowledge: History, Philosophy and Logic. Essays Dedicated to Göran Sundholm* (London: College Publications, Tribute series), 151–68.

Ramsey, F. (1931) 'Facts and Propositions', *The Foundations of Mathematics and other Logical Essays* (London: Kegan Paul), 138–55.

Tarski, A. (1956) 'On the Concept of Logical Consequence', *Logic, Semantics, and Meta-Mathematics* (Oxford: Clarendon Press), 409–20.

—— (1986) 'What are Logical Notions?', *History and Philosophy of Logic*, 7, 143–54. (Transcript of a 1966 talk, J. Corcoran (ed.))

—— (1987) 'A Philosophical Letter of Alfred Tarski', *Journal of Philosophy*, 84, 28–32. (Letter to Morton White, written in 1944.)

Wittgenstein, L. (1979) *Notebooks 1914–1916* (Oxford: Blackwell).

—— (1984) *Wittgenstein und der Wiener Kreis* (Frankfurt: Suhrkamp).

—— (1998) *Logisch-philosophische Abhandlung. Tractatus logico-philosophicus, Kritische Edition*, in B. McGuinness, and J. Schulte (eds) (Frankfurt: Suhrkamp).

—— (2003) *The Voices of Wittgenstein. The Vienna Circle, Ludwig Wittgenstein and Friedrich Waismann*, in G. Baker (ed.) (London: Routledge).

Part III

Ontology, Mereology and the Philosophy of Mathematics

10
Arithmetic in Leśniewski's Ontology

Peter Simons

For Jan

10.1 Introduction

One of the most entertaining aspects of friendship with Jan Woleński is the opportunity to learn from and relish his almost inexhaustible supply of anecdotes about the work, lives and foibles of the figures of Polish logic and philosophy, and it has been my pleasure and privilege to have enjoyed this entertainment for many years, since I first met Jan. He was at that time working on what was to become the definitive history of the Lwów-Warsaw School, first in Polish, and later, with a modicum of linguistic guidance from myself, in English.[1] In many ways the most fascinating of the many striking figures of that group of brilliant thinkers was Stanisław Leśniewski (1886–1939), a man whose many idiosyncrasies, both doctrinal and personal, make him the most entertaining object of all of Jan's anecdotes. Many's the time Jan and I have met again at a conference and he has said, 'I have a new story about Big Stan', that appellation being our private name for the great man.[2]

My knowledge of and interest in Leśniewski however predates my acquaintance with Jan by some years. As a graduate student in Manchester in the 1970s I knew and occasionally heard lectures by two admirers of Big Stan. One was Desmond Paul Henry, whose wonderful book *Medieval Logic and Metaphysics*[3] first made me aware of Leśniewski's ontology and mereology. Desmond, for whose introductory course I gave tutorials, relished applying Leśniewskian concepts to *outré* (his word) disputes of medieval logic and metaphysics. The other was the person who had first introduced Henry to Leśniewski, namely Czesław Lejewski. Czesław had studied with Leśniewski in Warsaw before the

227

war and having settled in England after it had dedicated himself to elucidating and continuing Leśniewski's work in a similar vein, an activity which earned him relatively scant respect among his colleagues and students, though it had impressed his erstwhile colleague and predecessor as professor in Manchester, Arthur Prior. Between the pair of them, Henry and Lejewski unwittingly set me off on what became my *Habilitationsprojekt*, namely the study of mereology.

Jan Woleński and I shared more than a merely antiquarian interest in Leśniewski however. Like myself, Jan also considered that Leśniewski had been somewhat poorly portrayed in subsequent literature, both by Leśniewski's often ignorant detractors and by his often equally uncritical admirers. We both take a somewhat more detached view than the latter, enabling us, as we think, to give a fair assessment of his important contributions. I am concerned to exhibit some of the striking advantages of doing more things Leśniewski's way even today, and this paper falls into that category.

10.2   Leśniewski on arithmetic

Leśniewski considered his life's work to be that of providing an acceptable antinomy-free foundation for mathematics, and his series of papers *O podstawach matematyki* advertises as much by its very title. One would therefore expect him to have considered the foundations of the arithmetic of natural numbers. And indeed Leśniewski gave courses of lectures in Warsaw with the following promising titles: 'Foundations of Arithmetic' (1920–3); 'Primitive Terms of Arithmetic' (1928/9); and 'Inductive Definitions' (1933/4). Of these, student lecture notes of the second and third have survived and were published in English translation in 1988.[4] The first, which has the most promising title, appears to have been irrevocably lost. On examination of the two extant series however, the principal reaction may be one of disappointment. In the last series it is shown how definitions by induction over the natural numbers can be converted into explicit definitions. This is unexceptionable. The second series however is deeply disappointing, in the sense that it uses very little of the resources of Leśniewski's own most important logical system, ontology. It consists simply in a formulation in Leśniewski's language of Peano's ideas about the natural numbers, taking these to be individuals of a certain kind, the natural numbers. Here are the axioms:

LP1 $1 \; \varepsilon \; \mathrm{nat}$

LP2 $A \; \varepsilon \; \mathrm{nat} \rightarrow Sq(A) \; \varepsilon \; \mathrm{nat}$

LP3 $A \; \varepsilon \; \text{nat} \rightarrow \sim (Sq(A) = 1)$

LP4 $A \; \varepsilon \; \text{nat} \wedge B \; \varepsilon \; \text{nat} \wedge Sq(A) = Sq(B) \rightarrow A = B$

LP5 $1 \; \varepsilon \; a \wedge \forall B^\lceil B \; \varepsilon \; \text{nat} \wedge B \; \varepsilon \; a \rightarrow Sq(B) \; \varepsilon \; a^\rceil \wedge A \; \varepsilon \; \text{nat} . \rightarrow A \; \varepsilon \; a$

LP6 $A \; \varepsilon \; \text{nat} \rightarrow A + 1 = Sq(A)$

LP7 $A \; \varepsilon \; \text{nat} \wedge B \; \varepsilon \; \text{nat} \rightarrow A + (B + 1) = (A + B) + 1$

LP8 $A \; \varepsilon \; \text{nat} \rightarrow A \times 1 = A$

LP9 $A \; \varepsilon \; \text{nat} \wedge B \; \varepsilon \; \text{nat} \rightarrow A \times (B + 1) = (A \times B) + A$

LP10 $A \; \varepsilon \; \text{nat} \wedge B \; \varepsilon \; \text{nat} \wedge A > B \rightarrow \exists C^\lceil C \; \varepsilon \; \text{nat} \wedge A = B + C^\rceil$

LP11 $A \; \varepsilon \; \text{nat} \wedge B \; \varepsilon \; \text{nat} \rightarrow A + B > A$

LP12 $A \; \varepsilon \; Sq(B) \rightarrow B \; \varepsilon \; \text{nat}$

Here 'nat' is a common name for (just) the natural numbers, '*Sq*' is a term for the successor function, and everything else is straightfoward. For those unfamiliar with Leśniewski's formalism let me however mention a few facts. Firstly, 'ε' is the singular inclusion functor, a sort of singular copula, best read as 'is one of (the)', so '$A \; \varepsilon \; \text{nat}$' means that A is one of the natural numbers, or, more briefly, that A is a natural number. The upper corners are quantifier scope delimiters, and Leśniewski employs the informal convention that upper-case letters (in variables like 'A' or functor expressions like 'Sq') signify individuals, while lower-case letters do not necessarily stand only for one individual. Hence LP5 is the principle of mathematical induction, since the variable 'a' ranges over not just single numbers but classes, multiplicities, or, as I shall say, *multitudes* of numbers. Having such plurally referring terms available enables Leśniewski to present his version of Peano as a first-order theory, quantifying only nominal variables, and without sets. Predicate quantification is not needed. Two final points should be mentioned. Firstly, the quantifiers do not *per se* carry existential or ontological import. This is now well enough known and I shall not belabour it. Secondly, in full dress, Leśniewski's axioms have no free variables, but all otherwise free variables are universally quantified with maximally wide scope. To avoid a sea of written initial universal quantifiers, I have elided these: they should be supplied in thought.

As ever, Leśniewski's axiomatization and treatment are logically exquisite and little more needs to be said. Note also that the addition and multiplication functors are in effect defined inductively, as they are in Peano. The later course provides the material for replacing this implicit inductive 'definition' by an explicit definition, using Frege's ancestral. The disappointment however centres on the treatment of the natural numbers as individuals, a view which not only runs counter to Leśniewski's own nominalism, but, more importantly, because, unlike Frege's treatment of arithmetic, it makes no attempt to connect these individuals to cardinality properties. There is simply no attempt to link say the number 3 with the logical property of being three-membered, or something similar. Frege, while regarding the natural numbers as (abstract) individuals, deliberately sets out to link them with concepts' having so and so many individuals falling under them. This is why he defines the number 3 as the class whose members are all and only three-membered classes, a class being for Frege the extension of a concept. Russell independently arrived at the same conception. In Leśniewski on the other hand, the members of nat simply form a sequence, generated by successive applications of the successor function. Any link to counting or cardinality is purely external, and for all the axioms tell us, nat could be any simple inductive sequence. No doubt some structuralists would be happy with this, but it would be very out of character for Leśniewski to be satisfied by it.

10.3   How to do arithmetic in ontology

Here is then how Leśniewski *should* have treated arithmetic – and for all we know might have treated it in the earlier and now lost lectures. Leśniewski's ontology, because it admits names that are not singular, but may be empty or plural, allows the numerical constants to figure as first-order predicates, whereas in Frege and Russell, where all names are singular, they are fundamentally second-order predicates. If the objects *a* are three in number, we can simply say this of them: 3(*a*). The key idea is expressible roughly by saying that the numbers form a certain kind of property, which we may call cardinality properties. A cardinality property n applies to objects *a* – we write 'n(*a*)' – if and only if there are precisely n objects *a*, where this can be defined logically. Describing the objects *a* as a 'multitude',[5] their number or cardinality is a non-distributive property of the multitude, meaning that it applies to or holds of the objects jointly but not severally, and indeed it applies to no proper submultitudes of the objects *a*, unless the *a* are infinite.

So here are a string of Leśniewski-style definitions which provide us with the relevant vocabulary. The treatment here is a sketch, and for the meanings of the symbols from Leśniewski's ontology, in my notation, I must for brevity refer elsewhere.[6] I will simply say that '*a ε b*' is true if and only if there exists exactly one *a*, and it is one of the one or more objects b, that 'E(a)' is true if and only if there exists at least one a, and 'E!(a)' is true if and only if there is exactly one a.

Definitions take the form of stipulated equivalences: we write ':: $\leftrightarrow$' to indicate that the equivalence of the left-hand side (*definiendum*) with the right (*definiens*) is stipulated. All constants are stipulated in a sentential context. We start by defining a necessarily empty or non-denoting name 'Λ':

$$A \, \varepsilon \, \Lambda ::\leftrightarrow A \, \varepsilon \, A \wedge \sim (A \, \varepsilon \, A)$$

then existence, non-existence, uniqueness and singularity predicates:

$$E(a) ::\leftrightarrow \exists A \ulcorner A \, \varepsilon \, a \urcorner$$

$$N(a) ::\leftrightarrow \, \sim \exists A \ulcorner A \, \varepsilon \, a \urcorner$$

$$!(a) ::\leftrightarrow \forall BC \ulcorner B \, \varepsilon \, a \wedge C \, \varepsilon \, a \to B \, \varepsilon \, C \urcorner$$

$$E!(A) ::\leftrightarrow E(A) \wedge !(A)$$

then inclusion (all a are b) and exclusion (no a are b):

$$a \subset b ::\leftrightarrow \forall A \ulcorner A \, \varepsilon \, a \to A \, \varepsilon \, b \urcorner$$

$$a | b ::\leftrightarrow \, \sim \exists A \ulcorner A \, \varepsilon \, a \wedge A \, \varepsilon \, b \urcorner$$

and general and singular identity:

$$a \equiv b ::\leftrightarrow \forall A \ulcorner A \, \varepsilon \, a \leftrightarrow A \, \varepsilon \, b \urcorner$$

$$A = B ::\leftrightarrow A \, \varepsilon \, B \wedge B \, \varepsilon \, A$$

then union, difference, and disjoint union:

$$A \, \varepsilon \, a \cup b ::\leftrightarrow A \, \varepsilon \, a \vee A \, \varepsilon \, b$$

$$A \, \varepsilon \, a - b ::\leftrightarrow A \, \varepsilon \, a \wedge \sim (A \, \varepsilon \, b)$$

$$A \, \varepsilon \, a \oplus b ::\leftrightarrow a | b \wedge (A \, \varepsilon \, a \vee A \, \varepsilon \, b)$$

We now define, in Cantor's fashion, equicardinality of two multitudes in terms of the existence of a one-one correlation between them.

$$Rel(R) ::\leftrightarrow \forall AB \ulcorner R(AB) \to A \, \varepsilon \, A \wedge B \, \varepsilon \, B \urcorner$$

that is, $\forall AB^\ulcorner R(AB) \rightarrow E!(A) \wedge E!(B)^\urcorner$

$OneOne(R) ::\leftrightarrow Rel(R) \wedge \forall ABCD^\ulcorner((R(AB) \wedge R(CB))$

$\rightarrow A = C) \wedge ((R(AB) \wedge R(AD)) \rightarrow B = D)^\urcorner$

$a\ eqcb ::\leftrightarrow \exists R^\ulcorner OneOne(R) \wedge \forall A^\ulcorner A\varepsilon a$

$\rightarrow \exists B^\ulcorner B\varepsilon b \wedge R(AB)^\urcorner^\urcorner \wedge \forall D^\ulcorner D\varepsilon b \rightarrow \exists C^\ulcorner C\varepsilon a \wedge R(CD)^\urcorner^\urcorner$

We say now that a one-place predicate is a cardinality predicate, or for short, a *cardinality*, when any multitudes falling under it are equicardinal:

$Card(f) ::\leftrightarrow \exists a^\ulcorner f(a)^\urcorner \wedge \forall a^\ulcorner f(a) \rightarrow \forall b^\ulcorner f(b) \leftrightarrow beqca^\urcorner^\urcorner$

We now define a successor functor applying to predicates (not necessarily just cardinalities):

$succ(f)(a) \leftrightarrow \exists Ab^\ulcorner E!(A) \wedge a \equiv A \oplus b \wedge f(b)^\urcorner$

This simply says that succ(f) applies to a if and only if f applies to a submultitude of a containing one member fewer. So for example, in addition to numerical predicates, suppose 'are located in my office' applies to a multitude consisting of a cupboard, a cup and a computer. Then 'succ(are located in my office)' applies almost trivially to the four-multitude consisting of these plus Barack Obama. It is not a practically interesting notion but it allows us to state as a more interesting and non-trivial theorem:

$Card(f) \leftrightarrow Card(succ(f))$

Note in passing that if f is a predicate true of infinitely many things but not every last thing, it is logically equivalent to succ(f).

Now let's introduce some numbers, or numerical predicates. The first two are trivial relabellings:

$0(a) ::\leftrightarrow N(a)$

$1(a) ::\leftrightarrow E!(a)$

It follows quickly that

$0(\Lambda)$

$A\varepsilon A \leftrightarrow 1(A)$

$Card(0)$

$Card(1)$

$1(a) \leftrightarrow succ(0)(a)$

then for a further couple of illustrations we can add

$2(a) \leftrightarrow succ(1)(a)$

$3(a) \leftrightarrow succ(2)(a)$

We can also introduce addition of predicates:

$(f+g)(a) ::\leftrightarrow \exists bc \lceil a \equiv b \oplus c \wedge f(b) \wedge g(c) \rceil$

Again this applies not just to cardinality-predicates: if f is 'is a cat' and g is 'is a dog' then '$f+g$' is 'is a cat or dog' (notice the 'or') and applies to any multitude consisting of at least one cat and/or at least one dog. It follows with little effort that:

$Card(f) \wedge Card(g) \rightarrow Card(f+g)$

The converse of this is not valid because of the 'absorptive' character of infinite cardinalities, but it does hold for finite cardinalities. So we should explain what we mean by 'infinite' and 'finite':

$inf(a) ::\leftrightarrow \exists b \lceil b \subset a \wedge \sim (a \subset b) \wedge b \; eqc \; a \rceil$

$fin(a) ::\leftrightarrow \sim (inf(a))$

So we get a few simple theorems:

$0(a) \rightarrow fin(a)$

$1(a) \rightarrow fin(a)$

But we want to describe *numbers* as finite or infinite, so define

$Inf(f) ::\leftrightarrow Card(f) \wedge \forall a \lceil f(a) \rightarrow inf(a) \rceil$

$Fin(f) ::\leftrightarrow Card(f) \wedge \sim Inf(f)$

and may derive as theorems

$Fin(f) \rightarrow Fin(succ(f))$

$Inf(f) \rightarrow Inf(succ(f))$

We also want to know when cardinalities (indeed all predicates) are "equal" or coextensive:

$$f \approx g ::\leftrightarrow \forall a \lceil f(a) \leftrightarrow g(a) \rceil$$

and then we can derive

$$Fin(f) \rightarrow \sim (f \approx succ(f))$$

$$Inf(f) \rightarrow f \approx succ(f)$$

10.4   Some tactical choices

The sentence 'n is a natural number' may be expressed via a predicate 'Nat' true of the predicate 'n' which is true of divers multitudes, so Nat(n), where n(a) for some a. In that case |Nat|, the semantic category of 'Nat', is $S\langle S\langle N \rangle \rangle$. Alternatively, we may introduce a new higher-order copula, a type-crossing 'higher epsilon' $\acute{\varepsilon}$ where $f \acute{\varepsilon} G$ is definable as

$$f \acute{\varepsilon} G ::\leftrightarrow \exists a \lceil f(a) \wedge G(f) \rceil$$

so that $|\acute{\varepsilon}| = S\langle S\langle N \rangle S\langle S\langle N \rangle \rangle \rangle$, then we will be able to say things like

2 $\acute{\varepsilon}$ Nat

There is no additional ontological commitment either way, but the second way is more reminiscent of the first-order Peanian form of expression we find in standard set theory and in Leśniewski's lecture notes.

So how are we to define the natural numbers as some of the cardinalities? The answer is basically Frege's Way: where R is a relation define the ancestral *R in the usual way (not given here) and use the relation folg defined as

$$m \text{ folg } n ::\leftrightarrow Card(m) \wedge Card(n) \wedge m \approx succ(n)$$

so

$$Nat(n) ::\leftrightarrow n \approx 0 \vee n^* \text{folg } 0$$

so we are treating natural numbers as those cardinals which are in the ancestral of succession to 0.

There are two standard definitions of the infinite:

(1) b are infinite iff a submultiple of b are equicardinal with the natural numbers

(2) some proper submultitude of *b* are equicardinal with *b* (Dedekind).

We opted for Dedekind's definition above. It is known that (2) ⊢ (1) and not (1) ⊢ (2), but (1) + Axiom of Choice ⊢ (2). So one question in the Leśniewskian framework is what to do about the Axiom of Choice. My own view is that the Axiom of Choice is a logical truth, and that we need have no scruples about adding it to the theses we hold to be true. A more delicate question is how to express it. My own preference is not available here, since it would involve using multitudes of higher order, something that Leśniewski does not acknowledge. But within the idiom we have adopted, let *f* be any predicate true only of non-empty multitudes. The Axiom of Choice says there is a function *F* such that for each *a* such that *f*(*a*), *F*(*a*) is an individual such that *F*(*a*)ε*a*:

AC $\quad \exists a^{\lceil} f(a)^{\rceil} \wedge \forall b^{\lceil} f(b) \to E(b)^{\rceil} \to \exists F^{\lceil} \forall a^{\lceil} f(a) \to F(a)\,\varepsilon\,a^{\rceil\rceil}$

With this in place, we can be satisfied that either choice of definition of the infinite is acceptable. However Nat is true not of multitudes but of predicates of multitudes, certain cardinality predicates. We can happily *say* that 0έNat, 1έNat, 2έNat, etc., and indeed

n $\acute{\varepsilon}$ Nat → succ(n) $\acute{\varepsilon}$ Nat

but in the vocabulary we have to date we cannot say that Nat is infinite, but have to 'raise' the notion of the infinite to the next higher type, adopt an Axiom of Choice of the next higher type, and so on. We are here trapped by the type-theoretic nature of Leśniewski's logical language. I say this while fully aware that Leśniewski did not subscribe to an *ontology* of types (apart from individuals), but he did retain the simple hierarchy of types of expressions that we find in Frege and later in Church, minus the associated platonistic ontology, and this still restricts our way of speaking. I will come back to this limitation at the end.

10.5 Ontological neutrality, infinity and logicism

Leśniewski's logic (LL) has no ontological implications. It is ontologically neutral, in that no existentially committing thesis follows from it: in particular, we do *not* have that

LL = Protothetic + Ontology ⊢ $\exists A a^{\lceil} A\,\varepsilon\,a^{\rceil}$

(there is some individual): defining

$$A\,\varepsilon\,V ::\leftrightarrow \exists a^\lceil A\,\varepsilon\,a^\rceil$$

(*A* is a thing iff *A* is something), we do *not* have

$$LL \vdash E(V)$$

We *do* have

$$LL \vdash \exists a^\lceil 0(a)^\rceil \quad (\text{since } \vdash 0(\wedge))$$

but *not* $LL \vdash \exists a^\lceil 1(a)^\rceil$

So we do *not* have as a theorem of Leśniewski's logic, together with the definitions we have added, that every natural number has a successor, since if there are more than m objects, the would-be natural number m is not a cardinality, since $Card(m) \to \exists a^\lceil m(a)^\rceil$. And if there are exactly m objects (m finite), that is, m(V), then for no *a* is succ(m)(*a*).

The first five Peano axioms, or rather their higher-order analogues in this language, are expressible as follows:

HP1 $0\,\acute{\varepsilon}\,Nat$

HP2 $n\,\acute{\varepsilon}\,Nat \to succ(n)\,\acute{\varepsilon}\,Nat$

HP3 $n\,\acute{\varepsilon}\,Nat \to \sim(succ(n) \approx 0)$

HP4 $n\,\acute{\varepsilon}\,Nat \wedge m\,\acute{\varepsilon}\,Nat \wedge succ(n) \approx succ(m) \to n \approx m$

HP5 $0\,\acute{\varepsilon}\,Nat \wedge \forall n^\lceil n\,\acute{\varepsilon}\,Nat \wedge n\,\acute{\varepsilon}\,G \to succ(n)\,\acute{\varepsilon}\,G^\rceil \wedge m\,\acute{\varepsilon}\,Nat. \to m\,\acute{\varepsilon}\,G$

I shall call the conjunction of these five 'PA'. Of these, only HP1 and HP3 are theses of LL, precisely because of the ontological neutrality of the logic. So:

$$LL \nvdash PA$$

However, if we adopt as an additional assumption Whitehead and Russell's Axiom of Infinity

AI   Inf(V)

then

$$LL + AI \vdash PA$$

However

$$LL + AI \nvdash PA1$$

where PA1 is Peano arithmetic formulated as a *first-order* theory about numbers as objects (individuals), as in Leśniewski's lectures, modulo the replacement of a 0-individual by a 1-individual in Leśniewski. To get from PA as we have it to PA1 would require us to interpolate a theory of abstraction to get us from cardinals *qua* predicates to cardinals *qua* singular terms (as in Frege, set theory, ...), and that is a temptation which can and probably should be resisted, for a number of reasons, into which I shall not go here. And in addition, we have no *a priori* reason to believe that AI is true. From this it follows that logicism, seen through Leśniewskian eyes, is simply false, something which he doubtless knew and of which he doubtless approved. The nearest we can get is to be able to assert the logical implication

$$AI \rightarrow PA$$

This does not advance us far beyond things as they stood in 1910 with the publication of Whitehead and Russell's *Principia Mathematica*, apart from the welcome replacement of ramified type theory by simple type theory, which eliminates worries about their Axiom of Reducibility. It may be that Leśniewski was satisfied enough with this not to follow through with a publication about arithmetic, or maybe he would have done so had he survived for longer: we don't know. But should we be satisfied? In the final section I will indicate why not.

10.6 Signposting the way ahead

There are two reasons to be *highly* satisfied with an approach to arithmetic via Leśniewski's logic along the lines we have indicated, or something similar. The first is that the ontological neutrality of logic is preserved. It is *not* a weakness of Leśniewski's approach that not all the assumptions of Peano arithmetic are logically true, but precisely a strength. Secondly, whereas nearly all modern treatments of arithmetic are based on some version of set theory, Leśniewski avoided sets as if they were the plague, and any nominalist must applaud his hard-headedness. Further, the kind of mathematics that is needed for the pursuit of natural science is going to be conditionally forthcoming from AI as a postulate, so with this reservation Leśniewski can be said to have done enough to show that arithmetic, and its extensions into analysis, are in no great danger as theoretical and applicable disciplines. However, the reliance on AI is irritating and worrying, since it may indeed be more probable that AI is actually false than that it is true. Current cosmology is tending towards the hypothesis that the universe is finite not only in extent but in the cardinality of its individuals. If we can weaken the assumptions while still remaining nominalistic, that would be welcome.

Consider how we teach small children arithmetic. First we teach them to count, reciting the number terms and then using these to enumerate groups of objects. Then we teach addition. The key fact about addition in application, not just as a pure symbolic exercise, is that it is used to give the overall number of two groups of objects which are disjoint, i.e. do not share an object. If I have 7 dogs and 5 cats, how many animals is that? Repeated addition repeats the procedure: if I have three groups of animals, 7 dogs, 5 cats and 8 mice, how many animals is that? Notice that we can count not just the individual animals but also the groups: two in the first case, three in the second. This comes into its own when we explain multiplication. Suppose I have six disjoint groups of animals, each consisting of five animals: how many animals do I have altogether? And suppose I have three of these groups of groups, again disjoint, then that gives $3 \times 6 \times 5 = 90$ animals altogether. And so on.

In the explanation of multiplication, we use one number word for groups of things, another for groups of groups of things, and perhaps more. Yet there is no obvious shift in meaning of the number words. The only difference is in the nature of what is counted: in one case individuals, in another groups of individuals, in a third groups of groups of individuals, and so on. This is not in principle any different from the fact that we can use the same number words for enumerating apples, pears, dogs, chickens, and so on. But if we look at Leśniewski's ontology, which has been our framework to date, we find it cannot represent the use of number terms in this simple way. If we say that a group a has five individuals, $5(a)$, we use a different constant from the one we use if we say there are are five groups of five individuals. This is forced on us by the language: the second 'five' has to be made a predicate of groups, and not a predicate of individuals. The situation changes again if we want to say there are five groups of five groups of five. In each case the word 'five' designates a different constant. So while Leśniewski's logical language does not lack the resources to count items at different levels, it forces us to do so by ascending the type-hierarchy, ruining the apparent univocity of the numerical terms and distorting the simple way we teach and learn arithmetic. Can we do better?

Yes we can. Here is a brief indication of how. It has to be brief, because to give a full treatment would involve nothing less than setting out a new logical language, going beyond that of Leśniewski. This would both go beyond our brief here, and if carried out in full would require a much longer paper setting out a much more expressively powerful logical system, comparable in power (though not in ontology) to a system of set theory. So the merest sketch will have to suffice here.

Leśniewski's ontology (the logical system) goes beyond standard Frege-Russell style predicate logic because it is more liberal with its names. Whereas standard predicate logic has only singular names, to which free logic adds empty names, ontology in addition allows plural names: names of several individuals. But it stops there. In a sense it doesn't take multitudes or groups (the objective counterparts of plurals) fully seriously, since a plural term is just one term for many individuals. We can have predicates true of multitudes, and with our numerical predicates we have seen how this allows us to bring arithmetic down a level by comparison with its standard (non-set-theoretic) treatment, where numerical terms are quantifiers, or predicates of predicates. And in Leśniewski's logic we have to use analogous numerical constants from ever higher types to continue the story. However, the illustration from kindergarten arithmetic shows that we naturally use the very same numerical terms for groups as for individuals, and indeed for groups of any desired level or order.

In order to facilitate this, we need to radically extend the names available. Not just names for individuals, and names for groups of individuals, but names for groups of groups of individuals, and so on, should be admitted. With this we open up a hierarchy of levels of name, and correlatively, a hierarchy of different orders of group or multitude. These may be naturally understood as being built up from below by individuals. If A, B and C are three individuals, then on their basis we have three groups of two individuals, writing them (by an obvious notation) AB, BC and CA, and one group of all three, ABC. These are four different groups or multitudes, since multitudes are the same if and only if they have the same members. Consider the two multitudes AB and BC. They are two despite sharing one individual, B. So now consider the different pair of multitudes BC and CA. These are a different pair, despite sharing the one group BC with the first. So the pairs of pairs are different: the pair AB|BC is different from the pair BC|CA, and both of these again different from AB|CA. And again each of these three pairs of pairs is different from the trio of pairs AB|BC|CA. At each level we unproblematically continue to count and distinguish groups, and there is no obvious logico-grammatical gear-change to different types when we do so. Hence it is just as legitimate to write '3(AB|BC|CA)' as it is to write '3(ABC)': what is different lies not in the predicate but in that to which it applies: first-order or second-order groups.

Once we acknowledge this, we need a radically new language, since we need a sense of membership in which AB is a member of AB|BC in the same way that A is a member of AB, and Leśniewski's 'ε' does not permit this.

The resulting language will inevitably have a lot in common with standard versions of set theory, though what principles it will need to fulfil is not yet fully settled. One alternative for example might be to mimic Zermelo's cumulative hierarchy without a universal multitude; another would be to stratify comprehension principles after the fashion of Quine's NF (New Foundations). Either way – and we are not deciding the matter here – the system of multitudes will needs differ from sets in three ways. Firstly, there exists no empty multitude. We will still have the empty *name* 'Λ', but by definition it does not stand for anything. Secondly, there can be no question of a distinction – at any level – between a multitude and the 'singleton' of just that multitude, as there is in standard set theory between a and {a} generally, and in Quine's NF for non-urelements. Finally – and this is perhaps the most contentious thought – admitting multitudes of higher order no more impugns nominalism than Leśniewski's admittance of first-order multitudes as subjects of predicates. It all comes down in the end to individuals for one thing, and for another, any multitude of particulars is itself a particular, albeit not an individual.

While the final form of such a theory is not yet certain, it holds out a number of promises. The first is that it will allow a type-free account of arithmetic that is still beyond Leśniewski. The second is that the principles of Peano arithmetic in this system will have a much weaker existential requirement for their truth: by dint of the ability to ramify multitudes up to any finite order, no more than two individuals are required to kick-start an infinite hierarchy and so ensure that every finite cardinality predicate is satisfied. This is not to give up on Leśniewskian ontological neutrality, since it remains a non-logical fact that even one individual exists, but we can be assured that in any universe we care about, in particular our own, Peano arithmetic will be true. And finally, the endless resources of such a system of multitudes holds out the promise of doing something that, following Quine, has been universally assumed to be impossible, namely to provide a nominalistically acceptable formal semantics for predicate logic, of first or higher order. These are heady prospects, and tasks for another time. But it all starts from the first step of Leśniewski having the courage and foresight to retain plurally referring names when all around him the logical world was rejecting them.

Notes

1. Woleński (1985; 1989).
2. I am pleased to say that some other historians of logic such as our mutual friend Göran Sundholm have adopted our informal moniker. (There is also a 'Little Stan', namely Stanisław Jaśkowski.)
3. Henry (1972).
4. Srzednicki and Stachniak (eds) (1988).
5. Instead of 'multitude' I will sometimes for euphony say 'group'. The terms are to be understood here as synonymous: in particular the latter is not to be thought of as in mathematical group theory.
6. Simons (1987), Chs. 1 and 2.

References

Simons, P. (1987) *Parts* (Oxford: Clarendon).
Srzednicki, J., and Stachniak, Z. (eds) (1988) *S. Leśniewski's Lecture Notes in Logic* (Dordrecht: Kluwer).
Woleński, J. (1985) *Filozoficzna szkoła lwowsko-warszawska* (Warsaw: PWN).
—— (1989) *Logic and Philosophy in the Lvov-Warsaw School* (Dordrecht: Kluwer).

11
Leśniewski, Tarski and the Axioms of Mereology

Arianna Betti

Alongside a respect for philosophically informed formal work and an interest in all things Polish, Jan Woleński and I share a profound admiration for Leśniewski's *oeuvre*. As Jan once told me, you can work on Leśniewski for your whole life. Indeed so. Eighteen years after I first met him, on a morning in late March at a bus stop in Sucha Bezkidzka, Southern Poland, here's a story about the axioms of Leśniewski's mereology, and Tarski's complicated role in it.

This story is for Jan.

11.1 A puzzle

Recently, Iris Loeb and I published a study of a short, frequently quoted paper by Alfred Tarski, *On the Foundations of the Geometry of Solids* (Betti & Loeb, 2012). In this paper, as is well known, Tarski is indebted to Leśniewski's Mereology.[1] While studying the paper, we have discovered a number of puzzling things. For instance, the original French version from 1929 (Tarski, 1929) and the 1956 English translation (done by Joseph Henry Woodger, and checked by Tarski himself and his pupil Richard Montague (Tarski, 1956)) show a number of remarkable differences, among which is this one: the 1956 English translation, but not the French 1929 original, includes the axioms of the system of mereology used by Tarski in his foundations. Why did Tarski include the axioms only in the English translation, but not in the French original? This is the question I will address in this paper.

The easiest answer would be the following: whereas in 1929 Poland Leśniewski's mereology was more or less a matter of common knowledge, in 1956 in the US it was not. Tarski therefore had to include the axioms in question in 1956 to enable understanding of his text, while in 1929 this was not necessary. But there is another, much less easy explanation for the change, and it is this one I will set out in this paper. I conjecture that the matter turned on subtle issues of priority and proper attribution, complicated by the specific non-Leśniewskian form that the axioms in question, if included, would have taken in Tarski's paper. This latter aspect – the non-Leśniewskian form of the axioms – may also be of significance for a broader aspect of research in the history of logic in Poland: namely for understanding Leśniewski's break with Tarski around 1929. In *Polish Axiomatics and its Truth* (Betti, 2008, pp. 51–6) I have indicated four specific facts that may have contributed to the break in Leśniewski and Tarski's professional relationship. Briefly stated, I submit that Tarski's deviant take on the axioms of mereology and issues of priority I discuss in my conjecture should be counted as a fifth fact.

Presenting my conjecture will require some digging into fine details. I should also warn the reader that my account will have a speculative character. We lack foolproof evidence to exclude more simple explanations, such as that Tarski simply may have lacked space to include the axioms in the 1929 volume. Yet historians are accustomed to a lack of foolproof evidence – and in this case the evidence is certainly conjectural, given the nature of the sources I rely upon. In any case, what I have to say will contribute to a more illuminating picture of the relationship between Leśniewski's and Tarski's work in this period. Here we go.

11.2 Tarski's 'simplification' of Leśniewski's axiom system

Two elements of Tarski (1929) are important here. First, Tarski's geometry of solids is not based on Mereology in the strict methodological sense in which a Leśniewskian theory can be said to be 'based' on another. Second, as Lejewski (1983) points out carefully, the axioms of mereology that appear in Tarski (1956) are not, strictly speaking, Leśniewski's axioms. As Betti and Loeb (2012) have argued, both elements of difference depend on the fact that Tarski worked with different frameworks in this paper. In particular, Tarski mixes topological notions, mereological notions and notions from Whitehead and Russell's *Principia*.

In the following I will extend the discussion of Lejewski (1983) and Betti and Loeb (2012) with an account of the historical aspects of the relation between Tarski 1956's axiom system of mereology and the corresponding system originally conceived by Leśniewski. It is relevant to my story that, barring differences and speaking loosely, the 1956 axioms *are* the axiom system for the specific system of Leśniewski's Mereology which Tarski uses in the paper, and yet strictly speaking they are *not*. Also relevant to my story is that Tarski (1929) is the text of a lecture given by Tarski two years earlier, in 1927, at the first conference of Polish mathematicians, and published two years later in the volume of proceedings of that conference.

This is the axiom system Tarski gives in 1956:

(Tarski)

Definition I *An individual X is called a proper part of an individual Y if X is a part of Y and X is not identical with Y.*

Definition II *An individual X is said to be disjoint from an individual Y if no individual Z is part of both X and Y.*

Definition III *An individual X is called a sum of all elements of a class α of individuals if every element of α is a part of X and if no part of X is disjoint from all elements of α.*

Axiom[2] I *If X is a part of Y and Y is a part of Z, then X is a part of Z.*

Axiom II *For every non-empty class α of individuals there exists exactly one individual X which is the sum of all elements of α.* (Tarski, 1956, p. 25)

This axiom system is based on the notion of (improper) *part* as primitive. It appeared for the first time in Appendix E to Woodger's *The Axiomatic Method in Biology* (Woodger, 1937) before appearing in Tarski (1956). On both occasions, Tarski says that the system is a 'simplification' of Leśniewski's (Tarski, 1956, p. 25, n. 1; Tarski, 1937a, p. 161, n. 1).[3] Although this system is not included in (Tarski, 1929), for the time being I will make the assumption that if Tarski had included an axiom system for mereology in that paper, it would have been this very system, or a very similar one: so similar as to make no difference for my purposes. I will return to the problem of giving evidence for this assumption in the following section.

The Leśniewskian axiom system of which Tarski's own is said to be 'a simplification' is the following:

(Leśniewski)

Axiom L1 *if P is an ingredient of Q and it is not the case that Q is P then Q is not an ingredient of P;*

Axiom L2 *if P is an ingredient of Q and Q is an ingredient of R then P is an ingredient of R;*

Axiom L3   *if (every a is an ingredient of P and an ingredient of Q and for all R, if R is an ingredient of P or R is an ingredient of Q then a certain ingredient of R is an ingredient of an a) then P is Q;*

Axiom L4   *if a certain object is an a then for some P ((for all Q, if Q is an a then Q is an ingredient of P) and for all Q, if Q is an ingredient of P then a certain ingredient of Q is an ingredient of a certain a).*

Definition L1   *P is a part of Q if and only if (P is an ingredient of Q and it is not the case that P is the same object as Q);*

Definition L2   *P is the class of as if and only if (P is an object, (for all Q, if Q is an a then Q is an ingredient of P) and for all Q if Q is an ingredient of P then a certain ingredient of Q is an ingredient of a certain a).* (Leśniewski, 1930a, p. 82; Leśniewski, 1991, p. 321 ff.)[4]

This system was constructed by Leśniewski in 1920 but published only in 1930, in the seventh chapter of *On the Foundations of Mathematics* (Leśniewski, 1930a). At the time when Tarski's solids paper was delivered in 1927, therefore, and even when it was published in 1929, Leśniewski's system had existed already for years, but was yet unpublished. One superficial element of terminological difference between (Tarski) and (Leśniewski) is that (Tarski) has *part, proper part, disjoint* and *sum* where (Leśniewski) has *ingredient, part, exterior* and *class*. The following differences are far more important:

(Tarski)	(Leśniewski)
definitions precede axioms	axioms precede definitions
axioms contain defined notions (that is, axiom II contains the notion of *sum* which is defined)	axioms do not contain defined notions
two notions of class are at work concurrently: distributive (*class*) and collective (*sum*). The Russellian notion of *individual* is also present	one notion of class is at work, *collective* class

These differences, especially the third, are strikingly non-Leśniewskian. The fact that Tarski allows axioms to be preceded by definitions and to contain defined notions (the first two differences) has already been identified as a non-Leśniewskian trait in Betti and Loeb (2012). Yet in that paper we did not take into account the Leśniewskian axiom system I am considering here. In this connection, note that Tarski is able to give an axiom system with *two* axioms instead of Leśniewski's *four* axioms. Recall that Tarski says that his system is a simplification of Leśniewski's. Two related things are interesting here. First, Leśniewski's axiom system is redundant, since the first axiom can be deduced from the remaining axioms,[5] whereas Tarski's is not. Redundancy in an axiom system is undesirable by usual standards, and certainly undesirable by Leśniewskian standards (cf. Sobociński, 1956), and turning an axiom system into a non-redundant one is, seemingly, at least one thing one would typically mean by 'simplification'. However, this is not the only meaning of the word: we might say that making an axiom system non-redundant is the minimal meaning of 'simplifying'. For, as noted by Tarski and Givant (1999) in another context (that is, concerning Tarski's own simplifications on axiom systems for geometry!), 'simplifying' may also mean removing defined notions from axioms and letting axioms contain only primitive notions:

> Another distinctive feature of Tarski's system is the formal simplicity of the axioms upon which the development is based. As opposed to Tarski's system, in all the systems of geometry known from the literature, at least some – and sometimes even most – axioms are not formulated directly in terms of primitive notions, but contain also other notions, previously defined. It is evident that the formal complexity of such an axiom set becomes apparent only if the axioms are reformulated exclusively in terms of primitive notions by eliminating all defined one. (Tarski & Givant, 1999, p. 192)

This notwithstanding, in the case of the axioms of mereology the shortening came at the expense of continuing to allow for defined notions in the axiom system. Therefore, Tarski's characterization of his axiom system of mereology as 'a simplification' of Leśniewski's can be meant only in the sense of shortening it by getting rid of redundant axioms. Tarski is explicit about this: 'The simplification consisted in eliminating one of Leśniewski's postulates [axioms, ab] by deriving it from the remaining ones.' (1956, p. 25, n. 2)

Having defined notions in the axioms does not seem to have caused much trouble for Tarski on this occasion. This brings me to the second interesting thing: even barring all other differences, it is doubtful that Leśniewski himself would have considered Tarski's result a simplification *of his own* axiom system since Tarski's axiom system includes, in addition to a primitive notion (improper part/ingredient), a *defined* one (sum/class). Compare Leśniewski's axiom system: here one finds just one primitive notion (improper part/ingredient). *Sum/class* is defined by means of this primitive notion.[6]

We come to the third difference. In the formulation of his axiom system of mereology, Tarski employs both the Russellian notion of distributive class, typical of set theory, and the Leśniewskian notion of collective class, typical of mereology. Moreover, Tarski adopts Russell's theory of *objective* types instead of Leśniewski's *linguistic* types, as is clear by his use of the term *individual*, signalling that he is formulating his axiom system of mereology with a restricted domain of quantification relative to a specific set of Russellian individuals taken as objects of the first type: namely, the values on which nominal variables range (for the crucial role of Russellian individuals in Tarski's paper, see Betti and Loeb, 2012, § 5). Now, first, this is not how quantifiers work in Leśniewski, where they are never restricted (nor is there any talk of values over which variables might be said to range); second, Leśniewski's opposition to set theory and dislike for the Cantorian notion of set – notably apparent in his fierce opposition to the notion of an empty set – is well known, as is that Leśniewski saw mereology as the only acceptable theory of multiplicities. Using, as Tarski does, both set theory and mereology (plus, in fact, topology), along with Russell's type theory – and refusing to use Leśniewski's Ontology as a background theory – places Tarski at a distance from Leśniewskian orthodoxy. Leśniewski's critical attitude towards set theory, and what this attitude meant for his relationship with Tarski, who had devoted himself more and more to set theory after his studies with Leśniewski, has already been discussed in Betti (2008). While writing the latter, I overlooked the hybrid nature of Tarski (1929), that is, Tarski's mixing of frameworks and use of notions from both set theory and mereology. This oversight became apparent when Iris Loeb and I studied it thoroughly. Yet the mix-up of theories in Tarski (1929) is quite relevant for judging Leśniewski's probable assessment of it, for there is, possibly, no other matter on which Leśniewski was so adamant as his opposition to set theory and belief in the unique adequacy of his own system of the foundations of mathematics.

To sum up: Tarski's '1929' axiom system simplifies Leśniewski's 1920 axiom system in the sense of merely reducing the number of axioms, but at the price of including defined notions in the axioms, thereby moving in a different direction with respect to Leśniewski's own results at that time. Moreover, the axioms include notions which were not only foreign to Leśniewski's systems, but derived from a rival system, namely the alternative formal framework of Whitehead and Russell's

Principia, both editions of which, on Leśniewski's assessment, possessed 'shocking defects' (Leśniewski, 1927, pp. 167–8, Eng. trans. Leśniewski, 1991, p. 179). From a Leśniewskian point of view, the extent to which Tarski's axiom system can be said genuinely to simplify Leśniewski's axiom system is doubtful.

11.3   The conjecture

I noted above that Leśniewski's axiom system – as presented in (Leśniewski) above – existed as early as 1920, but was yet unpublished when Tarski's talk was given in 1927 and his paper published in 1929. I also claimed that we can assume that if Tarski had included an axiom system for mereology in his 1929 paper, it would have been that of 1956 – (Tarski) above. But could Tarski have included the 1956 axiom system in the 1929 paper? Three specific issues arise here: first, how do we know that Tarski in 1927 knew Leśniewski's axiom system? Second, what reasons do we have to think that if Tarski had included an axiom system for mereology in the 1929 paper, it would have been exactly that which he included in 1956? Third, had Tarski already found the axiom system in question in 1927?

To answer the first question is easy: Tarski knew Leśniewski's system, if not earlier or by other means then at least from attending Leśniewski's 1921–4 lectures on Euclidean geometry based on mereology (cf. Betti & Loeb, 2012, § 2; Tarski studied with Leśniewski from 1919 to 1923). To answer the second and third questions is more difficult; both answers must be qualified. There is no evidence that Tarski had found *exactly this* axiom system by 1927, so we can't say simply that (Tarski) was a simple integration to the 1956 paper that could have been included in the 1927 paper. But support can be given for the claim that if Tarski had added an axiom system to the 1927 paper, it would have been an axiom system – call it (Tarski*) – *similar in some salient non-Leśniewskian aspects* to (Tarski). This claim carries a bit further the conclusions of Iris Loeb's paper in this volume (Loeb, 2013). Loeb (2013) provides support for two things, to a certain extent. First, including (Tarski*) in the 1956 paper was important for Tarski because of the system's connection with Tarski's results in Boolean algebra, and because of the connections among mereology, Boolean algebra and regular open sets (Loeb, 2013, 12.7). For this, (Tarski*) had at least to apply a (non-Leśniewskian) restriction of the domain to Russellian individuals. The reason for this restriction is the following. Tarski maintains in footnotes to the English translation of Tarski (1935), that (Tarski*) has a model in the family of all regular open sets of a Euclidean space equipped with the relation of set-theoretical inclusion. But this requires that the domain of (Tarski*) is restricted to the family of all regular open sets. The (Russellian) *individuals* are interpreted as (Kuratowski's) regular open sets (without the empty set), *part of* as set-theoretical inclusion, while *sum* is borrowed, crucially, from Leśniewski's mereology (Loeb, 2013, 12.5–7).[7] This structure, Tarski shows, is a Boolean algebra. Secondly, Loeb convincingly suggests that a footnote in Tarski (1937b) was meant by Tarski to show that all elements of the connection between regular open sets and Boolean algebra were already present in the 1927 talk on the geometry of solids, which was, as we know, published later as Tarski (1929) (cf. Loeb, 2013, fn 17.) Tarski's axiom system of mereology – explicitly restricted, in his own formulation and unlike Leśniewski's, to a specific domain of individuals – was fundamental for this connection.

If the above is correct, then it is all the more curious that an axiom system of mereology with the characteristics just noted – (Tarski*) – is absent in Tarski (1929). For I'd venture that, historically, the notion of a domain of Russellian individuals has been instrumental to the development of the connection between regular open sets and Boolean algebra via mereology. But if (Tarski*) was so important, why did Tarski not include it in 1927? Here is where my conjecture comes into play. Briefly stated, the reason why Tarski did not include (Tarski*) in the 1929 paper is that the original axiom system by Leśniewski – (Leśniewski) – was unpublished at that time, and Tarski waited to publish it until Leśniewski had done so first. For imagine what would have happened if Tarski's had included his unorthodox axiom system in the 1929 paper: Leśniewski's Mereology, at a time when it finally had a mature formulation and, in Leśniewski's view, a satisfying axiomatization from the methodological point of view, would have been presented for the first time ever by Leśniewski's own pupil and only student in an utterly deviant fashion. This would not have pleased Leśniewski in the least. When Tarski does finally publish his axiom system in 1937, Leśniewski's axioms had been public already for seven years (not to mention that by that time the relationship between the two had deteriorated). But in 1929, the problem was not only that this particular axiom system by Leśniewski had not been published but that *no* axiomatization of mereology satisfactory by Leśniewski's mature standard was available in print at that time.[8] The axiom system in question – that is, (Leśniewski), which Tarski simplified as described above – was the first that Leśniewski ever set up for a system of mereology based on the notion of ingredient (improper part), and that did not contain defined notions in the axioms (exactly the opposite of Tarski's 'simplification'.[9])

If this conjecture is right, then the story of the missing axiom system in Tarski (1929) links up with another story in an interesting way: namely, with the story of the rather remarkable form in which Leśniewski published his results between 1927 and 1931 in his *On the Foundations of Mathematics* (*O podstawach matematyki*).

11.4   Suum cuique

To understand these matters better, some context is needed. Leśniewski's first axiomatization of mereology (though, to his own mind, a deficient one) is given in Leśniewski (1916). The latter is considered a kind of work of transition between Leśniewski's early works and his period of mature production for which he is best known (when he is known at all): 'Leśniewski's systems'. One striking element here is that Leśniewski published nothing between this first booklet from 1916 and 1927, when *On the Foundations of Mathematics* (*O podstawach matematyki*) started appearing in instalments in *Przegląd filozoficzny* (Leśniewski, 1927; 1928; 1929a; 1930a; 1931). In the meantime, in 1919, Leśniewski had started lecturing in Warsaw. It was in the period from 1916 to 1927 that Leśniewski obtained his major results; if one speaks of Leśniewski's systems, one is basically speaking of his work in that period. This means that, until 1927, Leśniewski's most important results circulated among the Warsaw logicians only as presented by him in class or in discussions (see also *Leśniewski 1* quoted below). With *On the Foundations of Mathematics*, Leśniewski aimed to remedy this situation by publishing all the results he had obtained between 1916 and 1927.

On the Foundations of Mathematics is a curious work. First, it has what Leśniewski himself termed an 'autobiographical-synoptic' character (Leśniewski, 1927, p. 165; Eng. trans. Leśniewski, 1991, p. 175), which explains why it is written in a regimented natural language despite the fact that at that time Leśniewski had devised his own formal symbolism, an extremely precise notation to which he attached great value. Second, the work was supposed to cover the whole system (Leśniewski, 1927, pp. 165–6; Eng. trans. Leśniewski, 1991, pp. 176–7), which should have meant the three systems of Prototetics, Ontology and Mereology, but almost all the results presented in that work are mereological. The work counts eleven chapters and its publication spanned five issues of *Przegląd filozoficzny* over four years (1927–31). Ten of the chapters (composed in 1927–30) are on Mereology. The last chapter (of 1931) is on Ontology. No chapter is devoted to Prototetics.[10]

That *On the Foundations of Mathematics* is a cumbersome, prolix and pedantic work should not much surprise, for this is characteristic of Leśniewski's style. But the choice of an 'autobiographical-synoptic' *genre* is remarkable. For one chooses a certain *genre* over another, one might assume, for specific reasons: what pushed Leśniewski, after nine years of silence, to choose exactly this 'autobiographical-synoptic' *genre* to present his mature results? Was doing so the most effective way to convey the greatest number of results in the shortest time? This could hardly be the case, despite Leśniewski's statement to the contrary. Yet the 'autobiographical-synoptic' strategy was, arguably, the most effective way to make clear *who did what when*.

There is clear evidence for this in a passage from the Introduction:

Leśniewski 1   Just as I reached some of my views and some scientific results under the influence of conversations with my colleagues and in connection with their still unpublished scientific results, so also my views and observations, which I had formulated during a number of years in university lectures and in numerous scientific discussions, have contributed to the formation of certain opinions and results of my colleagues, who, out of an admirable loyalty towards me, have withheld from publication until now a number of their scientific results, until my own related results are published. (Leśniewski, 1927, pp. 164–5; Eng. trans. Leśniewski, 1991, p. 175)

A passage in chapter 5 (Leśniewski, 1929a) also supports this claim, where we read:

Leśniewski 2   All results known to me in the field of my 'general theory of sets' (Mereology, ab) which derive from this period (through 1920, inclusive, ab) were my own work. The situation began to change in this respect only from the year 1921. Putting various theorems and definitions into a separate whole according to the historical criterium of the date of their discovery, I have attempted to achieve here, as in other analogous cases, an easy satisfaction of the requirements of the chronologic principle '*suum cuique*' in presenting my own and other people's scientific contributions. (Leśniewski, 1929a, p. 60, Eng. trans. Leśniewski, 1991, p. 264) [adjusted by me]

So one may easily conjecture that Leśniewski chose the 'autobio-graphical-synoptic' *genre* mainly for reasons of primacy and proper attri-bution, and, I would argue, possibly from a particular concern for Tarski's role ('colleagues [...] have withheld from publication [...] a number of their scientific results, until my own related results are pub-lished'). The fact that the first instalment of this work appeared in 1927 suggests that a major motivation for Leśniewski to finally start publish-ing might have been Tarski: namely, that the latter had started to present work purportedly based on Mereology, but which in fact was neither exactly Mereology nor the kind of work on formal systems in general approved by his teacher.[11]

Let's examine the general structure of Leśniewski's book, and consider the context of some salient passages. The first instal-ment (Leśniewski, 1927) contains an introduction and chapters 1–3. Chapter 4 (Leśniewski, 1928), published in the second instalment of the work, presents the first published axiomatization of mereology from 1916. In chapter 5 (Leśniewski, 1929a), Leśniewski presents his unpub-lished results in mereology from the period between 1916 and 1920. Chapter 5 contains only theorems and definitions. Chapter 6 (1930) contains unpublished results in mereology from 1918, along with an axiomatization: this is the first system of mereology based on *proper part* as primitive; its axiom system contains no defined notions. Chapter 7 (1930) contains unpublished results in mereology from 1920, and also an axiomatization: this is the system based on ingredient (improper part) that concerns us here; the one that, in 1927 and in 1956, Tarski claims to have simplified. Chapter 8 (1930) describes which results exactly were obtained by Tarski and Kuratowski in 1921, and includes only theorems; no new axiomatization is presented. Chapter 9 (1930) contains further theorems from 1921–3 obtained by Leśniewski alone. Chapter 10 (Leśniewski, 1931) contains an axiomatization from 1921 for a system of mereology based on the only primitive *exterior to*. Chapter 11 interrupts the discussion of mereology due to unspecified 'technical and editorial problems', and is devoted instead to Ontology.[12]

Something strange happens in chapter 4. At its very end, Leśniewski adds a discussion of the relation between Mereology and Whitehead's theory of events (Leśniewski, 1928, pp. 286–91; Leśniewski, 1991, pp. 258–63). The issue discussed here was raised by Tarski in 1926, and concerns a possible defect in Whitehead's axiomatization of the theory of events (Whitehead, 1919, pp. 61, 62; pp. 68–81; pp. 101–3) which, as is well known, serves as a foundation of solid geometry. The issue is cen-tral to Tarski (1929). The footnote is evidence that Tarski and Leśniewski had discussed together the relationship between Whitehead's theory and Mereology and the possible defect in question (see also Betti & Loeb, 2012, § 3). The strange thing is that this information is appended here: the discussion in question had taken place in 1926 – a full ten years after Leśniewski discovered the results he is now reporting in his 'autobiographical-synoptic' way. As noted above, it is only in the third instalment, in chapter 5 (Leśniewski, 1929a), that Leśniewski starts to publish the *unpublished* results in mereology between 1916 and 1920. The reason is clear, then, why the report on the Whitehead discussion is not given earlier, for Leśniewski could not have done this before having introduced some basics of Mereology, as he does in chapters 3 and 4. This point of the text is therefore, in fact, the earliest point at which Leśniewski could have reported the discussion with Tarski. There is a link between Whitehead's axiomatization of the theory of events and two theorems from 1916 that immediately precede the footnote (Leśniewski, 1928, p. 288; Leśniewski, 1991, p. 261). The reason why it was in any case very convenient to append the footnote exactly at this point might be that Leśniewski in 1927 wanted to make public the discussion with Tarski as soon as possible, because that discussion concerned issues central to Tarski's talk on the geometry of solids presented that very same year (and which would appear in print two years later). In particular, Leśniewski showed – with very careful wording – that indeed, as Tarski conjectured, Whitehead's axiomatization was not sufficient to derive the two theses that Whitehead claimed were derivable, in this way offering 'steps towards a solution of Tarski's problem'. What Tarski does in his paper is provide a better axiomatization for the geometry of solids – that is, in Whiteheadian terms, events – and do so much more in the spirit of Whitehead than of Leśniewski since, among other differences, Tarski's axiomatization is, like Whitehead's, *atomless* (instead of being neither atomistic nor atomless, like Leśniewski's).[13]

One additional important reason for Leśniewski to include the footnote could very well have been the non-Leśniewskian character of Tarski (1929). As noted, in that paper Tarski mixes Leśniewskian results with results derived from at least two other, different frameworks, including Whitehead and Russell's *Principia Mathematica*. By appending the footnote, Leśniewski made clear first of all that it was not only Tarski who was at work in 1926 on Mereology and Whitehead's theory of events. Second, he takes the opportunity to tell the story from *his* point of view – including *his* criticism of Whitehead's work. For at this point in *On the Foundations of Mathematics*, the context is clear: the preceding chapters make clear that Leśniewski's mereology is very different from set theory, and moreover contain an attack on both the *Principia Mathematica* (chapter 1) and set theory (chapter 3).

11.5   Conclusion

I have advanced a conjecture as to why Tarski did not include an axiom system of mereology in Tarski (1929), but did in Tarski (1956), despite the fact that publishing a (saliently similar) axiom system right away would have been, arguably, important for Tarski's own work. At that time, Leśniewski, creator of the theory and Tarski's teacher, had not yet published the axiom system of mereology of which Tarski says his system is a simplification. Nor had Leśniewski published any other axiomatization of mereology that was, at that time, genuinely acceptable to him, nor was the general formal architecture of the systems available in print – certainly not in the orthodox form Leśniewski favoured. Publishing the axioms on Tarski's part would have created an awkward situation, the more so since Tarski's axiomatization deviates from Leśniewski's in at least three important ways. One of these, the fact that Tarski mixes notions from the set-theoretical apparatus and the type theory of Whitehead and Russell's *Principia* with notions from Leśniewski's systems, was at that time an extremely sensitive point between Leśniewski and Tarski, because Leśniewski was critical of *Principia*, disapproved openly and vehemently of set theory, and was involved in a fight with Tarski's supervisor in mathematics, Sierpiński, on exactly these matters. This suggests that the publication of Tarski (1929), and in general work on his Whitehead-like geometry of solids between 1926 and 1929, might be added as a fifth possible reason to the list of the four that, according to Betti (2008), led to the break between Leśniewski and Tarski around 1928–9.

To support this claim, I have considered the structure and some curious aspects of Leśniewski's major work *On the Foundations of Mathematics* (1927–31), which has an 'autobiographical-synoptic' character and clearly shows an extreme concern for the priority and paternity of results obtained in Mereology during the period when the theory was in full development and, also, Tarski was studying with Leśniewski in Warsaw.

Notes

1. I will write 'Mereology' with a capital 'M' when talking specifically about Leśniewski's system of mereology.

2. Woodger has 'postulate' instead of the 'axiom' of the original (Tarski, 1929, p. 230; Tarski, 1956, p. 25). Tarski was unhappy with this, and I write here 'axiom' (see Betti & Loeb, 2012, n. 16).

3. Here's how the axioms look in (Tarski, 1937a). First Tarski gives the axioms for mereology, which in this Appendix are three:

   1.11 $\mathbf{P}\varepsilon\,\mathrm{Trans}$ (Axiom)

   1.12 $y\Sigma\iota'x.\supset.x=y$ (Axiom)

   1.13 $\alpha \neq \Lambda.\alpha \supset.\vec{\Sigma}\alpha \neq \Lambda$ (Axiom)

   Then Tarski says that the following thesis can be substituted for the second and the third axiom (1.12 and 1.13):

   1.26 $\alpha \neq \Lambda\alpha \supset.E!\Sigma'\alpha$

   The two-axiom system (1.11 and 1.26) – the first axiom fixing the transitivity of parthood and the second fixing at once the uniqueness of composition and universal mereological composition – is the one Tarski presents in 1956. The reason why Tarski adopts the three axioms above as his axiom system in (Tarski, 1937a), is that taking the two theses 1.11 and 1.26 as the two axioms would make 'the first deductions somewhat more complicated'. It is interesting to note that the theoretical setup of Woodger's axiomatic foundations for biology is quite similar to the setup of Tarski's foundations of the geometry of solids: axioms and definitions for specific biological (Woodger) and geometrical (Tarski) notions (*precedes*, *organism*, etc. in Woodger; *sphere* in Tarski) are added to a layer of a formal system mixing the language and the set-theoretical tools from *Principia Mathematica* on the one hand with axioms and definitions from Leśniewski's mereology on the other. These traits are non-Leśniewskian. It can well be argued, I think, that Tarski's non-Leśniewskian way of operating with mereology has made a lasting mark on present-day investigations in metaphysics dealing with mereological issues, and had some awkward consequences for those investigations. The matter would deserve an investigation of its own, but doing so would bring me too far from my aims here.

4. This is the system in the formal notation used in NDJFL (Lejewski, 1983, p. 65), where A, B, C are Axioms L1, L2 and L3, and D and E are Definitions L1 and L2:

   A. $[AB] : A \in el(B).\sim (B \in A).\supset.B \in\sim \big(el(A)\big)$.

   B. $[ABC] : A \in el(B).B \in el(C).\supset.A \in el(C)$.

   C. $[ABCa] :.: C \in a[D] \therefore D \in a.\supset.D \in el(A).D \in el(B) :: [D]$

   $\therefore D \in el(A).\vee.D \in el(B) :\supset.[\exists EF].E \in a.\supset.F \in el(D).F \in el(E) ::\supset.A \in B$.

   D. $[Aa] :: A \in a.\supset\therefore [\exists B] \therefore [C] : C \in a.\supset.C \in el(B) \therefore [C] : C \in el(B).[\exists DE]$.

   $D \in a.E \in el(C).E \in el(D)$.

E. $[AB] : A \in pr(B) . \equiv . A \in A . A \in el(B) . \sim (A \in B)$.

F. $[Aa] :: A \in Kl(a) . \equiv . \therefore A \in A . \therefore [B] : B \in a . \supset . B \in el(A)$

  $\therefore [B] : B \in el(A) . \supset . [\exists CD] . C \in a . D \in el(B) . D \in el(C)$.

5. This is shown in Lejewski (1983, pp. 65–6).
6. For more information on the history of the axioms of mereology see (Sobociński, 1955; LeBlanc, 1983).
7. Thanks to Iris Loeb for discussion on this point.
8. At least, as it seems, if Leśniewski (1929a) (which did contain a proper axiomatization of mereology in this sense, based on *proper part*) was not yet published at that point. If, furthermore, Leśniewski (1929b) wasn't yet published either, then Leśniewski's strict methodological standards were also nowhere to be seen in practice: for the latter paper contains the most informative and formally orthodox and first-ever published presentation of the architecture of one of Leśniewski's systems (Prototethics). In any case, Tarski (1929) quotes neither, while he quotes Leśniewski (1927) and Leśniewski (1928).
9. This might be too quick: for note that whereas, as I have argued, there is some evidence *a posteriori* that (Tarski*) would have had the Russellian elements of (Tarski), it is more difficult to give grounds to say that (Tarski*) would have had, like (Tarski), defined notions in the axioms.
10. In the meantime, however, Leśniewski had published an eighty-page paper on Prototethics in *Fundamenta Mathematicae* (Leśniewski, 1929b) and another on Ontology, (Leśniewski, 1930b) the two very different in character.
11. Note that Tarski (1923), a fundamental contribution to Leśniewski's system, was instead orthodox Leśniewskian work, and there was accordingly no need for Leśniewski to respond by making clear, so to say, who did what when.
12. Although Leśniewski says that he will return once more to the subjects in that field in 'later chapters' (Leśniewski, 1931, p. 153; Leśniewski, 1991, p. 364), no further chapters of *On the Foundations of Mathematics* were published.
13. Leśniewski's footnote is also interesting because it seems to be the only published place where Leśniewski mentions 'the well-known method of "interpretation"' (Leśniewski, 1928, p. 288; Leśniewski, 1991, p. 261) and 'interpretation proofs of independence' (Leśniewski, 1928, p. 291; Leśniewski, 1991, p. 263).

References

Betti, A. (2008) 'Polish Axiomatics and its Truth: On Tarski's Leśniewskian Background and the Ajdukiewicz Connection', in D. Patterson (ed.) *Alfred Tarski: Philosophical Background, Development and Influence* (Oxford: Oxford University Press), 44–72.

Betti, A., and Loeb, I. (2012) 'On Tarski's Foundations of the Geometry of Solids', *Bulletin of Symbolic Logic*, 18, 230–60.

LeBlanc, A. O. V. (1983), *A Study of the Axiomatic Foundation of Mereology*, (University of Manchester: MA thesis).

Lejewski, C. (1983) 'A Note on Leśniewski's Axiom System for the Mereological Notion of Ingredient or Element', *Topoi*, 2, 63–71.

Leśniewski, S. (1916) 'Podstawy ogólnej teorji mnogości. I (Część. Ingredyens. Mnogość. Klasa. Element. Podmnogość. Niektóre ciekawe rodzaje klas)', in *Prace Polskiego Koła Naukowego w Moskwie, Sekcja matematyczno-przyrodnicza* (Moscow).

—— (1927) 'O podstawach matematyki' (Introduction and chapters I–III), *Przegląd Filozoficzny*, XXX, 164–206.

—— (1928) 'O podstawach matematyki' (chapter IV), *Przegląd Filozoficzny*, XXXI, 261–91.

—— (1929a) 'O podstawach matematyki' (chapter V), *Przegląd Filozoficzny*, XXXI, 60–102.

—— (1929b) 'Grundzüge eines neuen Systems der Grundlagen der Mathematik', *Fundamenta Mathematicae*, 14, 1–81.

—— (1930a) 'O podstawach matematyki' (chapters VI–IX), *Przegląd Filozoficzny*, XXXII, 77–105.

—— (1930b) 'Über die Grundlagen der Ontologie', *Sprawozdania z posiedzeń Towarzystwa Naukowego Warszawskiego, Wydział III, Nauk Matematyczno-fizycznych*, (Comptes Rendus des Séances de la Société des Sciences et des Lettres de Varsovie, Classe III, Science Mathématiques et Physiques) 22, 111–32.

—— (1931) 'O podstawach matematyki' (chapters X–XI), *Przegląd Filozoficzny*, XXXIV, 142–70.

—— (1991) *Stanisław Leśniewski: Collected Works*, in S. J. Surma *et al.* (eds), vol. I (Dordrecht and Warsaw: Kluwer and PWN).

Loeb, I. (2013) 'From Mereology to Boolean Algebra: The Role of Regular Open Sets in Alfred Tarski's Work'. [This volume].

Sobociński, B. (1955) 'Studies in Leśniewski's Mereology', *Rocznik Polskiego Towarzystwa Naukowego na Obczyźnie*, 5, 34–43.

—— (1956) 'On Well-Constructed Axiom Systems', *Rocznik Polskiego Towarzystwa Naukowego na Obczyźnie*, 6, 54–65.

Tarski, A. (1923) 'O wyrazie pierwotnym logistyki', *Przegląd Filozoficzny*, 26, 68–89.

—— (1929) 'Les fondements de la géométrie des corps' (résumé), *Annales de la Société Polonaise de Mathématiques/Księga pamiątkowa pierwszego polskiego zjazdu matematycznego, Lwów, 7-10.IX.1927* (Krakow), 29–33.

—— (1935) 'Zur Grundlegung der Boole'schen Algebras' I, *Fundamenta Mathematicae*, 24, 177–98.

—— (1937a) 'Appendix E', in J. H. Woodger, (ed.) *The Axiomatic Method in Biology* (Cambridge: Cambridge University Press), 161–72.

—— (1937b) 'Über additive und multiplikative Mengenkörper und Mengenfunktionen', *Sprawozdania z Posiedzeń Towarzystwa Naukowego Warszawskiego, Wydział III, Nauk Matematyczno-fizycznych* (Comptes Rendus des Séances de la Société des Sciences et des Lettres de Varsovie, Classe III, Science Mathématiques et Physiques), 30, 151–81.

—— (1956) 'Foundations of the Geometry of Solids', in *Logic, Semantics, Metamathematics* (Oxford: Oxford University Press), 24–9.

Tarski, A., Givant, S. (1999) 'Tarski's Systems of Geometry', *The Bulletin of Symbolic Logic*, 5, 175–214.

Whitehead, A. N. (1919) *An Enquiry Concerning the Principles of Natural Knowledge* (Cambridge: Cambridge University Press).

Woodger, J. H. (1937) *The Axiomatic Method in Biology* (Cambridge: Cambridge University Press).

12
From Mereology to Boolean Algebra: The Role of Regular Open Sets in Alfred Tarski's Work

Iris Loeb

To Jan Woleński on the occasion of his 70th birthday

12.1 Introduction

Can the content of a whole paper be explained by a single footnote? In 1956 Alfred Tarski made an effort in this direction in his chapter 'On the Foundations of Boolean Algebra' (Tarski, 1956c). This chapter had originally appeared as the paper 'Zur Grundlegung der Boole'schen Algebra, I' (Tarski, 1935), the sequel of which had never been published. This sequel should have contained, among other things, the atomless system of Boolean algebra, as Tarski points out (Tarski, 1956c, p. 341, fn. 2). In that same footnote Tarski also gives a model for atomless Boolean algebra, consisting of the family of so-called regular open sets of a Euclidean space and the relation of set-inclusion. He then refers back to another chapter in Tarski (1956b): *Foundations of the Geometry of Solids* (Tarski, 1956a). This reference and the model of atomless Boolean algebra appear only in the 1956 edition (Tarski, 1956c); they are absent in Tarski (1935).

These facts form the starting point for the research on which we will report in the current paper. We will trace back the origin of regular open sets to Kuratowski (1922), and argue that when these sets came into contact with Leśniewski's mereology, a bridge was created that made possible a connection to Boolean algebra. Both mereology and regular open sets play an important role in 'Foundations of the Geometry of Solids', which originally appeared as Tarski (1929) and which will thus play a vital role in our understanding of this particular history. It will show the fruitfulness of Tarski's practice of connecting various fields (see also Sinaceur, 2001; Betti & Loeb, 2012), free from any particular methodological constraint.

This story thus emphatically concerns the emergence and the development of the connection between mereology, regular open sets, and Boolean algebra in Tarski's work, and not that connection *per se*. The relation between regular open sets and Boolean algebra can be found in most textbooks on Boolean algebra or set theory (see, for example, Levi, 2002); the relation between (Leśniewski's) mereology and Boolean algebra has been subject to several systematic studies as well, as in Sobociński (1971) and Clay (1974).

These systematic studies on the relation between mereology and Boolean algebra set themselves the goal to make more explicit the observation that 'in a certain sense mereology is closely related to Boolean algebra' (Sobociński, 1971, p. 90). This is done, for example, by showing that:

(1) Mereology is a complete Boolean algebra with zero deleted.
(2) Complete Boolean algebra with zero deleted is a mereology.

(Clay, 1974, p. 639)

Although it may be fair to say that Tarski hinted at this observation in his footnote, the connection between mereology and Boolean algebra as mentioned above leaves out the crucial role of regular open sets, and the merger of the ideas of two great Poles: the mathematician Kazimierz Kuratowski (1896–1980) and the philosopher Stanisław Leśniewski (1886–1939).

The structure of this paper is as follows. In Section 2 we see the context in which Tarski (1929) appeared, and briefly point out the roles played in it by Leśniewski's mereology and Kuratowski's regular open sets.

Section 3, then, discusses Leśniewski's views and the specific formulation of the postulate system of mereology that can be found in Tarski (1956a).

Then the core of Tarski (1929; 1956a), Tarski's postulate system of the geometry of solids, is given in Section 4. This also shows the use of regular open sets as a formal notion of the theory.

We go deeper into regular open sets in Section 5. We argue that the intuitive connection between solids and regular open sets can be understood if we take into account Leśniewski's influence, especially through his notion of summation.

In Section 6 we discuss Tarski's model for the geometry of solids, the domain of which consists of the class of regular open sets without the empty set. This is also a model for the postulate system of Leśniewski's mereology that Tarski gives in 1956a.

Finally, in Section 7, we take all of this together. It is shown how mereology, regular open sets, and Boolean algebra are connected in Tarski (1956c). We can then fully understand Tarski's footnote that gives a model for atomless Boolean algebra.

12.2   Setting the stage

The so-called geometry of solids, which avoids the presupposition of ideal terms like points, enjoyed substantial interest in the beginning of the twentieth century (see for example Huntington, 1913; Leśniewski, 1928; Tarski, 1929). This was due at least partially to the success of Whitehead's 'theory of events' (Whitehead, 1919; 1920), which avoids the presupposition of moments, and thus can be seen as a temporal counterpart to the geometry of solids.

The paper (Tarski, 1929) offers a sketch of a foundation of the geometry of solids, starting from the primitive term *sphere*. It is the published abstract of a talk that Tarski gave two years earlier during the First Polish Mathematical Congress, and this fact may explain its somewhat sketchy and concise nature. At first reading it is unclear what framework Tarski has in mind as the 'background logic' for this formalization: Notions either of Leśniewski's mereology (of which we will see more in Section 3) or of Russell's type theory are frequently alluded to in Tarski (1929).[1] Here we will leave aside Russell's philosophical influence, and go deeper into the technicalities of Leśniewski's system, because we will see that we need his notion of *sum* (or *collection*, in Leśniewski's terminology) for our current purposes.

Although the question 'What framework was Tarski really working in?' thus does not seem especially relevant in this context, let us remark that even as late as 1930 Tarski wrote that he subscribed to Leśniewski's view (Tarski, 1930, see also Sundholm, 2003), but whether he actually meant it is another story (see Betti, 2008). Also in Tarski (1929) we see clearly Leśniewski's (technical) influences. Apart from Tarski's acknowledgment that his research has been conducted in reaction to a problem posed by Leśniewski (Tarski, 1929, p. 227), also explicit reference is made to Leśniewski's Mereology in the theoretical body of the paper. In the 1929 version we read:

> I will suppose here as known the deductive system founded by Leśniewski [...] and called by him *mereology*; I will use, in particular, the *relation of part to whole* as a known notion [...].[2] (Tarski, 1929, pp. 227–8)

Although one might argue that this leaves some room for the interpretation that the mention of Leśniewski's mereology here may only indicate a motivational influence upon Tarski's foundations of the geometry of solids, we can exclude this in light of the 1956 version, where Tarski's statement is more forceful: 'The deductive theory founded by S. Leśniewski and called by him *mereology* [...] will be essentially involved in our exposition.' (Tarski, 1956a, p. 24)

Furthermore in (1956a) Tarski gives postulates for mereology, which we will see in Section 3.

Subsequently, in a context disconnected from Leśniewski's mereology and Russell's simple type theory, Tarski introduces the notion of *regular open set*, which he takes from Kuratowski (1922) (see Tarski, 1929, p. 227). We will come back to this in Section 5. The role of regular open sets within Tarski (1929) and Tarski (1956a) is threefold. Regular open sets are brought to stage as:

1. The intuitive correlates of solids;[3]
2. A formal, defined notion of the theory;[4]

Furthermore we find in (Tarski, 1956a) regular open sets in the following role:

3. The interpretation of *solids* within the model of three-dimensional Euclidean geometry for the postulate system of the geometry of solids.[5]

The fact that regular open sets are both seen as intuitive correlates (1) and also play a more formal role (2) may not come as a surprise. However, being both a formal notion in the theory (2) and the interpretation of solids (3) seems rather awkward. To explain this at least partially it should be realized, first, that (3) is a rather new addition which may witness a development in mathematical view and practice; second, that the notion of model that Tarski takes here – despite being a later addition – may not be exactly what the modern reader has in mind (see also note 13); and third, that Tarski needed this explicit model for Tarski (1956c), as we will argue.

We will see the intuitive relation between regular open sets and solids in Section 5, and regular open sets as a formal notion of the theory in Section 4. Let's first go somewhat deeper into Leśniewski's mereology.

12.3 Leśniewski's mereology

Mereology refers to any theory of parts and wholes, and can contain as an axiom or theorem, for example, the transitivity of the 'part of'-relation ('If X is a part of Y, and Y is a part of Z, then X is a part of Z').[6] Stanisław Leśniewski's system of mereology is a formal and very well-developed one. Its intention can be understood only if we take into consideration Leśniewski's idiosyncratic view on formal systems in general. His conception is called *inscriptionism* (Simons, 2008) or *intuitionistic formalism* ('intuitionistischer Formalismus') (Tarski, 1930). It is a nominalistic view, in which systems are taken to be collections of concrete marks.

This means that, properly speaking, his system of mereology is also not a *completed* system, but can refer to any concrete construction in time and space made by someone following the rules, called 'directives'. Thus there are many systems of mereology: not only do various people make different decisions on what to build, but also similar constructions can be made in different spots (say, on different pieces of paper) or in different moments in time which would qualify them as different as well. At any moment there are only various systems in various stages.

Tarski (1956a, p. 25) gives the following definitions and postulates of Leśniewski's mereology, which are absent in Tarski (1929). In the definitions the notions 'proper part', 'disjoint' and 'sum' are defined, whereas the postulates take care of the transitivity of the 'part of'-relation and the existence of the sum of a class of individuals:

Definition I	*An individual X is called a* proper part *of an individual Y if X is part of Y and X is not identical with Y.*
Definition II	*An individual X is said to be* disjoint *from an individual Y if no individual Z is part of both X and Y.*
Definition III	*An individual X is called a sum of all elements of a class α of individuals [. . .] if every element of α is a part of X and if no part of X is disjoint from all elements of α.*
Postulate I	*If X is a part of Y and Y is a part of Z, then X is a part of Z.*
Postulate II	*For every non–empty class α of individuals there exists exactly one individual X which is a sum of all elements of α.*

The formulation of these definitions and postulates show, however, a marked disagreement with Leśniewski's view. First, the postulates follow the definitions, which does not agree with Leśniewski's methodology;

second, the formulation using 'individuals' imply a Russellian background logic (see also Betti & Loeb, 2012).

After giving these foundations for mereology, Tarski (1956a) continues with the core part of the paper: the foundations of the geometry of solids.

12.4   The Geometry of Solids

The core of Tarski (1929) and Tarski (1956a) consists of a system with the primitive notion *sphere*, nine definitions and four axioms. One of the defined notions is *point*, for example, which is defined as all spheres that are concentric to some given sphere. However, the definition that is most important for the current paper is that of *solid*:

Definition 8   A solid *is an arbitrary sum of spheres.*

(Tarski, 1956a, p. 27)

It is this summation of spheres (which we will study in Section 5), which forms, together with the 'part of' relation, one of the connections between mereology (through the geometry of solids), regular open sets, and Boolean algebra.

Although there are only four axioms, the first one – without going deep into the matter – arguably contains in abbreviated form a whole set of other axioms:

Postulate 1   *The notions of point and of equidistance of two points from a third satisfy all the postulates of ordinary Euclidean geometry of three dimensions.*

(Tarski, 1956a, p. 27)

Instead of listing all postulates of ordinary Euclidean geometry separately, they are taken together somewhat sloppily, it seems, using the notion of *satisfaction*.

The other three, according to Tarski, render the system categorical, and additionally establish exact connections between notions of the geometry of solids (of *solid* and of the *part of* relation) with the corresponding pair of ordinary point geometry (*regular open set* and the *inclusion*-relation) (Tarski, 1956a, p. 28). It is here, then, that we see *regular open set* used as a defined notion in the formal theory, although an explicit definition is not given:

Postulate 2   *If A is a solid, the class α of all interior points of A is a non–empty regular open set.*

Postulate 3   *If the class α of points is a non–empty regular open set, there exists a solid A such that α is the class of all its interior points.*

Postulate 4   *If A and B are solids, and all the interior points of A are at the same time interior to B, then A is a part of B.*

(Tarski, 1956a, p. 28)

But what exactly are these regular open sets? We will now address this question, starting from Kuratowski (1922) to explain their intuitive connection to the geometry of solids.

12.5   Regular open sets

Regular open sets spring indeed from Euclidean geometry, as Tarski also points out, though via topology rather than directly. In Kuratowski (1922) the starting point is an n dimensional Euclidean space, in which letters like A refer to a set of points in that space, $1 - A$ (or A^1) its complement, and $\overline{A}$ (or A^-) to the closure of A.[7] Other notions taken are that of equality ($=$), that of *sum* (set union, $+$), that of non-strict set inclusion ($\subset$), and that of the empty set (0). Kuratowski does not explicitly explain his notation as such, because he approaches it in a more axiomatic way.

He puts forward, that is, the following statements,[8] which, he says, 'can be easily seen to hold' (Kuratowski, 1922, p. 182):[9]

I. $\overline{A + B} = \overline{A} + B$ [sic]

II. $A \subset \overline{A}$

III. $\overline{0} = 0$

IV. $\overline{\overline{A}} = \overline{A}$

In the remainder of Kuratowski (1922) the above statements are taken as axioms and the operations in them as primitive. We can thus say that Euclidean geometry, or – to be more precise – Euclidean *spaces* were a motivation for Kuratowski's axioms, but that his treatment, and especially his axioms in which points, for example, play no role, are far removed from the more classical approach.

It is in this context that Kuratowski introduces both regular closed and regular open sets as follows:

I call a set A a *regular closed set*, when[10]

$$A^{1-1-} = A.$$

(Kuratowski, 1922, p. 192)

We will say that A is a regular open set, when[11]

$$A = A^{-1-1}.$$

(Kuratowski, 1922, p. 194)

In other words, if for a set X we call X^{1-1} its *interior* (Kuratowski, 1922, p. 187), then we can say that a set is a regular closed set if it coincides with the closure of its interior, and that it is a regular open set if it coincides with the interior of its closure.[12]

Let's see a few simple one-dimensional examples. The open interval $(0,1)$ considered in the space of real numbers ($\mathbb{R}$) is a regular open set, because:

$$(0,1)^{-1-1} = [0,1]^{1-1}$$
$$= ((-\infty,0) \cup (1,\infty))^{-1}$$
$$= ((-\infty,0] \cup [1,\infty))^{1}$$
$$= (0,1).$$

On the other hand, it is not regular closed, because:

$$(0,1)^{1-1-} = ((-\infty,0] \cup [1,\infty))^{-1-}$$
$$= ((-\infty,0] \cup [1,\infty))^{1-}$$
$$= (0,1)^{-}$$
$$= [0,1] \neq (0,1).$$

The closed interval $[0,1]$ is an example of a set that is regular closed, and the union of two open intervals $(0,1) \cup (1,2)$ is – although open – not regular open (and not regular closed).

After giving these definitions, Kuratowski proves several theorems about regular open and regular closed sets. For example, he shows that the sum of two regular closed sets is again regular closed, which implies that the sum of any finite number of regular closed sets is regular closed as well.

For us it is more relevant to understand which properties regular open and regular closed sets do *not* have. Although regular closed sets are thus closed under finite summation, they are not closed under *infinite* summation. Regular open sets, on the other hand, are not closed under summation at all: The union of two disjoint open intervals already shows that the sum of a finite number of regular open sets need not be regular open itself. (In higher dimensions we can find similar counterexamples.)

The lack of exactly these properties makes the intuitive correspondence between solids and Kuratowski's regular open (or closed) sets less obvious. And there is (or should be), Tarski writes, an intuitive correspondence: '[...] solids – the intuitive correlates of open (or closed) regular sets of three-dimensional Euclidean geometry.' (Tarski, 1956a, p. 24)

It is essential for our argument that Tarski identifies this *intuitive* correspondence, over and above the use of regular open sets as a *formal* notion of the system (see Section 4). Whereas a formal notion of the system may be defined in such a way as to have any property that Tarski desires, it is the intuitive connection which shows that Tarski wants his regular open sets and regular closed sets to have those properties of *Kuratowski's* regular open and, respectively, closed sets.

Looking again at Definition 8, we may now identify the difficulty:

Definition 8   A solid *is an arbitrary sum of spheres.*

(Tarski, 1956a, p. 27)

Thinking of spheres as open spheres – and thus as a special kind of regular open sets – an arbitrary sum of those would be open, but in general not regular open; a fact that seemingly destroys an intuitive correspondence between regular open sets and solids. Interpreting spheres as closed spheres, on the other hand, would imply that an infinite sum of those is not necessarily regular closed, which would likewise forbid any straightforward correspondence between solids and regular closed sets. Note that the difficulty just mentioned appears under the assumption that Tarski and Kuratowski have the same notion of summation in mind, that is, set union.

This issue – that Kuratowski's notion of regular open set does not correspond to Tarski's, using Kuratowski's notion of 'sum' – is dealt with only indirectly in Tarski (1929) and Tarski (1956a), where it appears that although the notion of regular closed (and regular open) sets are taken from Kuratowski, it is not *his* notion of sum that Tarski takes, but rather Leśniewski's. Here we will see that by combining notions from several fields, Tarski was able to obtain innovative results.

Let's look again at the definition of sum that Tarski gives in (1956a):

Definition III   *An individual X is called a* sum *of all elements of a class α of individuals [...] if every element of α is a part of X and if no part of X is disjoint from all elements of α.*

(Tarski, 1956a, p. 25)

To be more specific: the sum of a class of regular open sets is defined as the smallest regular open set that contains all elements of that class.

This is a notion of sum that we can give explicitly in terms of set-union, closure and interior. For two sets (in modern notation):

$$A + B := (\overline{A} \cup \overline{B})^{o};$$

or more generally for a family of sets:

$$\sum_{i} A_i := (\bigcup_{i} \overline{A_i})^{o}.$$

And it is well-known nowadays that with this notion of summation, the regular open sets form a Boolean algebra (see, for example, Levi, 2002, p. 264). Tarski may have been the first one to realize this fact. (See also note 17.)

12.6   A model for the geometry of solids

Having now seen that Tarski used regular open sets as a formal notion in his axioms (Section 4), and having understood the intuitive correspondence between solids and regular open sets, let's look briefly at how they appear in the description of a model[13] of the geometry of solids. In Tarski (1956a) we find the following theorem:[14]

> Theorem B   *The postulate system of the geometry of solids, with the postulates of mereology included, has a model in three–dimensional geometry. To obtain such a model we interpret spheres as interiors of Euclidean spheres, and the relation of part and whole as the inclusion relation restricted to non–empty regular open sets [. . .].*
>
> (Tarski, 1956a, p. 29)

On a first reading of this theorem, what seems to be missing is the domain of the model. It is given only implicitly by the restriction of the inclusion relation to non-empty regular open sets. Because the relation of part and whole is defined on every individual of the geometry of solids, its interpretation has to be defined on the whole domain of the model as well. Because the inclusion relation is restricted to non-empty regular open sets and needs to be defined everywhere on the domain, the domain consists of the class of non-empty regular open sets. (Note the non-emptiness requirement here; we will come back to this in Section 7.) We, the readers, are now able to fill another gap here as well by the following reasoning: It follows that the class of individuals[15] of the background logic consists of the class of all solids.[16]

It is, among other things, this theorem (Theorem B) that Tarski refers to in the footnote of Tarski (1956c, p. 341, fn. 2) to explain the content of the never-published paper on atomless Boolean algebra. As I argue that the chief connection that Tarski made is between Boolean algebra, regular open sets and *mereology* (and not the geometry of solids), a sincere worry may be the seemingly heavy involvement here of the geometry of solids. In what way, we may ask, are the geometry of solids and its particular geometrical postulates essential for this theorem and its further use?

Of course, the fact that the non-empty regular open sets together with the inclusion relation is a model for the *geometry of solids* is important in the context in which the theorem has been given, which is a paper on the foundations of the geometry of solids (Tarski, 1956a). It is, in addition to this, also important for the conclusion that Tarski draws immediately from it: The postulate system of the geometry of solids and that of three-dimensional Euclidean geometry are equiconsistent.

For the further use of Theorem B, though, it seems safe to leave aside the context of the geometry of solids. The full geometrical context seems no longer to play any role. Instead one need take into account only the fact that this structure is a model for the postulates of mereology. What we should remember is that the domain, the class of individuals, of mereology (in the formulation given by Tarski) is interpreted by the class of regular open sets without the empty set. The fact that in some extension of the postulate system of mereology it becomes a postulate system of geometry, and that this is even a model for that extended system, is not relevant for its further use.

12.7   Foundations of Boolean algebra

We have now collected all ingredients necessary to grasp Tarski's connection between mereology, regular open sets, and Boolean algebra. Especially, we have seen in the previous section that the class of Kuratowski's regular open sets (without the empty set) forms a model for mereology, though such a notion of model and interpretation may be far from Leśniewski's intentions. We will come to this later.

We will now see how Tarski claims that mereology has influenced his axiomatization of Boolean algebra, and, when we understand this relation, it will also be clear how the class of regular open sets, a model for mereology, became a model for (atomless) Boolean algebra.

In Tarski (1935; 1956c) Tarski gives two postulate systems for Boolean algebra, and some variation on those, and proves them to be equivalent.

The first set of postulates is labeled '$\mathfrak{A}$' and the second '$\mathfrak{B}$'. See the appendix for full details. He then goes on to study properties of *atomistic* Boolean algebra.

The first postulate system contains all 'usual' primitive notions, like 'B' (the universe of discourse (Tarski, 1956c, p. 320)), '$<$' (the inclusion relation), '$=$' (equality), '$+$' (sum), '$.$' (product), '0' (zero), '1' (unit element), "$'$" (complement), and some more. The second set, the $\mathfrak{B}$s, on the other hand, gets by with only three of them: 'B', '$=$' (which can be defined), and '$<$'.

This is a clear connection to the postulates for mereology that Tarski gives in (Tarski, 1956a), and that we have seen in Section 3. In fact, it suffices to compare Postulate I (mereology) to Postulate $\mathfrak{B}_2$ (Boolean algebra), and Definition III (mereology) to Definitions $\mathfrak{B}_5$(a) and $\mathfrak{B}_6$(a). The only puzzling element may be what to take as the counterpart in Tarski's formulation of the postulate system of mereology for the universe of discourse 'B' that we encounter in that of Boolean algebra. The answer is that, in Tarski's formulation of the former, the domain of discourse is encoded in the formal theory by way of Russellian individuals (instead of by a constant in the language, as in the postulate system for Boolean algebra). (More on this in Betti & Loeb, 2012.)

There are two footnotes in Tarski (1956c) that refer to Tarski (1956a). The first one starts as follows, where it should be noted that II is Tarski (1956a):

> The formulation of Posts. $\mathfrak{B}_4$ and $\mathfrak{B}_4^*$ as well as some fragments of the proofs of Ths. 1 and 2 have been influenced by the researches of S. Leśniewski. The extended system of Boolean algebra is closely related to the deductive theory developed by S. Leśniewski and called by him *mereology*. The foundations of mereology have been briefly discussed in article II, where bibliographical references to the relevant works of Leśniewski will also be found. The relation of part to the whole, which can be regarded as the only primitive notion of mereology, is the correlate of Boolean-algebraic inclusion. The postulate system $[\mathfrak{B}_2, \mathfrak{B}_4^*]$ has been obtained by a slight modification of the postulate system for mereology (Posts. I and II) suggested in II; regarding the relation of the latter system to the original postulate system of Leśniewski see II, p. 25, footnote 2. (Tarski, 1956c, p. 333)

Tarski's need to connect Boolean algebra explicitly to the postulates of mereology may be one of the reasons that he has added these postulates to the 1956 version of Tarski (1929).[17]

The footnote then continues to compare models for mereology and Boolean algebra, which we may already connect to Theorem B, and which sets the stage for understanding the later footnote on the model of regular open sets for atomless Boolean algebra:

> The formal difference between mereology and the extended system of Boolean algebra reduces to one point: the axioms of mereology imply (under the assumption of the existence of at least two different individuals) that there is no individual corresponding to the Boolean-algebraic zero, i.e. an individual which is part of every other individual. If a set *B* of elements (together with the relation of inclusion) constitutes a model of the extended system of Boolean algebra, then, by removing the zero element from *B*, we obtain a model for mereology; if, conversely, a set *C* is a model for mereology, then, by adding a new element to *C* and by postulating that this element is in the relation of inclusion to every element of *C*, we obtain a model for the extended system of Boolean algebra. (Tarski, 1956c, pp. 333–4)

In the final part of that footnote, Tarski shows awareness that his approach to mereology is far removed from that of Leśniewski, which we have briefly met in Section 3:

> Apart from these formal differences and similarities, it should be emphasized that mereology, as it was conceived by its author, is not to be regarded as a formal theory where primitive notions may admit many different interpretation. [...] (Tarski, 1956c, p. 334)

We are now almost through, and end here with the footnote which formed also the starting point of our research and which addresses a model for atomless Boolean algebra that explains a part of the content of the never published sequel of Tarski (1956a):[18]

> This article has been conceived as the first part of a more comprehensive paper. The second part (which was never published) included, among other things, a discussion of the so-called atomless system of Boolean algebra, i.e. the system in which Post. $\mathfrak{D}$ is replaced by a postulate stating that there are no atoms. The atomless system of Boolean algebra can, of course, be constructed as a system in which 'B' and '$x < y$' occur as the only primitive expressions. A model for such a system is provided by the family of all regular open sets of a Euclidean space and the relation of set-theoretical inclusion. In view of the remarks made above in p. 333, footnote, this follows immediately from the discussion in II, in particular, from Th. B. (Tarski, 1956c, p. 341)

Looking back, we can now understand also why in Theorem B (and in other places in the 1956 edition of the paper on the foundations of the geometry of solids) the requirement of being non-empty is present so prominently; a very non-Leśniewskian addition, by the way. The reason why the non-emptiness requirement is explicitly stated, I argue, is that it is the empty set that should be added to this model of mereology to get a model for atomless Boolean algebra: It is the interpretation of the zero element, as explained in the construction in the footnote on page 333, and which we have cited above.

12.8   Conclusion

We have seen how Tarski in Tarski (1929) connected Leśniewski's notion of summation to Kuratowski's notion of regular open sets. By doing so, Tarski was able to construct a model for the postulate system of the geometry of solids, which – more importantly – is also a model for the postulate system of mereology. Through the connection that Tarski made between mereology and Boolean algebra, he was able not only to propose a new postulate system for this latter field, but also to make clear that the class of regular open sets with the inclusion relation forms a model for *atomless* Boolean algebra, even though a paper on atomless Boolean algebra by Tarski was never published.

This is the story that Tarski tells us through footnotes in Tarski (1956c).

Acknowledgments

The author is supported through ERC Starting Grant TRANH Project No 203194. She also thanks Arianna Betti for discussions on an earlier version of this paper.

Appendix: Tarski's postulate system for atomistic Boolean algebra

In this appendix a part of the several postulate systems for Boolean algebra are given that can be found in Tarski (1956c). Basically, there are two sets of them: those labelled 𝔄, and those labelled 𝔅. Definition ℭ and Postulate 𝔇 make the system one for *atomistic* Boolean algebra.

For more postulates, which play no role in the current paper, and details on the formal language we refer to Tarski (1956c).

Postulate $\mathfrak{A}_1$ *(a) If $x \in B$, then $x < x$; (b) if $x, y, z \in B, x < y$, and $y < z$, then $x < z$.*

Postulate $\mathfrak{A}_2$ *If $x, y \in B$, then $x = y$ if and only if both $x < y$ and $y < x$.*

Postulate $\mathfrak{A}_3$ *If $x, y \in B$, then*

    (a) $x + y \in B$;

    (b) $x < x + y$ and $y < x + y$;

    (c)   if, moreover, $z \in B, x < z$ and $y < z$, then $x + y < z$.

Postulate $\mathfrak{A}_4$ *If $x, y \in B$, then*

    (a) $x.y \in B$;

    (b) $x.y < x$ and $x.y < y$;

    (c)   if, moreover, $z \in B, z < x$ and $z < y$, then $z < x.y$.

Postulate $\mathfrak{A}_5$ *If $x, y, z \in B$, then*

    (a) $x.(y + z) = x.y + x.z$;

    (b) $x + y.z = (x + y).(x + z)$.

Postulate $\mathfrak{A}_6$ *(a) $0, 1 \in B$; (b) if $x \in B$, then $0 < x$ and $x < 1$.*

Postulate $\mathfrak{A}_7$ *If $x \in B$, then*

    (a) $x' \in B$;

    (b) $x.x' = 0$;

    (c) $x + x' = 1$.

Postulate $\mathfrak{A}_8$ *If $X \subseteq B$, then*

    (a) $\sum_{y \in X} y \in B$;

    (b) $x < \sum_{y \in X} y$ for every $x \in X$;

    (c)   if, moreover, $z \in B$ and $x < z$ for every $x \in X$, then
    $\sum_{y \in X} y < z$.

Postulate $\mathfrak{A}_9$ *If $X \subseteq B$, then*

    (a) $\prod_{y \in X} y \in B$;

    (b) $\prod_{y \in X} y < x$ for every $x \in X$;

    (c)   if, moreover, $z \in B$ and $z < x$ for every $x \in X$, then
    $z < \prod_{y \in X} y$.

Postulate $\mathfrak{A}_{10}$ *If $x \in B$ and $X \subseteq B$, then*

    (a) $x.\sum_{y \in X} y = \sum_{y \in X} (x.y)$;

    (b) $x + \prod_{y \in X} y = \prod_{y \in X} (x + y)$.

Postulate $\mathfrak{B}_1$ *If $x, y \in B, x < y$, and $y < x$, then $x = y$.*

Postulate $\mathfrak{B}_2$ *If $x, y \in B, x < y$, and $y < z$, then $x < z$.*

Postulate $\mathfrak{B}_3$ *If $x, y \in B$ and x non-$<$ y, then there is an element $z \in B$ such that $z < x$, and z non-$<$ y, and the formulas $u \in B, u < y$, and $u < z$ always imply $u < v$, for every $v \in B$.*

Postulate $\mathfrak{B}_4$ *If $X \subseteq B$, then there exists an element $x \in B$ which satisfies the following conditions: (1) $y < x$ for every $y \in X$; (2) if $z \in B$, $z < x$, and if, for every $y \in X$, the formulas $u \in B, u < y$, and $u < z$ always imply $u < v$ for every $v \in B$, then also $z < v$ for every $v \in B$.*

Definition $\mathfrak{B}_5$ *(a)   For all $x, y, z \in B, z = x + y$ if and only if z is the only element which satisfies the conditions:*
      (1) $x < z$ and $y < z$,
      (2)   for every $u \in B$, if $x < u$ and $y < u$, then $z < u$,
   (b)   for all $x, y, z \in B, z = x.y$ if and only if z is the only element which satisfies the conditions:
      (1) $z < x$ and $z < y$,
      (2)   for every $u \in B$, if $u < x$ and $u < y$, then $u < z$;
   (c)   for every $x \in B, x = 0$ if and only if x is the only element which satisfies the condition: $x < v$ for every $v \in B$;
   (d)   for every $x \in B, x = 1$ if and only if x is the only element which satisfies the condition: $v < x$ for every $v \in B$;
   (e)   for every $x, y \in B, y = x'$ if and only if y is the only element which satisfies the conditions:
      (1)   the formulas $u \in B, x < u$, and $u < y$ always imply $u < v$ for every $v \in B$;
      (2)   the formulas $u \in B, x < u$, and $y < u$ always imply $v < u$ for every $v \in B$.

Definition $\mathfrak{B}_6$ *If $x \in B$ and $X \subseteq B$, then*
   (a) $x = \sum_{y \in X} y$ if and only if x is the only element which satisfies the conditions:
      (1) $y < x$ for every $y \in X$;
      (2)   if $z \in B$ and $y < z$ for every $y \in X$, then $x < z$;
   (b) $x = \prod_{x \in X} y$ [sic] if and only if x is the only element which satisfies the conditions:
      (1) $x < y$ for every $y \in X$;
      (2)   if $z \in B$ and $z < y$ for every $y \in X$, then $z < x$.

Postulate $\mathfrak{B}_4^*$ *If $X \subseteq B$, then there is exactly one element $x \in B$ which satisfies the conditions (1) and (2) of Post. $\mathfrak{B}_4$.*

Definition $\mathfrak{C}$ *$x \in At$ (1) if and only if $x \in B$ and $x \neq 0$, and (2), for every element $y \in B$, the formulas $y < x$ and $y \neq 0$ imply $y = x$.*

Postulate $\mathfrak{D}$ *If $x \in B$ and $x \neq 0$, then there is an element $y \in At$ such that $y < x$.*

Notes

1. For more on the formal and methodological particularities of Tarski (1929), (see Betti & Loeb, 2012).
2. Je supposerai ici comme connu le système déductif fondé par M. Leśniewski [...] et appelé par lui *meréologie*; je vais me servir, en particulier, de la *relation de partie au tout* comme d'une notion connue [...].
3. Or rather the other way around: Solids are introduced as the intuitive correlates of regular open sets.
4. Although the precise definition of the formal notion is lacking.
5. This role is implied by the interpretation of spheres as interiors of Euclidean spheres; more on this in Section 6.
6. From this example it may already be seen that the 'part of' relation in mereology is in some sense closer to the inclusion relation of set–theory than to the 'element of' relation.
7. Although not all of this notation is in use anymore nowadays, we will stick to it in the current paper.
8. There is an error in the first statement, which should have been:

$$\overline{A+B} = \overline{A} + \overline{B}.$$

9. On montre aisément que les énoncés suivants subsistent: [...]
10. J'appelle l'ensemble A un *ensemble fermé régulier*, lorsque

$$A^{1-1-} = A.$$

11. Nous dirons que A est un ensemble régulier ouvert, lorsque

$$A = A^{-1-1}.$$

12. These are the 'informal' explanations of regular open and regular closed sets that Tarski gives in Tarski (1956a, p. 24). They are absent in Tarski (1929).
13. It would lead too far astray to go into Tarski's notion of model here. Although this theorem is thus a relatively new addition, it seems that Tarski had in mind here a notion of model similar to Weyl's (1927), in which one formal system has a model in another one. Tarski, namely, continues Theorem B by stating: 'Conversely, the postulate system of three–dimensional Euclidean geometry has a model in the geometry of solids.'
14. This theorem as such is not included in Tarski (1929).
15. The term 'individual' appears in Tarski's formulation of the postulates of Leśniewski's mereology.
16. This interpretation is consistent with what we have argued in Betti and Loeb (2012).
17. Another, quite important reason for Tarski to clarify that all ingredients for the connection between regular open sets and Boolean algebra were already present in Tarski (1929), may be a priority issue with Stone's work (1934; 1935a; 1935b; 1936) and Tarski (1937, p. 321):

Proposition 7.23 has been known to me since 1927; in its application to Euclidean spaces it is implied in an essay that I have published in *Księga Pamiątkowa pierwszego Polskiego Zjazdu Matematycznego. Lwów, 7–10.IX. 1927* (Proceedings of the First Polish Conference), appendix of the Ann Soc. Pol. Math., Kraków (1929, p. 29 ff.)

Satz 7.23 ist mir seit 1927 bekannt; in seiner Anwendung auf Euklidische Räume steckt er implizite in einem Aufsatz, den ich im *Księga Pamiątkowa pierwszego Polskiego Zjazdu Matematycznego. Lwów, 7–10.IX. 1927* (Andenkenbuch des I. Poln. Math. Kongr.), Annex zu Ann Soc. Pol. Math., Kraków (1929, S 29 ff.) veröffentlicht habe[.]

In Schlimm (2009) the emergence of Boolean algebra is discussed, focusing on the work of Boole, Stone, and Tarski. However, Tarski's development is not traced back to mereology or Tarski (1929), nor is a possible priority issue between Stone and Tarski hinted at.

18. The footnote on p. 333 – referred to in this passage – is the footnote that we have cited above.

References

Betti, A. (2008) 'Polish Axiomatics and its Truth: On Tarski's Leśniewskian Background and the Ajdukiewicz Connection', in D. E. Patterson (ed.) *Alfred Tarski: Philosophical Background, Development and Influence* (Oxford: Oxford University Press), 44–71.

Betti, A., and Loeb, I. (2012) 'On Tarski's Foundations of the Geometry of Solids', *The Bulletin of Symbolic Logic*, 18(2), 230–60.

Clay, R. E. (1974) 'Relation of Leśniewski's Mereology to Boolean Algebra', *Journal of Symbolic Logic*, 39(4), 638–48.

Givant, S. R., and Mackenzie, R. (eds) (1986) *Alfred Tarski: Collected Papers*, vol. 1 (Basel: Birkhäuser).

Huntington, E. (1913) 'A Set of Postulates for Abstract Geometry, Expressed in Terms of the Simple Relation of Inclusion', *Mathematische Annalen*, 73(4), 522–59.

Kuratowski, C. (1922) 'Sur l'opération $\overline{A}$ de l'Analysis Situs', *Fundamenta Mathematicae*, 3, 182–99.

Leśniewski, S. (1928) 'O podstawach matematyki (On the foundations of mathematics)' *Przegląd Filozoficzny*, XXXI, 261–91. Translated and reprinted in Surma *et al.* (1991).

Levi, A. (2002) *Basic Set Theory* (Mineola, NY: Dover Publications). 1st edn (Springer 1979).

Schlimm, D. (2009) 'Bridging Theories with Axioms: Boole, Stone, and Tarski', in B. van Kerkhove (ed.) *New Perspectives on Mathematical Practices: Essays in Philosophy and History of Mathematics* (Singapore: World Scientific Printers), 222–35.

Simons, P. (2008) 'Stanisław Leśniewski', in Zalta E. N. (ed.) *The Stanford Encyclopedia of Philosophy*, http://plato.stanford.edu/entries/lesniewski/, date accessed 20 October 2012.

Sinaceur, H. (2001) 'Alfred Tarski: Semantic Shift, Heuristic Shift in Metamathe-matics', *Synthese*, 126, 49–65.

Sobociński, B. (1971) 'Atomistic Mereology I', *Notre Dame Journal of Symbolic Logic*, XII, 89–103.

Stone, M. H. (1934) 'Boolean Algebras and Their Application to Topology', *Proceedings of The National Academy of Sciences of the U.S.A.*, 20(3), 197–202.

—— (1935a) 'Postulates for Boolean Algebras and Generalized Boolean Algebras', *American Journal of Mathematics*, 57(4), 703–32.

—— (1935b) 'Subsumptions of the Theory of Boolean Algebras', *Proceedings of The National Academy of Sciences of the U.S.A.*, 21(2), 103–5.

—— (1936) 'The Theory of Representations for Boolean Algebras', *Transactions of the American Mathematical Society*, 40(1), 37–111.

Sundholm, G. (2003) 'Tarski and Leśniewski on Languages with Meaning Ver-sus Languages without Use', in J. Hintikka *et al.* (eds) *In Search of the Polish Tradition-Essays in Honor of Jan Woleński on the Occasion of his 60th Birthday*, (Dordrecht: Kluwer Academic Publishers), 109–27.

Surma, S. *et al.* (eds) (1991) *Stanisław Leśniewski: Collected Works*, vol. I (Warsaw and Dordrecht: PWN and Kluwer).

Tarski, A. (1929) 'Les fondements de la géométrie des corps', *Annales de la Société Polonaise de Mathématiques*, 29–34. Reprinted in Givant and Mackenzie (1986).

—— (1930) 'Fundamentale Begriffe der Methodologie der deduktiven Wis-senschaften, I', *Monatshefte für Mathematik und Physik*, 37, 361–404.

—— (1935) 'Zur Grundlegung der Boole'schen Algebra, I', *Fundamenta Mathemat-icae*, 24, 177–98.

—— (1937) 'Über additive und multiplikative Mengenkörper und Mengen-funktionen', *Sprawozdania z Posiedzeń Towarzystwa Naukowego Warszawskiego, Wydział III, Nauk Matematyczno-fizycznych (Comptes Rendus des Séances de la Société des Sciences et des Lettres de Varsovie, Classe III, Science Mathématiques et Physiques*, 30, 151–81.

—— (1956a) 'Foundations of the Geometry of Solids', in Tarski (1956b), 24–9.

—— (1956b) *Logic, Semantics, Metamathematics* (Oxford: Oxford University Press).

—— (1956c) 'On the foundations of Boolean algebra', in Tarski (1956b), 320–41.

Weyl, H. (1927) 'Philosophie der Mathematik und Naturwissenschaft', in *Hand-buch der Philosophie* (Munich: Oldenbourg Verlag).

Whitehead, A. N. (1919) *An Enquiry Concerning the Principles of Natural Knowledge* (Cambridge: Cambridge University Press).

—— (1920) *The Concept of Nature* (Cambridge: Cambridge University Press).

13
The Philosophy of Mathematics and Logic in Cracow between the Wars[1]

Roman Murawski

The aim of the paper is to present a philosophical reflection on the logic and mathematics in Cracow in the inter-war period. This was a time of the intensive development of mathematics (Polish Mathematical School) and of logic (Warsaw Logical School, Lvov-Warsaw Philosophical School) in Poland. One can ask whether this development was accompanied by philosophical and methodological reflection. We shall present and analyse philosophical concepts concerning the logic and mathematics of Jan Sleszyński, Stanisław Zaremba, Zygmunt Zawirski, Witold Wilkosz and Leon Chwistek who were active (mainly) in Cracow. Their ideas will be confronted and compared with the philosophical concepts of the logicians and mathematicians who were members of the Warsaw School.

The connections of the indicated scholars with Cracow were of a varying nature and degree of intensity. There is no doubt that Sleszyński, Zaremba and Wilkosz should be associated with Cracow. The situation is no longer so simple in the case of Zawirski and Chwistek. Zawirski obtained his *Habilitation* at the Jagiellonian University in Cracow in 1924 and then between 1924–8 he was professor at the Technical University in Lvov and then, subsequently, from 1928 until 1937 he was at the University in Poznań and from 1937, professor of the Jagiellonian University in Cracow. Therefore he is sometimes treated as a member of the Lvov-Warsaw philosophical school. Chwistek, on the other hand, obtained his doctorate and his *Habilitation* in Cracow and was active there until 1930 when he became professor of the Faculty of Mathematics and Natural Sciences of the Jan Kazimierz University in Lvov. He is sometimes associated with the Lvov mathematical school but, nevertheless, we have decided to include Zawirski and Chwistek in the group of philosophers associated with Cracow.

13.1 Jan Sleszyński

Jan Sleszyński (1854–1931), who worked in mathematical analysis and in the calculus of variations as well as in number theory, can – in a certain sense – be treated as a pioneer of mathematical logic in Poland. He stressed the meaning and role of it for mathematics. He saw in it a tool which enabled mathematical reasoning to be made clear and precise. It is important because, as he wrote: 'Everything that is obscure and complicated has no value' (cf. 1925–9, vol. II, p. 212).

Sleszyński distinguished in mathematics between the context of discovery and the context of justification. In the first one, intuition plays an important role. Its results should be made precise and clear using methods of logic. Hence the need to construct complete proofs of mathematical theorems – they are necessary in foundational studies. 'Normal' mathematicians do not usually give such proofs – they are satisfied by sketches of proofs. One should elaborate methods that will help to avoid mistakes in such reasonings. And this is the task of logic understood by Sleszyński as a theory of proof. He realized this task in his lectures devoted alternately to the methodology of mathematics and to mathematical logic. The notes to those lectures (prepared by his students) were published as a book *Teorja dowodu* [Proof Theory] (two volumes, 1925 and 1929). It contains general considerations on the concept of a deductive system understood as a collection of sentences, some of which have been assumed without proof (axioms and definitions) and others which have been deduced from them. Furthermore, there is some historical information on the development of logic and an analysis of the concept of a deductive proof in mathematics. The second volume, describing the contribution of several thinkers to the development of modern formal logic, stresses the passage from a logical calculus which serves as a tool to solve logical problems to calculus as a tool for the justification of theorems. It also contains an exposition of a system of propositional calculus and an analysis of several examples showing how rules and laws of logic can be applied to construct complete proofs of mathematical theorems.

One of the greatest merits of Sleszyński was the attempt to formulate and realize a program of the reconstruction of the real process of proving theorems in mathematics. It was later undertaken by Stanisław Jaśkowski in the form of a system of natural deduction. It should be stressed here that Sleszyński rejected psychologism and that he treated a deductive system as a hypothetical system and emphasized that mathematical theorems are in fact statements about connections between an antecedent and a consequent.

Sleszyński saw the role and meaning of symbolism in logic and mathematics but simultaneously warned against – as Kazimierz Twardowski called it – 'symbolomania and pragmatophobia'. Symbols serve only as a tool to make proofs simpler and more transparent.

Sleszyński – although he avoided ontological considerations – was in fact an anti-fictionalist. He treated the fictions, especially inconsistent fictions, which had been introduced into mathematics as being destructive – they usually appear at those places where the used concepts are not precise enough.

The most important contribution of Sleszyński to the development of the philosophical and methodological reflection on mathematics was his conviction of the role and meaning of logic and of formal methods for mathematics and its methodology. But one should stress that he did not treat logic as an independent and autonomous discipline – just the opposite, for him it was an auxiliary discipline with respect to mathematics. Consequently he considered and studied it exclusively in the context of its possible applications in mathematics and not for its own sake. This view was characteristic and typical for scientists in Cracow – we shall see similar views in the case of Zaremba. This distinguished them from the Warsaw School, where logic was treated as an independent and autonomous discipline which formed the foundations and methodology of mathematics.

13.2   Stanisław Zaremba

In this way we come to the next important figure in Cracow, namely Stanisław Zaremba (1863–1942). He was interested in modern mathematical logic and, although he did not work himself in it, he saw and appreciated its role for mathematics. He was interested not in logic itself but rather in the foundations of mathematics and the analysis of the logical structure of mathematics. One can see here the influence of his studies in France (where he obtained his doctorate in 1899). His views towards logic and its role in mathematics corresponded to the attitude in this respect to that of most French mathematicians.

Zaremba was a versatile and multi-faceted mathematician. He worked mainly in mathematical analysis, in particular in the theory of partial differential equations of the second order. It could be a consequence of his conviction that mathematics should not be a goal for itself, that the ultimate aim of mathematics is its applications, in particular applications in natural sciences. Nevertheless he also appreciated the meaning and the beauty of pure mathematics.

Considering the independence of the fifth postulate of Euclid (concerning the parallels) Zaremba (1911) distinguished the provability of certain propositions on the basis of accepted axioms and their truth.

Being interested in the applications of mathematics, he considered relations between mathematics and physics (cf. Zaremba, 1923; 1938). He treated this problem as very important since its solution can help us to understand both disciplines better. It is clear that mathematics is an indispensable tool for physics but also mathematics needs physics – various physical problems stimulated research in mathematics and contributed to the development and evolution of mathematics, for example, the theory of vibrations of a string or the problems connected with the theory of heat.

Though the deductive method – which is characteristic of mathematics – is used in physics, one should stress that deductive reasoning does not justify the claim that the thesis which has been proved is true. It depends on the truth of accepted assumptions used in the proof. Additionally it is usually the case that the postulates of any deductive theory can be interpreted in various ways. Hence those theories do not concern a particular kind of concrete objects. Mathematics provides tools for the deduction of consequences from assumed empirical hypotheses. Those consequences are then checked by experiments and in this way one can confirm or reject those hypotheses. Hence, on the one hand, mathematics plays an auxiliary role towards physics providing it with formal methods and, on the other, it was developed (among others) by attempts to solve the problems suggested by physics.

As said above, Zaremba was interested in mathematical logic. He treated it as an auxiliary discipline with respect to classical fields of mathematics, as a tool in the didactic of mathematics. Consequently he was not interested in the theoretical inner problems of logic itself. According to him, logic should be 'in mathematics' (*en mathématique*), should be *ancilla mathematicae* as he wrote in his work *La logique des mathématiques* (Zaremba, 1926). His views were here in full accordance with the views of French mathematicians and in contrast with the views of the members of the Warsaw school of logic.

This divergence of opinion was demonstrated in particular in the polemics between Zaremba and Jan Łukasiewicz and which were described extensively by Jan Woleński in his book *Szkoła Lwowsko–Warszawska w polemikach* [Lvov-Warsaw School in Polemics] (Woleński, 1997).

In 1916, Łukasiewicz devoted one of his courses at Warsaw University to the methodology of deductive sciences. During his lectures he discussed the book *Arytmetyka teoretyczna* [Theoretical Arithmetic] by Zaremba (1912) and analysed it from the methodological point of view, challenging some principles adopted by Zaremba as well as his definition of a magnitude (in particular he criticized the usage of sentences with no contents). Łukasiewicz published his remarks in a paper (Łukasiewicz, 1916). This was the beginning of a dispute in which several persons took part, including among others Kazimierz Kuratowski, Tadeusz Czeżowski, Leon Chwistek and, of course, Zaremba. The essence of the dispute concerned in fact not the very concept of a magnitude – this problem can be solved by relativization of the formalism to a certain given interpretation or model – but the role of logic in mathematics. Zaremba was of the opinion that logic should play an auxiliary role in mathematics – it should help us to construct correct mathematical reasonings. Hence logic belongs in fact to the propaedeutics of mathematics and cannot be itself a subject of study. Consequently one cannot speak about a priority of logic towards mathematics. The requirements to provide complete proofs and the idea that definitions are in fact superfluous (since they can be eliminated) – as postulated by the new mathematical logic – are, according to Zaremba, only ballast and they rather hinder the process of understanding and communicating mathematical results.

The opinion of Łukasiewicz was different. He was convinced that incomplete proofs are not only didactically imperfect but they can be also a source of errors. Science is not a collection of true statements – it should be an edifice in which every element is connected with the whole. Only logic helps and enables us to uncover those connections, logic provides tools to discover such connections. The Warsaw Logical School shared the opinion of Łukasiewicz and treated logic as an autonomous discipline, as a discipline that belongs to the core of mathematics. This view corresponded with the programme of Warsaw Mathematical School (Sierpiński, Janiszewski, Mazurkiewicz) where stress was put on set theory, foundations of mathematics and just logic (cf. Murawski, 2010).

13.3   Zygmunt Zawirski

Zygmunt Zawirski (1882–1948) was mainly interested in the philosophy of science, in particular in the methodological, epistemological and ontological problems connected with physics. He was also interested in the problems of time. He represented moderate realism, criticized neo-Kantianism and Empirio-criticism. He was also interested in the problem of applications of results of formal disciplines in natural sciences.

The philosophical problems of logic and mathematics appear rather at the margins of his considerations. One can distinguish here two circles of problems that interested him: the connections between logic and mathematics and the meaning and role of non-classical logics. The main aim of his papers in this domain was to inform the philosophical circle about new achievements in the world and to correlate them with research undertaken in Poland. Consequently he rather rarely formulated his own opinions.

He stressed that mathematics is older than logic – the ancient Greeks formulated correct mathematical proofs long before systematic investigations on the essence of logical reasoning started. The deductive sciences investigate formal objects whereas in the natural sciences one is interested in the examination of the real world. One searches there not for *ens* (as in logic and mathematics) but *ens existens* – and this can be done only with the help of empirical methods. However, this does not mean that logic and mathematics have no meaning for natural sciences. For mathematical theories can be interpreted. In this way, mathematical constructions become the components of physical theories and mathematical theories interpreted in such a way can be checked empirically.

Zawirski was interested in particular in the connections between physics and geometry. According to a classical approach, that is, before the development of non-Euclidean geometries – both investigate space. Thanks to axiomatization one can claim that the differences between physics and geometry disappear: physics becomes interpreted geometry (Schlick) and geometry becomes a natural science (Einstein, Born). According to Zawirski, physics and geometry are distinct disciplines. Geometry constructs its subject independently of experience and of the existing physical reality. It justifies its theses exclusively by deduction. Physics, on the other hand, investigates objects given in experience and formulates its theses usually using inductive methods. In a developed physical theory one can justify particular theses by deducing them from accepted axioms – this can be the case only in the context of justification. However, accepted axioms must have a certain empirical justification. In geometry, as in the whole of mathematics, certain statements and laws can have empirical sources but they can be accepted only after deducing them from the axioms. Experience is no justification in mathematics.

How did Zawirski understand logic? According to him, logic is a discipline that considers forms of reasoning which are used in any inference or proof. It reflects the common structure of justification used in all disciplines. Hence Zawirski understood logic in a broad sense – not only as a formal system (or a collection of such systems) but also as a discipline about any reasoning. One should add that such opinion was rather popular in Poland at the time.

Let us come now to the second point, that is, to the problem of nonclassical logics. Zawirski was interested mainly in intuitionistic and many-valued logics. In particular he was interested in the problem of the possible applications of many-valued logics to the solution of problems of quantum mechanics or of problems generated by the introduction to physics of statistical laws. He claimed that the new many-valued logic was the only way to help understand the phenomena of the microworld. Using some of the ideas of Łukasiewicz and Post he attempted to develop a system of logic that would be appropriate both to the problems of modern physics as well as to probability calculus. He described the certain parallelism between expressions of the probability calculus and formulas of the many-valued logic of Łukasiewicz and Post. He claimed that both should be treated as different systems providing an empirical base for the other. Such an approach would make possible the application of many-valued logic in quantum mechanics. This idea was developed further in particular by P. Suppes. In this way Zawirski can be treated as a precursor of quantum logic.

13.4   Witold Wilkosz

Under the influence of Sleszyński, some young mathematicians in Cracow became interested in logic – among them was Witold Wilkosz (1891–1941).

He did not write any separate papers devoted to the philosophy of logic or mathematics, nevertheless in his works one finds several remarks of a philosophical nature. They contain neither revolutionary ideas nor form a compact system yet still they indicate tendencies and philosophical sympathies and show that Polish mathematicians were interested in philosophical problems.

Let us start with his understanding of the origin of the concept of number. In a popular work *Liczę i myślę. Jak powstała liczba* [I count and I think. How Number Came About] (Wilkosz, 1938) he accepts the logico-set theoretical conception (as opposed to the intuitionistic one) according to which the concept of a number and the whole of arithmetic comes from more general intuitions connected with equipollency. He justifies this thesis with reference to many examples of a psychological and ethnological nature.

Logic is – according to Wilkosz – only a tool for developing a system on the basis of accepted axioms and definitions. Hence it is connected with deductive systems. Consequently it can be successfully applied in mathematics but attempts to apply it in other disciplines have been rather ineffective. The reason is that the axiomatic methods can be applied in situations when the starting and fundamental postulates are constant and fixed. But this is not the case in many disciplines. On the other hand, he was of the opinion that strict methods based on the application of formal logic should also be used in the humanities and in the social sciences as well as even in theology.[2]

Formal logic does not fully fit 'natural' methods and the ways of reasoning used, for example, by mathematicians. In fact, it is suitable rather to the reconstruction of real mathematical reasoning, hence in the context of justification and not directly in the context of discovery.

Wilkosz also tried to apply formal logical methods to explain some philosophical problems, in particular the problem of abstraction. The problem is: what is an abstract object? What should it be identified with and how can it best be described? In trying to answer this, Wilkosz used equivalence relations and equivalence classes. He suggested applying the method of representation and to identify the abstract either with the appropriate equivalence class or with one element of such a class. But he also saw another solution: instead of deciding what is the abstract object, one can explain what sentences about it in fact mean.

13.5 Leon Chwistek

The most important and best known figure among the Cracow logicians and mathematicians interested in the philosophy of logic and mathematics was Leon Chwistek (1884–1944).[3] He is known mainly for his logical works, in particular for his simplification of Whitehead and Russell's theory of types. His logical investigations, however, were connected with his philosophical ideas concerning logic and mathematics.[4] They were motivated by those ideas and, in this sense, he was an exception among Polish logicians and mathematicians. His aim was not only to solve particular fragmentary problems but (as it was also the case by Leśniewski) he attempted to construct a system containing the whole of mathematics.

Chwistek (1924; 1925) formulated a pure theory of logical types – a theory of constructive types. In this theory, the nonconstructive objects are rejected but the price for that is the greater formal complication of the system.

Those investigations led Chwistek to the construction of a full theory of expressions and of so-called rational metamathematics. The latter should be a system more fundamental than logic. It should enable the reconstruction of a classical logical calculus and of Cantor's set theory. According to Chwistek, it should be based on nominalistic assumptions – hence in particular it should be free of any existential axioms, first of all of the reduction axiom and the axiom of choice. The basic assumption was that theorems of the system being constructed, and consequently theorems of classical logic and of set theory, refer only to expressions/inscriptions that can be obtained in a finite number of steps by a rule of construction fixed in advance and not to the meaning of those expressions. Moreover, those expressions/inscriptions were understood as physical objects.

All of those ideas were developed by Chwistek later as part of his system of philosophy of logic and mathematics, in particular as a part of his methodology of deductive sciences. He developed it mainly in the book *Granice nauki. Zarys logiki i metodologii nauk ścisłych* from 1935 – English translation *The Limits of Science. Outline of Logic and of the Methodology of the Exact Science* appeared in 1948.

According to Chwistek, human knowledge is neither complete nor absolute. It cannot be complete because statements concerning the totality of objects lead to inconsistencies. On the other hand, it cannot be absolute because there exists no absolute reality. In *The Limits of Science* he wrote:

> It follows from these considerations that the principle of contradiction does not permit complete knowledge, i.e., knowledge which includes the answer to all questions. The attempts to secure such knowledge will sooner or later conflict with sound reason. (Chwistek, 1948, p. 42)

And common sense is – according to Chwistek – a factor common to all correct cognitive processes. It functions beside the admission of experience as a fundamental source of knowledge and of the necessity of schematization of experienced objects and phenomena. Common sense consists in rejecting all assumptions that cannot be experimentally checked or are inconsistent with experiments or are not based on reliable and certain statements concerning simple facts or cannot be logically reduced to such statements. Both empirical and deductive knowledge are relative. Empirical knowledge is relative because there are various types of experiments corresponding to various realities, and so is deductive knowledge – because it depends on the accepted system of concepts. Chwistek is talking here about rational relativism.

Chwistek was decidedly against irrationalism – he accepted the principle of the rationalism of knowledge. Rationalism consists in accepting only two sources of knowledge, namely experience and strict reasoning. It concerns not only mathematics and the exact sciences but experimental sciences and philosophy as well. Chwistek wrote in *The Limits of Science*: '[...] the point of departure in constructing a world view should be not a confused metaphysics, but simple and clear truths based upon experience and exact reasoning'. (1948, p. 3)

Consequently he was against irrationalism, metaphysics and idealism in philosophy and mathematics.[5] He sharply criticized Plato, Hegel, Husserl and Bergson. Despite the defects of positivism, he appreciated its epistemological conceptions. He also greatly appreciated dialectical materialism, seeing in fact almost no fundamental conflict between it and positivism. His own epistemological conceptions he described as critical rationalism and set it against dogmatic rationalism.[6]

A way out of the difficulties caused by irrationalism and simultaneously a weapon in a struggle against it is formal logic, in particular the rational metamathematics founded by him. Chwistek began his book *The Limits of Sciences* by writing in the first sentence (cf. 1948, p. 1): 'We are living in a period of unparalleled growth of anti-rationalism.' And he finishes the Introduction by writing (cf. 1948, p. 23): 'History teaches that ultimately victory has always been the destiny of societies who employ the principles of exact reasoning.'

In the Introduction he also wrote:

> When this new system [that is, the system of rational metamathematics – R.M.] is completely worked out, we will be able to say, that we have at our disposal an infallible apparatus which sets off exact thought from other forms of thought. (cf. Chwistek, 1948, p. 22)

The epistemological views of Chwistek were closely connected with those of Neo-positivism. He claimed that an object of a scientific knowledge can only be what is or can be given in experience, hence only what can be seen or experienced by the senses, perhaps assisted by instruments. He wrote in Chwistek (1935): 'Talking about reality we do not think about an ideal object but about those schemes we have to do with in a given case.'

Chwistek appreciated constructive methods. He was of the opinion that they should be used both in science and in philosophy. This method was explained by him in the paper 'Zastosowanie metody konstrukcyjnej do teorii poznania' [Application of a Constructive Method to Epistemology] (Chwistek, 1923). According to him, it can be applied in a full form mainly in the deductive sciences but nevertheless it can be used in empirical sciences and in philosophy as well. The method is based on the analysis of the intuitive concepts used in a given discipline. This permits the separation of the primitive notions characterized in axioms. On the basis of axioms new theorems can be obtained by laws of (formal) logic. Later Chwistek came to the conclusion that attempts to construct deductive systems in philosophy are in fact useless – the reason is the fact that philosophical investigations are too complicated.

As indicated above, Chwistek claimed that an object of cognition can only be what is given in an experience. There are, however, various types of experience. In this way we come to the best-known original philosophical conception of Chwistek, namely to his theory of the plurality of realities. It was explained by him for the first time in the paper 'Trzy odczyty odnoszące się do pojęcia istnienia' [Three Lectures Concerning the Concept of Existence] (Chwistek, 1917). Chwistek (1917, p. 145) claimed there that: 'the intuitive belief in one reality seems to be a superstition' and saw the concept of many realities already in Pascal and Mach (cf. Chwistek, 1917, pp. 149–50). The conception of the plurality of realities was developed in his book *Wielość rzeczywistości* [Plurality of Realities] (1921). The final version can be found in *Granice nauki* (Chwistek, 1935). Its foundations were explained once again in the English version of this book (cf. Chwistek, 1948) but this in fact brought nothing new.

In the first period of his scientific activity, that is up until 1925, Chwistek distinguished between the concepts 'reality' and 'existence'. The latter has – according to him – a more general character because it can concern not only objects of reality but also abstract objects such as the objects of mathematics. He wrote: 'If we assumed that everything that exists is in fact real, then we should accept as real all mathematical relations together with elements of experience'. (Chwistek, 1917, p. 145)

Chwistek (1917) distinguished three possible positions concerning the problem of existence: nominalism, realism and hyper-realism. According to him 'nominalists demand descriptions by words excluding inconsistencies' (Chwistek, 1917, p. 126), realists do not demand descriptions by words but they 'exclude inconsistent objects' (Chwistek,

1917), and 'hyper-realists do without descriptions by words and do not exclude inconsistent objects' (Chwistek, 1917).

In the beginning he accepted only two types of reality and attempted to formalize his theory. In the book *Granice nauki* he gave up his attempts at formalization and accepted four types of reality corresponding to possible types of experience. Hence he distinguished the reality of impressions, the reality of images, the reality of things (the reality of everyday life) and physical reality (that which is constructed in the exact sciences). Independent existence and full theoretical equality of rights were attributed to all particular kinds of reality.

Let us turn now to Chwistek's views connected directly with the philosophy of mathematics. The characteristic feature of them is nominalism.

Chwistek claimed that the object of deductive sciences, hence in particular of mathematics, are not any abstract ideal entities but in fact expressions constructed according to accepted rules of construction. Hence the objects of mathematics are not points, lines, numbers, sets, and so on, but expressions, physical objects given to us in experience. Those expressions can be transformed according to accepted rules – in every given system such rules as well as some expressions that play the role of axioms and form the basis on which one deduces theorems are accepted. The rules of transformation and axioms are chosen in such a way that the expressions can be interpreted as descriptions of considered states of things. To apply deductive theories to particular disciplines and generally to get to know particular domains of the reality, elements of the latter should be schematized.

Chwistek claimed that geometry is an experimental discipline. In Chapter VIII of *The Limits of Science* he wrote:

> Geometry is an experimental science. It depends upon the measurement of segments, angles and areas. The Egyptians conceived it in this way and it has remained essentially the same up to this very day. Today what is generally regarded as geometry, i.e., what is included in textbooks, is the peculiar mixture of experimental geometry and the geometrical metaphysics which was inherited from the Greeks as Euclid's *Elements*. (Chwistek, 1948, p. 170)

In Chwistek's opinion the development of systems of non-Euclidean geometry of Bolyai, Gauss and Lobachevsky[7] in the nineteenth century rejected Kantian idealism. Those geometries have shown that, for example, the concept of a line has no objective character but depends on adopted axioms. This can suggest that conventionalism should be a proper philosophy for geometry. Indeed, in his first papers, for example in the paper 'Trzy odczyty odnoszące się do pojęcia istnienia' [Three Lectures Concerning the Concept of Existence] (Chwistek, 1917) Chwistek stated explicitly that the existence of consistent systems of non-Euclidean geometries refutes the thesis about the objective character of geometry. Nevertheless he never stated explicitly that he would be ready to accept conventionalism. He wrote:

> Both systems [that is, the system of Euclidean geometry and systems of non-Euclidean geometries – R.M.] are free of inconsistencies – in fact they can be reduced to analytic geometry. Hence there are almost no fundamental differences between them from the theoretical point of view. Intuition easily accepts Lobachevsky's theorems that only at first glance seem to be paradoxical [...]. So we come to the conclusion that both geometries are equally true, each of them refers to different lines; the differences between those two types of lines can be caught neither by experimental means nor by intuitive ones, hence a segment of a line we draw or think can serve as an illustration of one or another type depending on our will. (Chwistek, 1917, pp. 144–5)

In *The Limits of Science*, however, Chwistek clearly and categorically rejected conventionalism. He claimed that geometry – like all other fundamental experimental sciences – should be based on a theory of expressions. The reason is that in fact conventionalism introduces hypothetical entities (this was the case already in J. S. Mill or later in Poincaré, the propagator of this tendency).[8] He wrote:

> It seems that it is impossible to attain the general concept of a geometry without using formulae. It is therefore clear that the conception of geometry as the science of ideal spatial constructions must be nullified. [...] To speak of [...] four-dimensional space-time it is necessary to employ five-dimensional space-time. It is clear that all this has only as much meaning as do mathematical formulae. (Chwistek, 1948, pp. 186–7)

One should also treat arithmetic, mathematical analysis and other mathematical theories just like geometry, obtaining in this way a nominalistic interpretation of all of them.

To sum up I would add that philosophical investigations of Chwistek had no systematic character and it seems that they were not treated by himself with a full sense of responsibility (cf. Chwistek, 1961 Preface, p. vii). He did not explain many of the concepts he used, his conceptions were 'earlier proclaimed than checked' (Chwistek, 1961).

He did not develop his systems in detail but satisfied himself by sketching them only. Chwistek's system of rational metamathematics could not be developed by his collaborators (among them were Jan Herzberg, Władysław Hetper and Jan Skarżyński) nor by his students and pupils (Wolf Ascherdorf, Celina Gildner, Kamila Kopelman, Abraham Melamid, Józef Pepis and Kamila Waltuch) because all of them were killed during the Second World War. His investigations were not in the main stream of the development of logic and philosophy of mathematics. Therefore his works did not generally find any interest amongst logicians and philosophers (with the exception of his version of type theory). Like Leśniewski (cf. Murawski, 2004), Chwistek worked on his own conceptions and ideas without any collaboration with other logicians, mathematicians or philosophers, he went his own way on his own. Although a professor of Lvov University he had in fact no strong contacts with the Lvov-Warsaw philosophical school (cf. Woleński, 1989). Only after 1945, together with the growing interest in nominalism in the philosophy of mathematics, did some of his ideas find recognition. One should also add that recently a reference to Chwistek's pluralism was made by the Australian philosopher R. Sylvan (1997).

13.6 Conclusions

As we see, mathematicians, logicians and philosophers in Cracow (or connected with Cracow) were interested in the philosophical problems of logic and mathematics but rather on the margins of their proper scientific work. The sole exception here was Chwistek, whose logical investigations were connected with his philosophical ideas concerning logic and mathematics and were in a sense motivated by them. What is characteristic for all of the Cracow scientists presented here is that they were interested in logic not as an autonomous discipline but treated it only as a useful tool that could help us to make mathematical reasoning more precise and clear. According to them, logic should provide practical tools for constructing proofs and justifying reasonings, it is an auxiliary discipline and can be reduced to a theory of proof or rather a theory of constructing (first of all mathematical) proofs. This is the difference with respect to the Warsaw School where logic was developed as an independent discipline. In fact the approach to mathematics represented by Cracow mathematicians was rather classical and not modern – as it was the case in Warsaw or Lvov. This was rightly characterized by the Russian mathematician Nikolai Lusin who in a letter to Arnold Denjoy wrote in 1926:

> It seems that the mathematical life in Poland is being developed along two different paths: one of them leans toward classical fields of mathematics, the other – toward set theory (theory of functions). Those tendencies in Poland exclude each other, each of them is hostile toward the other and a stout battle between them goes on. Both parties are very vigorous but, as seems to me, their strengths are not equal. [...] The classical side is represented nowadays only by the old [...] Cracow University. [...] The most inflexible follower of this tendency among Polish mathematicians is professor Zaremba. Other adherents are close to him. [...] However the classical tendency came to the end in many cities [...] where it has been replaced by the tendency of the school of Mr Sierpiński. (Lusin, 1983, p. 66)

Notes

1. The financial support of the National Center for Science [Narodowe Centrum Nauki] (grant no NN101 136940) is acknowledged.
2. In fact such attempts have been undertaken, for example, by the so-called Cracow Circle to which belonged I. M. Bocheński, F. Drewnowski, and J. Salamucha – they have tried to apply logico-axiomatic methods to philosophical and theological problems.
3. For Chwistek's philosophy of logic and mathematics see also Murawski (2011a).
4. Similar interconnections between philosophical ideas and research in logic can also be seen in the case of Stanisław Leśniewski (1886–1939) – cf. Murawski (2004) and (2011b).
5. It is worth noting here that Chwistek was against irrationalism and idealism not only because they are – in his opinion – incorrect philosophical theories but also because they are the source of human sufferings, social injustice, cruel excesses and wars.
6. A certain difficulty in interpreting Chwistek's philosophical views should be noted. In fact he often used classical philosophical notions but gave them a special meaning which he never explained or explained in an insufficient way.
7. It is worth noting that Chwistek considered them to be the most important achievement in the exact sciences.
8. It is worth adding here that Chwistek rejected conventionalism not only for this reason. He claimed also that conventionalism became a source of reactionary social views and tendencies by reducing truth to efficiency and leading in this way to the reinforcement of ruling classes. In (1948, p. 234) he wrote: "It should be observed that idealism clothed in the feathers of conventionalism became a very dangerous instrument in the hands of those who were reacting against the old dogmatic idealism."

References

Chwistek, L. (1917) 'Trzy odczyty odnoszące się do pojęcia istnienia' [Three Lectures Concerning the Concept of Existence], *Przegląd Filozoficzny*, 20, 122–51. Reprinted in Chwistek (1961), 3–29.

—— (1921) *Wielość rzeczywistości* [Plurality of Realities] (Cracow). Reprinted in Chwistek (1961), 30–105.

—— (1923) 'Zastosowanie metody konstrukcyjnej do teorii poznania' [Application of a Constructive Method to Epistemology], *Przegląd Filozoficzny*, 26, 175–87.

—— (1924) 'The Theory of Constructive Types (Principles of Logic and Mathematics). Part I: General Principles of Logic: Theory of Classes and Relations', *Annales de la Société Polonaise de Mathématique*, II, 9–48.

—— (1925) 'The Theory of Constructive Types (Principles of Logic and Mathematics). Part II: Cardinal Arithmetic', *Annales de la Société Polonaise de Mathématique*, III, 92–141.

—— (1933) *Zagadnienia kultury duchowej w Polsce* [Problems of a Spiritual Culture in Poland]. (Warsaw: Gebethner and Wolff). Reprinted in Chwistek (1961), 149–277.

—— (1935) *Granice nauki. Zarys logiki i metodologii nauk ścisłych* (Lvov and Warsaw: Książnica Atlas). Reprinted in Chwistek (1963), 1–232. English translation: Chwistek (1948).

—— (1948) *The Limits of Science. Outline of Logic and of the Methodology of the Exact Sciences*. English translation by H. C. Brodie, and A. P. Coleman (New York and London: Kegan Paul, Trench Trubner & Co Ltd). Reprinted by Routledge, London 2000.

—— (1961) *Pisma filozoficzne i logiczne* [Philosophical and Logical Writings], vol. I (Warszawa: PWN).

—— (1963) *Pisma filozoficzne i logiczne* [Philosophical and Logical Writings], vol. II (Warszawa: PWN).

Lusin, N. (1983) 'List do Arnolda Denjoy z 1926 r'. [Letter to Arnold Denjoy from 1926], *Roczniki Polskiego Towarzystwa Matematycznego. Seria II: Wiadomości Matematyczne*, 25, 65–8.

Łukasiewicz, J. (1916) 'O pojęciu wielkości (Z powodu dzieła Stanisława Zaremby)', *Przegląd Filozoficzny*, 19, 1–70. Published in English: Łukasiewicz, J. (1970) 'On the Concept of Magnitude. In Connection with Stanisław Zaremba's Work' in Łukasiewicz, J. *Selected Works*, L. Borkowski (ed.) (Amsterdam and Warszawa: North-Holland Publishing Company and PWN), 16–83.

Murawski, R. (2004) 'Philosophical Reflection on Mathematics in Poland in the Interwar Period', *Annals of Pure and Applied Logic*, 127, 325–37.

—— (2010) 'Philosophy of mathematics in the Warsaw Mathematical School', *Axiomathes*, 20, 279–93.

—— (2011a) 'On Chwistek's Philosophy of Mathematics', in N. Griffin, B. Linsky, and K. Blackwell (eds) *Principia Mathematica at 100* (Hamilton, Ontario: The Bertrand Russell Centre), 121–30. Reprinted in *Russell: The Journal of Bertrand Russell Studies*, 31, 121–30.

—— (2011b) *Filozofia matematyki i logiki w Polsce międzywojennej* [Philosophy of Mathematics and Logic in the Interwar Poland] (Toruń: Wydawnictwo Naukowe Uniwersytetu Mikołaja Kopernika).

Sleszyński, J. (1925–9) *Teorja dowodu* [Proof Theory] [Wykłady uniwersyteckie oprac. S. K. Zaremba], vols. I–II (Kraków: Kółko Matematyczno-Fizyczne Uczniów UJ).

Sylvan, R. (1997) *Transcendental Metaphysics* (Cambridge: The White Horse Press).

Whitehead, A. N., and Russell, B. (1925–27) *Principia Mathematica*, 2nd edn, vols. I–III (Cambridge: The University Press).

Wilkosz, W. (1938) *Liczę i myślę. Jak powstała liczba* [I count and I think. How Number Came About] (Kraków: Księgarnia Powszechna). 2nd edn: (1951) (Warszawa: Państwowe Zakłady Wydawnictw Szkolnych).

Woleński, J. (1989) *Logic and Philosophy in the Lvov-Warsaw School* (Dordrecht: Kluwer).

—— (1997) *Szkoła Lwowsko-Warszawska w polemikach* [Lvov-Warsaw School in Polemics] (Warszawa: Wydawnictwo Naukowe Scholar).

Zaremba, S. (1911) 'Pogląd na te kierunki w badaniach matematycznych, które mają znaczenie teoretyczno-poznawcze' [Remarks on Those Trends in Mathematical Investigations Which have Epistemological Meaning], *Wiadomości Matematyczne*, 15, 217–23.

—— (1912) *Arytmetyka teoretyczna* [Theoretical Arithmetic] (Kraków: Polska Akademia Umiejętności).

—— (1923) 'O stosunku wzajemnym fizyki i matematyki' [On Connections Between Physics and Mathematics] in *Poradnik dla samouków*, vol. III: *Matematyka. Uzupełnienia do tomu pierwszego* (Warszawa: A. Hefler and St. Michalski), 131–67.

—— (1926) 'La logique des mathématiques', *Mémorial des Sciences Mathématiques*, XV (Paris: Gauthier-Villars), 1–52.

—— (1938) 'Uwagi o metodzie w matematyce i fizyce' [Remarks on the Methods of Mathematics and Physics], *Przegląd Filozoficzny*, 41, 31–6.

Jan Woleński, Photo: Piotr Falkowski

A Selection of Jan Woleński's Publications

This bibliography contains a selection of Jan Woleński's publications on the history and philosophy of Polish logic, as well as his most important contributions to philosophical logic, epistemology, ontology, philosophy of logic, the applications of logic to philosophy, analytic philosophy of law, the philosophy of religion and philosophy of the Holocaust (represented by (1985b), (1987b), (1989a), (1995c), (1997), (2003c), (2004a), (2004c), (2004e), (2004f), (2006), (2007a), (2008b), (2009d), (2009e), (2009f), (2010a), (2010b), (2011c), (2012a)). Jan Woleński has published 25 books, over 600 research papers and has edited 30 collections. A more extensive list of his publications can be found at http://www.iphils.uj.edu.pl/~j.wolenski/publications.pdf

Books

(1985) *Filozoficzna Szkoła Lwowsko-Waszawska* (Warsaw: PWN).

(1989) *Logic and Philosophy in the Lvov-Warsaw School* (Dordrecht: Kluwer); English version of (1985).

(1999) *Essays in the History of Logic and Logical Philosophy* (Cracow: Jagiellonian University Press).

(2004) *Lvovsko-warszawskaja filosofskaja szkoła*, W. N. Porus (tr.) (Moskva: Rossijskaja Politichieskaja Enciklopedia); Russian translation of (1985).

(2011a) *Essays on Logic and Its Applications in Philosophy* (Frankfurt am Main: Peter Lang).

(2011b) *L'école de Lvov-Varsovie: Philosophie et logique en Pologne (1895–1939)*, A. Zielińska (tr.) (Paris: Vrin); French version of (1985).

(2013) *Historico-Philosophical Essays* (Cracow: Copernicus Center Press).

Books Edited

(1988) *Logische Rationalismus. Philosophische Schriften der Lemberg-Warschauer Schule* (Frankfurt am Main: Athenäum) [with D. Pearce].

(1990) *Kotarbiński: Logic, Semantics and Ontology* (Dordrecht: Kluwer).

(1993) *Polish Scientific Philosophy: The Lvov-Warsaw School* (Amsterdam: Rodopi) [with F. Coniglione and R. Poli].

(1994a) *Philosophical Logic in Poland* (Dordrecht: Kluwer).

(1994b) *60 Years of Tarski's Definition of Truth* (Cracow: Philed) [with B. Twardowski].

(1995) T*he Heritage of Kazimierz Ajdukiewicz* (Amsterdam: Rodopi) [with V. Sinisi].

(1998) *The Lvov-Warsaw School and Contemporary Philosophy* (Dordrecht: Kluwer) [with K. Kijania-Placek].

(1999a) *Alfred Tarski and the Vienna Circle* (Dordrecht: Kluwer) [with E. Köhler].

(2004a) *Handbook of Epistemology* (Dodrecht: Kluwer) [with I. Niiniluoto and M. Sintonen].

(2006) *Church's Thesis After 70 Years* (Frankfurt: Ontos) [with A. Olszewski and R. Janusz].

(2009) *The Golden Age of Polish Philosophy. Kazimierz Twardowski's Philosophical Legacy* (Berlin: Springer) [with S. Lapointe, M. Marion, and W. Miśkiewicz].

Papers

(1973) 'Wajsberg on the First-Order Predicate Calculus for Finite Domains', *Bulletin of the Section of Logic*, II, 107–13.

(1985a) 'Ajdukiewicz and Quine', *Science of Science*, 1–2, 83–98 [with H. Jakubiec]; reprinted in (1999), 85–94.

(1985b) 'Deontic Logic and Consequence Operations', in G. Holmström, and A. Jones (eds) *Action, Logic and Social Theory* (Helsinki: Societas Philosophica Fennica), 314–26.

(1986) 'Reism and Leśniewski's Ontology', *History and Philosophy of Logic*, VII, 167–76; reprinted in (1999), 13–21.

(1987a) 'Suszko's Analysis of the Development of Knowledge', in J. Perzanowski (ed.) *Essays on Philosophy and Logic* (Cracow: Wydawnictwo Uniwersytetu Jagiellońskiego), 181–5; reprinted in (1999), 237–40.

(1987b) 'Is, Ought and Logic', *Archiv für Rechts-und-Sozialphilosophie*, XXVIII, 373–85 [with K. Opałek].

(1989a) 'On Comparison of Theories by Their Contents', *Studia Logica*, XLVIII, 617–22.

(1989b) 'The Logical Works of Jerzy Słupecki', *Studia Logica*, XLVIII, 401–11 [with J. Zygmunt].

(1989c) 'De Veritate: Austro-Polish Contributions to the Theory of Truth from Brentano to Tarski', in K. Szaniawski (ed.) *The Vienna Circle and the Lvov-Warsaw School* (Dordrecht: Kluwer), 391–442 [with P. Simons].

(1989d) 'The Lvov-Warsaw School and the Vienna Circle', in K. Szaniawski (ed.) *The Vienna Circle and the Lvov-Warsaw School* (Dordrecht: Kluwer), 443–53; reprinted in (1999), 45–51.

(1990a) 'Kotarbiński, Many-Valued Logic and Truth', in (1990), 190–7; reprinted in (1999), 115–20.

(1991a) 'Theories of Reasonings in the Lvov-Warsaw School', in L. Albertazzi and R. Poli (eds) *Topics in Philosophy and Artificial Intelligence*, (Bozen: Instituto Mitteleuropeo di Cultura), 101–12; reprinted in (1999), 52–8.

(1991b) 'Polish Logic', in H. Burkhardt and B. Smith (eds) *Handbook of Metaphysics and Ontology* (München: Philosophia Verlag), 462–6.

(1991c) 'Philosophical Aspects of Tarski's Truth Theory', *Warsaw Scientific Society – Logic Group Bulletin*, 1, 27–30.

(1993a) 'Gödel, Tarski and the Undefinability of Truth', *The Yearbook of Kurt Gödel Society 1991* (Wien), 97–108; reprinted in (1999), 134–8.

(1993b) 'Tarski as a Philosopher', in (1993), 319–38.

(1993c) 'The Lvov-Warsaw School: Its Origin, Development and Ideas', in R. Stachowski (ed.) *Roots of Polish Psychology* (Poznań: Instytut Psychologii UAM), 13–23.

(1993d) 'Two Concepts of Correspondence', *From the Logical Point of View*, II, 3, 42–55.

(1994c) 'Some Polish Contributions to Fallibilism', in G. Debrock and M. Hulswitt (eds) *Living Doubt* (Dordrecht: Kluwer), 187–95; reprinted in (1999), 227–32.

(1994d) 'Is Tarski's Conception of Truth Relativistic?', in (1994b), 96–112.

(1994e) 'Jan Łukasiewicz on the Liar Paradox, Logical Consequence, Truth, and Induction', *Modern Logic*, 4, 392–400; reprinted in (1999), 121–5.

(1995a) 'On Ajdukiewicz's Refutation of Scepticism', in 1995, 353–6; reprinted in (1999), 204–6.

(1995b) 'On Tarski's Background', in J. Hintikka (ed.) *From Dedekind to Gödel* (Dordrecht: Kluwer), 331–41; reprinted in (1999), 126–33.

(1995c) 'Logic and Falsity', in T. Childers and O. Majer (eds) *Logica'94* (Praha: Filosofia), 95–105.

(1995d) 'Mathematical Logic in Poland 1900–1939: People, Circles, Institutions, Ideas', *Modern Logic*, 5, 363–405; reprinted in (1999), 59–84.

(1995e) 'Leśniewski's Logic and the Concept of Being', in D. Miéville and D. Vernant (eds) *Stanisław Leśniewski Aujourd'hui* (Neuchâtel: Centre de Recherches Sémiologiques), 93–101.

(1997) 'Two Theories of Transcendentals', *Axiomathes*, VIII, 367–80.

(1998a) 'Ajdukiewicz, Kazimierz', in E. Craig (ed.) *Routledge Encyclopedia of Philosophy*, vol. 1 (London: Routledge), 135–8.

(1998b) 'Leśniewski, Stanisław', in E. Craig (ed.) *Routledge Encyclopedia of Philosophy*, vol. 5 (London: Routledge), 570–4.

(1998c) 'Łukasiewicz, Jan', in E. Craig (ed.) *Routledge Encyclopedia of Philosophy*, vol. 5 (London: Routledge), 860–3.

(1998d) 'The Reception of the Lvov-Warsaw School', in (1998), 3–19.

(1998e) 'Reichenbach's Probability Logic and the Lvov-Warsaw School', in H. Poser and U. Dirks (eds) *Hans Reichenbach Philosophie im Umkreis der Physik* (Berlin: Akademie Verlag), 89–95; reprinted in (1999), 214–19.

(1999b) 'Semantic Conception of Truth as a Philosophical Theory', in J. Peregrin (ed.) *The Nature of Truth (If Any)* (Dordrecht: Kluwer), 51–66; translated into Spanish as: 'La concepción semántica de la verdad como teoría filosófica', in *Lógica y filosofía* (Madrid: Publicaciones de la Faculdad de Teologia "San Dámaso") [with P. Domínguez], 2005, 13–37.

(1999c) 'Semantic Revolution: Rudolf Carnap, Kurt Gödel, Alfred Tarski', in (1999a), 1–15.

(2000) 'Jan Łukasiewicz und die Satz vom Widerspruch', in N. Öffenberger and N. Skarica (eds) *Beiträge zum Satz vom Widerspruch und zur Aristotelischen Prädikationtheorie* (Hildesheim: Olms), 1–42.

(2001a) 'In Defence of the Semantic Definition of Truth', *Synthese*, 126, 1–2, 67–90.

(2001b) 'The Rise of Many-Valued Logic in Poland', in M. Stöckler (ed.) *Zwischen traditioneller und moderner Logik. Nichtklassische Ansätze* (Paderborn: Mentis), 193–204; reprinted in (2013), 37–50.

(2002) 'From Intentionality to Formal Semantics (From Twardowski to Tarski)', *Erkenntnis*, 56, 9–27.

(2003a) 'Polish Attempts to Modernize Thomism by Logic (Bocheński and Salamucha)', *Studies in East European Thought*, 55, 299–313; reprinted in (2011a), 31–42.

(2003b) 'The Achievements of Polish School of Logic', in Th. Baldwin (ed.) *The Cambridge History of Philosophy 1870–1945* (Cambridge: Cambridge University Press), 401–16.

(2003c) 'Psychologism and Metalogic', *Synthese*, 137, 179–93; reprinted in (2011a), 31–42.

(2004b) 'The Reception of Frege in Poland', *History and Philosophy of Logic*, 25, 37–51; reprinted in M. Beaney and E. Reck (eds) *Gottlob Frege: Critical Assessments of Leading Philosophers*, vol. I (London: Routledge), 2006, 290–309.

(2004c) 'Analytic vs. Synthetic and A Priori vs. A Posteriori', in (2004a), 781–839.

(2004d) 'Gödel, Tarski and Truth', *Revue Internationale de Philosophie*, 59, 459–90; reprinted in (2011a), 101–23.

(2004e) 'First-Order Logic: (Philosophical) Pro and Contra', in V. Hendricks *et al. First-Order Logic Revisited* (Berlin: λογος), 369–99; reprinted in (2011a), 61–80.

(2004f) 'What is Formal in Formal Semantics?', *Dialectica*, 58, 427–36; reprinted in (2011a), 81–9.

(2006a) 'Tarskian and Post-Tarskian Truth', in R. E. Auxier and L. E. Hahn (eds) *The Philosophy of Jaakko Hintikka* (La Salle: Open Court), 647–72; reprinted in (2011a), 163–82.

(2006b) 'Les paradoxes logiques et la logique en Pologne', in R. Pouivet and M. Rebusch (eds) *La philosophie en Pologne 1918–1939* (Paris: Vrin), 113–34.

(2007a) 'The Cognitive Relation in a Formal Setting', *Studia Logica*, 86, 479–97; reprinted in (2011a), 197–211.

(2007b) 'Andrzej Grzegorczyk: Logic and Philosophy', in S. Krajewski *et al.* (eds) *Topics in Logic, Philosophy and Foundations of Mathematics and Computer Science. In Recognition of Professor Andrzej Grzegorczyk (A Special Issue of Fundamenta Informaticae 81)* (Amsterdam: IOS Press and Polish Mathematical Society), 1–17 [with S. Krajewski].

(2008a) 'The Character of T-Equivalences', in C. Dégremont, L. Keiff, and H. Rückert (eds) *Dialogues, Logics and Other Strange Things. Essays in Honour of Shahid Rahman* (London: College Publications), 511–21; reprinted in (2011a), 231–41.

(2008b) 'Applications of Squares of Oppositions and Their Generalizations in Philosophical Analysis', *Logica Universalis*, 2(1), 13–29; reprinted in (2011a), 255–69.

(2008c) 'Chwistek-Tarski Competition in Lvov: A Contribution to Social History of Logic', in P. Bernhard and V. Peckhaus (eds) *Methodisches Denken in Kontext. Festschrift für Christian Thiel* (Paderborn: Mentis Verlag), 229–37.

(2008d) 'Mathematical Logic in Warsaw: 1918–1939', in A. Ehrenfeucht, V. W. Marek, and M. Srebrny (eds) *Andrzej Mostowski and Foundational Studies* (Amsterdam: IOS Press), 30–46.

(2008e) 'Andrzej Mostowski on the Foundations of Mathematics', in A. Ehrenfeucht, V. W. Marek, and M. Srebrny (eds) *Andrzej Mostowski and Foundational Studies* (Amsterdam: IOS Press), 324–37 [with R. Murawski].

(2008f) 'Tarski and His Predecessors on Truth', in D. Patterson (ed.) *New Essays on Tarski and Philosophy* (Oxford: Oxford University Press), 21–43 [with R. Murawski].

(2009a) 'The Rise and Development of Logical Semantics in Poland', in (2009), 43–59.

(2009b) 'Logic and Foundations of Mathematics in Lvov (1900–1939)', in B. Bojarski, J. Ławrynowicz, and Y. G. Prytula (eds) *Lvov Mathematical School in the Period 1915–1945 as Seen Today* (Warsaw: Instutute of Mathematics, Polish Academy of Sciences, Banach Center Publications), vol. 87, 27–44.

(2009c) 'The Principle of Bivalence and Suszko Thesis', *Bulletin of the Section of Logic*, XXXVII(3/4), 99–110; reprinted in (2011a), 285–92.

(2009d) 'Induction and Metalogic', in S. Philström, P. Raatikainen, and M. Sintonen (eds) *Approaching Truth. Essays in Honour of Ilkka Niiniluoto* (London: College Publications), 241–50.

(2009e) 'Theism, Fideism, Atheism, Agnosticism', in L. G. Johannson, J. Österberg, and R. Śliwiński (eds) *Logic, Ethics, and All That Jazz. Essays in Honour of Jordan Howard Sobel* (Uppsala: Uppsala Universitet), 512–21; reprinted in (2011a), 271–83.

(2009f) 'Executioners, Victims and Bystanders', in J. Ambrosewicz-Jacobs (ed.) *The Holocaust. Voices of Scholars* (Cracow: Jagiellonian University, Centre for Holocaust Studies, Auschwitz-Birkenau State Museum), 267–78.

(2010a) 'Truth and Consistency', *Axiomathes*, 20(2), 347–55; reprinted in (2011a), 303–11.

(2010b) 'Meaningfulness, Meaninglessness and Language-Hierarchies: Some Lessons from Ingarden's Criticism of the Verifiability Principle', *Polish Journal of Philosophy*, IV(2), 35–47; reprinted in (2011a), 313–23.

(2010c) 'Łukasiewicz and Popper on Induction', *History and Philosophy of Logic*, 31, 285–8 [with J. Agassi].

(2011a) 'Logic, Formal Methodology and Semantics in Works of Ryszard Wójcicki', *Studia Logica*, 99, 7–30 [with G. Malinowski].

(2011b) 'Jan Łukasiewicz sur la logique Stoïciens', *Philosophie Antique. Problèmes, Renaissances, Usages*, 11, 9–13.

(2011c) 'Remarks on Axiomatization within Logic', in (2011a), 325–30.

(2012a) 'Logic as Calculus Versus Logic as Universal Medium, and Syntax Versus Semantics', *Logica Universalis*, 6(3), 587–96.

(2012b) 'Truth is Eternal if and only if It is Sempiternal', in E. Tegtmeier (ed.) *Studies in the Philosophy of Herbert Hochberg* (Frankfurt am Main: Ontos Verlag), 223–30.

Index for Names

Printed in the United States
by Baker & Taylor Publisher Services